New Horizons in Astronomy

SECOND EDITION

New Horizons in Astronomy

SECOND EDITION

John C. Brandt

NASA-Goddard Space Flight Center
Greenbelt, Maryland
and University of Maryland
College Park, Maryland

Stephen P. Maran

NASA-Goddard Space Flight Center
Greenbelt, Maryland

W. H. Freeman and Company
San Francisco

Sponsoring Editor: Arthur C. Bartlett

Design and Production: Rick Chafian

Copyeditor: Larry Olsen

Compositor: Graphic Typesetting Service

Printer and Binder: Kingsport Press

Indication of the authors' affiliation with the National Aeronautics and Space Administration on the title page of this book does not constitute endorsement of any of the material contained herein by NASA. The authors are solely responsible for any views expressed.

Library of Congress Cataloging in Publication Data

Brandt, John C.
   New horizons in astronomy

   Bibliography: p.
   Includes index.
   1. Astronomy. I. Maran, Stephen P., joint author.
II. Title.
QB43.2.B7 1979   520   78-11717
ISBN 0-7167-1043-9

Frontispiece: A portion of the great cluster of galaxies in the constellation Coma Berenices. This cluster contains more than 1,000 galaxies, each of which is an "island universe" of billions of stars. Many of these stars perhaps resemble our own sun. (Kitt Peak National Observatory.)

Cover: Under construction on the Plains of San Agustin in New Mexico, the Very Large Array is already in use to receive radio waves from distant parts of space. When completed, this system will be in many respects that most powerful radio telescope ever built. Only a small portion is shown here. The full system will contain 27 dish-shaped antennas, each 82 feet in diameter. The antennas will be arranged in a Y-shaped pattern, each arm of the Y extending 13 miles, and its base extending almost 12 miles. (National Radio Astronomy Observatory.)

Contents

Preface

Countless examples throughout history and in our own time show that science is not an isolated activity of scholars but rather a vital force in human affairs. In fact, the impact of science on society has never been regarded so critically, nor discussed by so many people, as it is today. In particular, astronomy, which sometimes seemed to have a largely academic character with limited applications, now, through the growth of "Big Science" and especially the space exploration programs, figures prominently in discussions of the allocation of national resources.

Teaching an introductory, one-semester course in astronomy for non-science majors, we found that no single text existed that satisfied the criteria we had set for the course. These included: avoidance of mathematics; emphasis on the relationship of life (and the individual) to the astronomical universe; the involvement of astronomy in fundamental questions that have occupied scientific philosophers; and the development of astronomical thought in the context of the struggle for knowledge, with the adversary being sometimes Nature, and sometimes Man and his institutions.

So we wrote our own text. *New Horizons in Astronomy* is intended to develop in the reader an understanding of key physical aspects of astronomy and space research. Where possible, this is done by means of a historical treatment that demonstrates the manner in which the present understanding of an astronomical phenomenon has been reached. In our attempt to make the book readable, we have been constrained to keep it much less than encyclopedic. Thus, we have chosen to emphasize some topics, such as the results of lunar and planetary exploration and new developments in stellar and extragalactic astronomy, at the expense of space that otherwise might have been devoted to subjects that other authors may find equally valid. Our aim here is not to emphasize novelty for its own sake but to portray astronomy as the living subject that it really is. Inspired by—not resting upon—the achievements of the past, astronomers today are making discoveries and reaching understandings perhaps as fundamental as those of any previous time.

In preparing the second edition, besides the obvious task of updating the individual chapters, we have added problem sets and rewritten the biology and geology material to relate it more closely to the astronomical context. We have added material inspired by the results of x-ray astronomy and other frontier areas, and we have deleted the discussion of the solar oblateness controversy in favor of two new sections on black holes and radio galaxies. We hope that students and teachers will find these sections to be useful and not merely obligatory. In the chapter on "Our Sun," we have added material on the case of "the missing sunspots" and on the neutrino problem. We have revised the discussion of the solar system to incorporate the implications of the discovery that Pluto has low density and a moon. The chapter on "Exploration of Space" has been expanded to match the progress since 1972. In the chapter on "The Milky Way Galaxy," we have added a brief discussion of the density-wave theory of spiral arms.

In the chapter on "Space Age Astronomy," there is now a description of the Kuiper Airborne Infrared Observatory, a discussion of the scientific results from the Copernicus satellite, and a section on solar research satellites. Many of the topics in this and other chapters are amplified in *The New Astronomy and Space Science Reader,* a collection of magazine and journal articles edited by us and published by W. H. Freeman and Company in 1977 in a paperbound edition intended for use as a source of supplemental reading assignments for introductory astronomy courses.

We have added a modern observational H-R Diagram to Chapter 10 and a symbolic H-R Diagram to the color plates. In the chapter on Birth and Death of Stars, we have added the fascinating stellar interior cross-section diagrams of Icko Iben.

We also added the brief paragraphs on Benjamin Banneker and Thomas Hariot to Chapter 2 (Chapter 4 in the first edition). We think that these are interesting footnotes to astronomical and social history.

J.C.B. has continued to benefit from the experience of teaching astronomy at the University of Maryland. S.P.M. brought a fresh approach to the preparation of this edition from his experience in teaching at UCLA while on sabbatical in 1976. We thank our students, and we wish to acknowledge the many colleagues who sent useful comments and criticism of the first edition of this text. Comments on this second edition will also be welcome and should be forwarded to us in the care of Astronomy Publisher, W. H. Freeman and Company, 660 Market Street, San Francisco 94104.

November 1978   *John C. Brandt*

Stephen P. Maran

Introduction

SCIENCE AND THE ULTIMATE PURPOSE OF ASTRONOMY

How did the universe originate, and how did it reach its present state? What formed the earth, and how did life arise? These are among the central questions that have concerned every human culture, and they obviously involve events buried in the distant past. How can we progress in understanding so as to choose among the various theories for the origin of the universe, or possibly formulate a better theory? Surely the way must lie in observing and determining as many of the properties of the universe as we can, and in looking for a common pattern in the evidence. This process of exploration is what we call astronomy, and the final purpose of astronomy is thus nothing less than to account for the origin and physical nature of the universe around us.

At every stage in history we have had a particular understanding or view of the cosmos, but as new information developed through further observations, experiments, and theoretical analyses, our picture has continually improved. This second edition of *New Horizons in Astronomy* was prepared after scientists revised our basic concepts of the solar system, thanks to discoveries made with spaceprobes sent to Mercury, Venus, Mars, and Jupiter, and as astronomers gathered new data indicating the likely presence of black holes in the universe.

Thus, many of the following chapters differ in content from the corresponding sections of books written just a few years ago, and we must accept as a certainty that future discoveries, the nature of which we cannot foresee, will in turn make this volume obsolete. Therefore, in introducing the principal concepts that underlie the present state of astronomy and some related fields, we ask you to keep in mind that "scientific truths" are often susceptible to improvement or revision, and currently accepted theories are as subject to the dictum that "this, too, shall pass away" as are human affairs.

ASTRONOMY AND HISTORY

Beyond the basic purpose of astronomy discussed above, astronomical discoveries have played a vital role in the history of ideas and indeed the development of intellectual freedom. Historians generally recognize three major conflicts between established authority and scientists, and astronomy relates to two of them.

The first confrontation involved the rejection of the old idea that the earth is stationary at the center of the universe (Chapter 2). The second concerned the age of the earth (Chapter 3), to which scientists assigned an increasingly greater age (now considered to be about 5 billion years) as geological evidence accumulated. This was in direct conflict with the age of several thousand years that followed from a literal interpretation of the Bible. The third of these ideological battles concerned the evolution of species theory proposed by Charles Darwin (Chapter 4). Almost all scientists today believe that the concept of evolution has been fully verified, but this particular debate has not vanished from the social arena. According to the law in some areas, biology teachers must teach the Biblical story of Creation on an equal basis with the theory of evolution.

At present, science is involved in a different, perhaps more fundamental crisis—the apparent disenchantment of large segments of the public with the achievements and results of science and technology. "Progress" of the sort that was welcomed in the past is being questioned now by many people who are concerned about the deterioration of our environment, the development of weapons of mass destruction, increased automation, and so on. Astronomy is not deeply involved in these matters, although a type of pollution—light pollution (Chapter 6)—is affecting our capabilities of observing the stars. On the other hand, there is a clear trend toward reducing the support of basic research in favor of other important national priorities, and this has serious consequences for employment of young astronomers and for some aspects of astronomy that require expensive equipment, such as radio telescopes and space vehicles.

HOW THE WORLD BEGAN

It may be useful at this point to summarize briefly the history of the universe and our planet. Some aspects of this summary are the subjects of considerable controversy among scientists, although others, like the theory of evolution, are supported by an extensive array of evidence. Some of the terms used here may be new to you, but they are explained in the chapters that follow. Proceeding through these chapters we will explore, among other things, the full range of sizes in nature, from the atomic nucleus at 10^{-14} meters or 10^{-12} centimeters, to the distance to the farthest observed galaxies at 10^{26} meters or 10^{28} centimeters. The hierarchy of physical sizes is illustrated in Figure 1-1. If you do not know what a meter or centimeter is, or if you do not understand the meaning of 10^{-12} and 10^{28}, turn to Appendixes 1 and 2, which summarize the units and numerical notation used throughout this book.

According to our view, about 20 billion years ago, there was no earth, no moon, no sun, no stars, no galaxies. Just one thing existed—the enormously hot and dense *Primeval Fireball* that contained all the matter and energy in the universe. The fireball exploded into a rapidly expanding and cooling gas of protons, neutrons, and electrons, immersed in an intense sea of radiation. At first, pressure of the radiation maintained the smooth expansion, but eventually the matter, now consisting mostly of hydrogen atoms with some helium, began to form clumps. The clumps or blobs of gas continued to fly apart from each other, although the matter in an individual blob tended to contract because of its own gravitation.

In time, turbulent motions within a contracting gas blob disrupted it. One of the smaller objects produced by this disruption spun faster and faster as it contracted gravitationally, becoming flatter but continuing to condense even as it broke up into yet smaller blobs. How did this process of condensation and fragmentation come to an end? Some of the objects exploded, returning their material to the surrounding space. But, in one particular (and rather typical) case, the condensation produced a gaseous sphere that was sufficiently compact to resist the tendency to break into smaller pieces. Heat was released as this sphere contracted, and the temperature near the center of the object rose to the point where nuclear reactions began to occur, and these reactions constitute a source of radiant energy that has continued up to the present day. *In this way the sun was born.* It lay at the center of a disk-shaped array of matter, left over from the contracting cloud. The earth and other planets formed in the disk through the accumulation and condensation of this celestial debris (Figure 1-2).

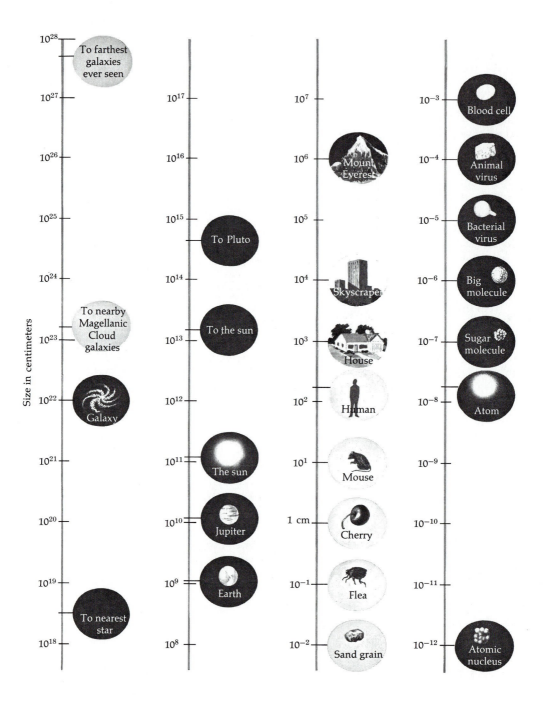

Figure 1-1. The scale of distances and dimensions in the universe is illustrated over the range from almost 10^{28} cm, the distance to the farthest known galaxies, to 10^{-12} cm, the diameter of an atomic nucleus. (After *Chemistry* by Linus Pauling and Peter Pauling. W. H. Freeman and Company. Copyright © 1975; the first column at the left is revised to take into account current values as of 1978.)

Figure 1-2. Five stages in the evolution of the expanding universe are sketched according to a popular theory. (a) The expansion was dominated by radiation (photons or particles of light are symbolized here by wavy arrows) during the first few minutes. The presence of matter is represented by the black dots (protons), circles (neutrons), and small circles (electrons). (b) Within 30 minutes, however (second stage), the effects of radiation had diminished to the point that particles of matter within the expanding gas could form more complex structures, such as the deuterium nucleus (neutron and proton) and helium⁴ nucleus (two neutrons and two protons). Later, most of the matter took the form of hydrogen atoms (each consists of a proton and an electron), and in the third stage (c), hundreds of millions of years after the expansion began, immense condensations formed in the gas—these eventually became the galaxies. (d) Within a galaxy (fourth stage), stars formed from smaller condensations and were sometimes accompanied by planets; our own solar system is about 4.6 billion years old. (e) After several more billion years elapsed, some of the matter on the surface of a planet reached the living state. The estimates of the times are very uncertain. (After *Modern Cosmology* by George Gamow. Copyright © 1954 by Scientific American, Inc. All rights reserved.)

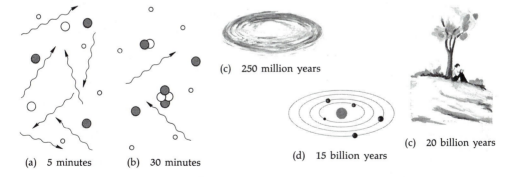

(a) 5 minutes   (b) 30 minutes

(c) 250 million years

(d) 15 billion years

(c) 20 billion years

The surface of the newly formed earth was bombarded by rocks left over from the formation of the planets. Gradually the oceans formed as volcanoes released water from the interior. The other basic geological processes of sedimentation, mountain building, and erosion began to operate. As the continents slowly drifted over the face of the earth, the world assumed its present general appearance.

At some time in the past, chemical compounds in the ocean water and the primitive atmosphere combined, and larger, more complex molecules were formed. Eventually, these molecules developed into the simplest forms of living organisms, which in turn evolved into more advanced marine life-forms. As evolution continued, some organisms adapted to life on the land and in the air. The numbers and complexity of animal and plant species increased, and competition for food and living space intensified. As new creatures arose by mutation and as climatic changes occurred, this competition led to the natural selection of those species best equipped to adapt and survive. The less efficient, less adaptable organisms were doomed to extinction.

Figure 1-3. Artist's impression of the map of the world drawn by John Speed in 1627. Note that the Mediterranean region is accurate, but errors occur elsewhere; for example, California is shown as an island.

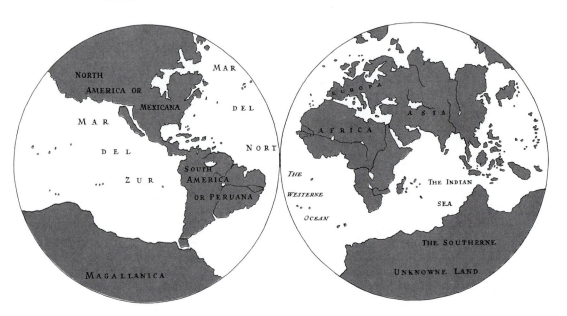

Figure 1-4. India and Ceylon, photographed from the manned spacecraft Gemini 11 at an altitude of 655 km (410 miles). (Courtesy of National Aeronautics and Space Administration.)

New species evolved, and an enormous, ever-changing variety of animal and plant life inhabited the earth. At one point, for example, dinosaurs dominated the planet for some 100 million years. Finally, perhaps 4 million years ago, man-apes appeared in Africa. As brain capacity and use of tools generally increased, these hunter–gatherers evolved toward *Homo sapiens,* passing through such intermediate stages as *Homo erectus* and Cro-Magnon Man. Modern *Homo sapiens* has been around for at least 35,000 years—the Kalahari Bushmen of South Africa remained hunter–gatherers until the 1960s, preserving their way of life as practiced over at least the past 11,000 years.

Homo sapiens became the dominant species on earth, and curiosity, often reinforced by cultural and economic factors, led to a continual expansion of human horizons. Cautious inspection of the regions immediately surrounding dwelling sites was the first step in this process, which led to the systematic exploration of the planet and has culminated in space travel.

The voyages of Marco Polo and Magellan are especially celebrated in the annals of Western civilization. In our own country we remember the nineteenth-century journeys of Lewis and Clark into the Pacific Northwest, and the expeditions of John Wesley Powell down the Grand Canyon of the Colorado River. These men traveled through parts of America that were *terra incognita* to the coastal residents. Their explorations were of tremendous value to the cartographer, for geographic information was hard to come by and was often inaccurate, as shown in Figure 1-3. Imagine how an early map-maker might have welcomed the information contained in a picture such as Figure 1-4, which shows India and Sri Lanka (Ceylon), photographed from space by an astronaut.

About 5,000 years ago, humans began to record information in an orderly and often permanent fashion. Thus, the knowledge available to succeeding generations no longer depended on direct experience or word of mouth. This stage began the cultural revolution that has accelerated so dramatically in the twentieth century. New information is being developed and published so rapidly that scholars are hard pressed to keep abreast of even the developments in their own fields, and some people believe that we are threatened by incomplete understanding and misuse of new facts and technologies.

Given the innate curiosity and constantly expanding horizons of human beings, it was inevitable that they would turn their attention to the heavens above as well as to the earth below.

Astronomy, the study of matter and radiation in space, enables us to understand the position and nature of the earth as an element of the cosmos and to search for an explanation of its origin. Astronomy played a basic role in the development of science and philosophy. In recent years certain astronomical subdisciplines and related fields, such as *general relativity*, have concentrated on the attempt to understand the nature of space itself.

2

Historical Astronomy— Sky and No Telescopes

On a clear, dark night, an observer located far from the lights and smog of the cities can perceive several thousand stars without optical aid. The people of ancient civilizations were familiar with the stars, many of which they named, and they recognized familiar patterns—the outlines of animals, people, ships—among them. These patterns are called *constellations*. Other points of light in the sky, called *planets*, were distinguished from the stars by the fact that they slowly moved across the backdrop of the constellations. (The word "planet" stems from the Greek *planetes*, meaning wanderer.) Five planets (Mercury, Venus, Mars, Jupiter, Saturn) were recognized before the telescope was introduced, but no one thought that they too were worlds like the earth.[1] The nightly rising and setting of the stars as they perform circular motions in the sky about the North and South Poles were ascribed to a rotation of the sky as a whole—most people did not believe that the earth itself turned. The Milky Way was perceived as simply a hazy band of light. Its appearance actually results from a great many faint stars spaced relatively close together in the sky, but this fact was not recognized until telescopes became available in the seventeenth century. The *Magellanic Clouds* (actually two relatively nearby galaxies; see Chapter 13) were familiar to inhabitants of the Southern Hemisphere and were recorded in the log books of sixteenth-century Por-

---

[1]Actually, the Greeks counted the sun and the moon as planets too, since both of them move with respect to the stars.

tuguese navigators. Resembling luminous patches of cloud, they were long regarded as isolated fragments of the Milky Way. Another hazy, dim light phenomenon is the *zodiacal light* (Figure 2-1). Now identified as the product of the scattering of sunlight by dust particles in the solar system, the zodiacal light was recorded as a cone- or pyramid-shaped glow, seen in the west just after dusk or in the east shortly before morning twilight.[2] Occasionally, comets were seen, but their celestial nature was not admitted, and they were generally considered to be cloud phenomena in the atmosphere. The celestial origin of meteors (often called "shooting stars") was also unrecognized; they were sometimes compared to lightning flashes.

THE PHASES OF THE MOON AND ECLIPSES

The most obvious thing about the moon is that the shape of its illuminated surface is different from one night to the next. Throughout the month we observe a systematic progression. The thin crescent seen low in the western sky just after sunset grows in illuminated area *(waxes)* and is found farther to the east each night after sunset, becoming a half-moon (referred to by astronomers as the *first quarter* stage) and continuing to grow until it reaches *full moon;* then the moon rises as the sun sets. The full moon, as the month continues, decreases in lit-up area *(wanes)*, passes through a half-moon stage (called *last quarter* or *third quarter* moon), becomes a thin crescent that rises in the east just before sunrise, vanishes briefly *(new moon)*, and finally becomes a thin crescent in the west at sunset, which signals the start of another monthly lunar cycle. This sequence is shown in Figure 2-2, where one can also see that the moon keeps the same face toward the earth throughout the month. The term *gibbous moon* applies whenever the illuminated area is larger than at quarter-phase, but less than at full moon.

The phases of the moon are a simple consequence of two facts: (1) the moon is not self-luminous, but shines by reflecting the light of the sun; (2) the moon orbits around the earth. Since the sun always illuminates the half of the moon that faces it at any given time (except during an eclipse of the moon), the phases depend solely on what fraction of the illuminated lunar hemisphere can be seen from the earth (Figure 2-3). As early as the fifth century A.D., Anaxagoras correctly explained the lunar phases by asserting that the moon shines by reflected sunlight, but for radical

---

[2]The zodiacal light is dim, and one needs to be in a dark location away from city lights in order to see it.

Figure 2-1. The zodiacal light as seen from Mount Chacaltaya, Bolivia. (Courtesy of D. E. Blackwell and M. F. Ingham.)

ideas such as this, he was convicted of impiety and banished from Athens.

Anaxagoras also understood that eclipses are simply shadow effects. The moon is eclipsed (which can occur only at full moon phase) when it moves within the shadow of the earth (see Figure 3-1). The sun is eclipsed (only possible at new moon phase) when the earth falls within the moon's shadow. Then why don't we see such eclipses every month at new moon and full moon? The explanation for this is also geometrical. There is a small tilt of about 5 degrees between the planes of the moon's orbit about the earth and the earth's orbit about the sun. At most new and full moon times the moon lies slightly above or below the sun-earth line, and thus does not eclipse the sun. But, if the moon happens to be in the plane of the earth's orbit (which is called the *plane of the ecliptic*, for obvious reasons) at new moon, then the sun, moon, and earth are in a straight line and a solar eclipse occurs. If the corresponding circumstances happen at the time of full moon, a lunar eclipse occurs.

The maximum number of eclipses possible in a year is seven (either four solar and three lunar, or five solar and two lunar); the minimum is

Figure 2-2. The moon as seen through its monthly cycle of phases. The first example (top left) is the thin crescent at age 4 days after new moon. The additional examples, going from left to right, from top row to bottom, are age 7 days (first quarter); age 10 days (gibbous); age 14 days (full moon); age 18 days (gibbous); age 20 days (gibbous); age 22 days (third quarter); age 24 days; age 26 days. Compare with the diagram in Figure 2-3. (Courtesy of the Lick Observatory.)

Figure 2-3. The geometry of the moon's appearance in the sky as seen looking down on the earth's North pole. As the moon moves in its orbit around the earth, different parts of the sunlit and shadowed lunar surface are visible from earth, and this situation gives rise to the appearance of new, crescent, quarter, gibbous, and full moon as shown. The same position on the moon is marked by an arrow; note that the moon keeps the same face pointed toward the earth. The outer, boxed moons show its appearance as seen from the earth. Compare these views with the photographs in Figure 2-2.

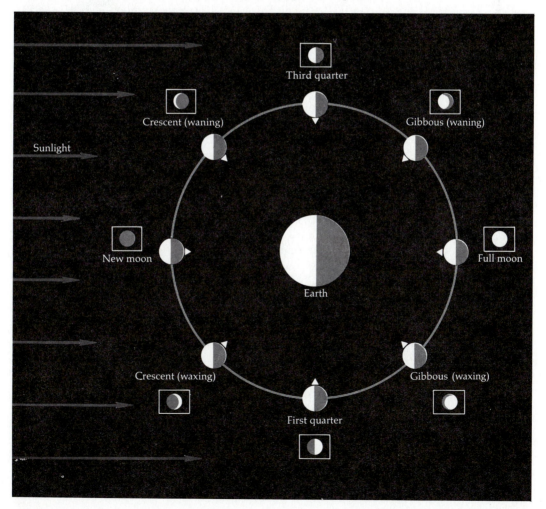

two (both solar); and the average is four. A lunar eclipse can be seen from any place on the night side of the earth. The shadow of the earth is seen to be circular and to move across the full moon. At the time of totality, the moon seems to have a dull red or copper color, due to rays of sunlight that, passing through the earth's atmosphere, are bent so they enter the shadow and strike the moon.

A total eclipse of the sun is a much more impressive phenomenon. A solar eclipse is seen as total when the observer is in the darker part of the moon's shadow, called the *umbra;* a partial eclipse is seen when the observer is in the lighter part of the shadow, called the *penumbra* (see Figure 2-4). When a total solar eclipse occurs, the *path of totality,* that is, the region on the earth that lies within the umbra, is typically only 120 kilometers (75 miles) wide. The apparent diameters of the sun and moon are nearly equal (they are each about one-half degree), and the narrow width of the path of totality is due to the fact that the umbra barely reaches to the earth (Figure 2-4). The umbra tracks across the earth at a typical speed of 1,600 to 3,200 km/hour (1,000 to 2,000 mph), and the duration of the total eclipse as seen from any given spot in the path of totality never exceeds eight minutes. A rather small fraction of the earth's population has witnessed a total solar eclipse; if you were to sit down at one point on the earth and wait for one, you would have a wait, on the average, of 400 years.

The duration of totality at a given spot depends primarily on the exact distance of the moon from the earth at the time of eclipse. Sometimes, because its orbit is slightly elliptical, the moon is far enough away so that the earth does not lie in the umbra. In other words, the moon is too far from the observer to block out the whole sun. In this case, a thin ring of sunlight is seen around the moon's edge (an *annular eclipse,* Figure 2-4b).

While a lunar eclipse or a partial or annular solar eclipse is an interesting event to watch, a total eclipse of the sun is a really spectacular phenomenon. As the moon moves across the sun, the observer's surroundings grow darker and colder until, at the moment of totality, the dark disk of the moon is suddenly surrounded by the glowing white *corona* or outer atmosphere of the sun. Animals of many types react to eclipses as though night had fallen, and humans have also had unusual responses. An ancient Chinese myth tells that the eclipsed sun is being consumed by a monster (Figure 2-5) so that one has to beat drums and gongs to drive the monster away and rescue the sun. So far, this procedure has always been successful.

If an opportunity to view a total solar eclipse arises, it is well worth taking and, in view of the scarcity of such events at a given location, it

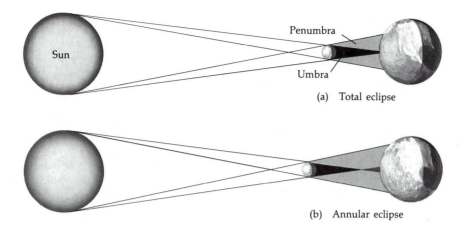

Figure 2-4. Geometry of solar eclipses. (a) In a total solar eclipse the umbra (from which no portion of the sun is visible) reaches to the earth's surface. Outside of the region of totality a partial eclipse occurs where the penumbra (from which part, but not all, of the sun is visible) strikes the earth's surface. (b) In an annular eclipse the moon's shadow does not reach to the earth's surface; the result is that even in the most total phase of the eclipse, part of the sun is still visible as a bright ring around the moon. (Not drawn to scale.)

may be a once-in-a-lifetime occasion. *Take care not to view the partially eclipsed sun directly,* as considerable infrared radiation remains and eye damage could result; it is safe to view the sun directly only during totality. When the total phase is about to end, small portions of the sun's disk emerge through the irregularities on the edge of the moon, thereby producing the *diamond ring effect* (Figure 2-6). This effect can also occur at the beginning of totality.

Many scientific investigations are made during eclipses of the sun. At these times the outer layers of the solar atmosphere can be studied without interference from the blinding glare of the sun's disk. Observers can search for comets and small planets near the sun, and an experimental test of the general theory of relativity can be made.[3] Eclipses have been used to date the events described in ancient records. If the location at

---

[3] According to this theory by Albert Einstein, a ray of light from a star will be bent slightly as it passes the sun. Without a solar eclipse, however, we cannot observe stars near the sun, and hence cannot check the theory.

Figure 2-5. The monster responsible for solar eclipses according to an old Chinese myth.

Figure 2-6. The sun emerging from the total eclipse of March 7, 1970, showing the diamond ring effect. (Photographed by J. C. Brandt and R. G. Roosen, Goddard Space Flight Center.)

which an eclipse was seen is recorded, it is frequently possible to calculate the date in terms of our modern calendar.

TIME AND THE CALENDAR

Three natural clocks existed in the sky for ancient people: the rising and setting, or *diurnal motion,* of the sun and stars, which measures the day and night; the phase cycle of the moon, which was used to define the month; and the apparent motion of the sun, which gave rise to the concept of the year (and which seemed closely related to the seasons).

The first of these clocks, which measures the day, is due to the rotation of the earth from west to east about the axis that passes through the North and South Poles. This produces the effect that the stars appear to move in the opposite direction, namely, from east to west. The *meridian* is an imaginary line in the sky that passes through the north and south points on the horizon and through the *zenith* point overhead. One way of defining the day is in terms of the interval between two successive crossings of the meridian by the sun *(solar day).* Another way is in terms of the interval between two successive crossings of the meridian by a star *(sidereal day).* The days defined by these two methods differ in length by about four minutes. We will return to this difference near the end of the chapter.

We should note in passing that, unlike the day, month, and year, there is no basis in celestial motions for the week. Nevertheless, in many languages, the days of the week are named after the seven bright, moving astronomical bodies that were known to the ancients, as shown in Table 2-1.

Table 2-1.   The Days of the Week

Sky Object	Day (English)	Day (Other Languages)
Sun	Sunday	—
Moon	Monday	—
Mars	Tuesday	Martes (Spanish)
Mercury	Wednesday	Mercoledi (Italian)
Jupiter	Thursday	Jeudi (French)
Venus	Friday	Venerdi (Italian)
Saturn	Saturday	—

The second of the three clocks, on which the month was based, is the motion of the moon in its orbit about the earth, which gives rise to the phase cycle shown in Figure 2-2. Similarly, the third clock, on which the year is based, is the motion of the earth about the sun. The sun appears projected against the background of different constellations as we view it from different points in the earth's orbit. Of course we don't see stars when the sun is in the sky, but we can, for example, note the point that is a given distance east of the sunset location at some fixed time after sunset, and find in this way that this point moves systematically through the constellations, returning to the same place among the stars after an interval of slightly more than 365 solar days.

A problem of considerable importance in the construction of a calendar is that the length of the year is not a whole number of days. As a practical matter, one does not want to have a fraction of a day left at the end of a calendar year, so a set of rules was formulated to give some calendar years an extra day and make the average of a large number of calendar years approximately equal to the true length of the earth's journey around the sun, 365.2422 solar days. Many of the ancient calendars were hopelessly inadequate, and when Julius Caesar came to power he found that the seasons were occurring in the "wrong" months.

Caesar extended the length of the year 46 B.C. to 445 days to make up for past errors. He also established the *Julian calendar* on the basis of advice from the Alexandrian astronomer Sosigenes, approximating the average length of the year as 365¼ or 365.2500 days. This average was obtained by adding an extra day every fourth year to a basic calendar containing 365 days. This is how the leap year originated.

While the Julian calendar was a great improvement, the calendar year and the astronomical year still differed by 365.2500 minus 365.2422 or 0.0078 days. This is a small difference, but given enough time a substantial discrepancy will accumulate. If the calendar were allowed to run for 1,600 years without adjustment, a difference of about twelve days would occur. In fact, this happened, and in the year 1582 Gregory XIII (on the advice of the astronomer Clavius) issued a papal bull designating the day after Thursday, October 4, as Friday, October 15. This proclamation established what we now call the *Gregorian calendar*. It omits as leap years the centurial years that are not evenly divisible by 400; thus, 1900 was not a leap year, but 2000 will be. On the Julian calendar there would be 100 leap years in 400 years; on the Gregorian calendar three of these are dropped. Hence, the average length of a Gregorian year is 365 plus $^{97}/_{400}$, or 365.2425 days. This value is so close to the astronomical one that the error we make in using the Gregorian calendar is only three days in 10,000 years.

The Gregorian calendar was adopted immediately by the Roman Catholic countries, but it was opposed initially in Protestant lands. Gradually it gained acceptance (Great Britain adopted it in 1752, despite riots over "lost days") and today it is the official or *civil calendar* in use over most of the globe. The calendar change leads to an interesting historical oddity. Modern history books list the birthday of George Washington as February 22. However, calendars hanging in the American colonies at the time of his birth showed the date as February 11.

THE SEASONS

The ancients noticed that the sun was lower in the sky during winter than in summer and thus deduced that the seasons have an astronomical cause. In fact, the seasons arise from the circumstance that the earth's axis is not perpendicular to the plane of the earth's orbit around the sun, but rather is tilted some 23.5 degrees with respect to the perpendicular. The axis points north to a fixed position in the sky (North Celestial Pole) and thus, as is clear from Figure 2-7, the Northern Hemisphere leans away from or toward the sun, depending on the season of the year. As perceived by an observer in, say, New York, the sun is higher in the sky at a given time of day (such as noon) in the summer than it is in winter. This is an effect caused by the Northern Hemisphere leaning toward the sun in summer and away from it in winter. When the sun is high in the sky, it heats the observer's surroundings more effectively than when it is low in the sky (see Figure 2-8). Also important is the length of time each day that the earth is heated by the sun as compared to the length of the night, when the earth loses heat. In the Northern Hemisphere the sun is highest in the sky and the day longest near June 22, but the warmest days do not occur at that time, because the earth still has not recovered from the cold of winter. It takes about a month more for the surface of the earth to reach its maximum temperature.

Note that the distance from the earth to the sun is not an important factor in determining the seasons. In fact, we are closest to the sun when winter occurs in the Northern Hemisphere; at that time it is summer in the Southern Hemisphere, which is then leaning toward the sun.

SYSTEMATIC DESCRIPTION OF THE SKY

When we study natural objects, whether they are geological formations, plant and animal species, or stars, one of the first things we want to know is: Where are they? The position of a given star can be described in terms of two quantities—its distance and its direction. We defer the discussion

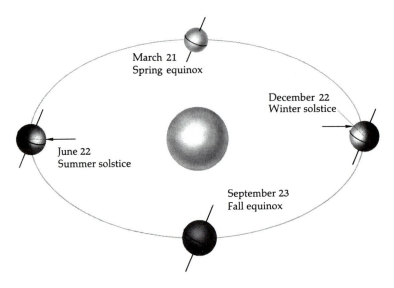

Figure 2-7. Tilt of the earth's axis and the seasons. The earth's axis of rotation points throughout the year in the same direction in space, a direction which is tilted 23½° with respect to a line at right angles to the plane of the earth's orbit. This effect causes the sun's rays to strike a given latitude on earth at different angles throughout the year. An example is indicated in the figure; at noon on June 22, the sun is overhead for latitude 23½° north and heating is efficient. At noon on December 22, the sunlight strikes the ground obliquely and heating is inefficient. See Figure 2-8.

Figure 2-8. Differences in ground heating between (a) summer and (b) winter. The solar intensity is equal in both cases (as indicated by the equal spacing), but because of the different angles of incidence, a given area of the ground receives much more energy in summer than in winter.

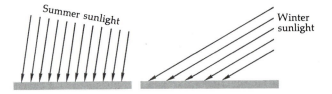

of star distances to Chapter 10, as the ability to measure them is a relatively recent development in the history of astronomy. On the other hand, the directions of the stars, as recorded on maps of the constellations, have been studied for thousands of years. They are conveniently specified by the *equatorial coordinate system*. Just as we can describe a location on the earth according to its *latitude* (angular distance north or south of the equator) and *longitude* (angular distance east or west of Greenwich, England), we use two coordinates, called *Right Ascension* and *Declination*, to identify points in the sky (Figure 2-9). The *Celestial Equator* is defined as the intersection of the earth's equatorial plane with the *celestial sphere*, an imaginary sphere in the sky, centered on the earth. The equatorial coordinate system can be visualized as a grid on this sphere, resembling the grid of latitude and longitude on a globe. Declination in the sky is analogous to latitude on the earth; it is measured in degrees north or south of the Celestial Equator. Right Ascension is measured eastward from one of the two points at which the earth's orbital plane (determined by plotting the sun's position on a sky map) intersects the Celestial Equator. This point, called the *Vernal Equinox*, is the location of the sun on the first day of spring[4] when the sun crosses the Celestial Equator, moving from the south celestial hemisphere to the north (that is, moving from southern to northern Declinations). The path of the sun across the background of stars is called the *ecliptic*; it represents the intersection of the plane of the earth's orbit with the celestial sphere.

The Age of Aquarius

We have mentioned the three celestial clocks used by ancient people. There is still another and slower clock that requires 26,000 years for a single revolution. Although the earth's axis appears to point toward the same direction in space, actually its orientation is slowly changing, making the axis trace a circle of radius 23.5 degrees on the sky. At the present time the location to which the axis points (North Celestial Pole) is near the star alpha Ursae Minoris, which we therefore call the North Star or Polaris, and the gradual motion of the axis, called *precession*, is moving the North Celestial Pole closer to this star. It will be closest around A.D. 2100 and will then move away toward the constellation Cepheus and on to the bright star Vega, returning to alpha Ursae Minoris after 26,000 years have elapsed. Precession (Figure 2-10) is due to the gravitational pull of the moon and sun on the earth's equatorial bulge and is analogous to the

---

[4]Spring in the Northern Hemisphere of the earth; fall in the Southern Hemisphere.

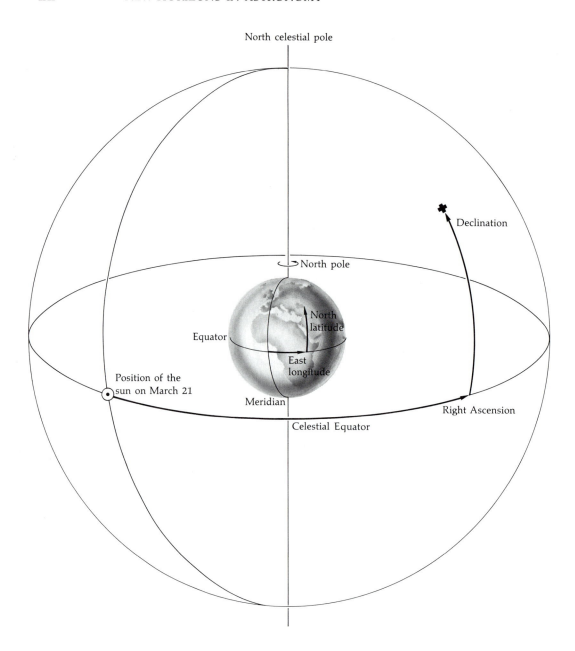

Figure 2-9.   Geographical and astronomical coordinate systems.

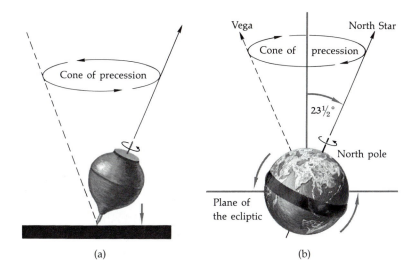

Figure 2-10. Precession. (a) The axis of a spinning top sweeps out a cone of precession because the top is pulled down by the earth's gravity. (b) Similarly, the moon and sun pull on the earth's equatorial bulge (marked by the dark band) and tend to pull the earth's equator into the plane of the ecliptic. Thus the earth's rotation axis also sweeps out a cone of precession, thereby tracing a circle on the sky. The north half of the axis presently points approximately to the star alpha Ursae Minoris, but in A.D. 14,000 it will point near the bright star Vega, and Vega will become the North Star.

wobbling motion of a spinning top or toy gyroscope. The existence of precession as a slow, continuous process affecting the equatorial coordinates of the stars was recognized by Hipparchus in the second century B.C. and was independently recorded in Chinese annals a few hundred years later.

If the direction in which the axis of the earth points is changing slowly, then as the equator (by definition) always lies in a plane at right angles to the axis, and as the intersection of this plane with the celestial sphere defines the celestial equator, the celestial equator must also be in gradual motion against the star background. The Vernal Equinox must therefore move through the constellations. This point, where the sun lies on the first day of spring, had great significance to the ancients; it was called the "First Point of Aries" because it lay in the constellation of Aries, the Ram. More recently, it has been in Pisces, the Fish, and now it is moving toward Aquarius, the Water Carrier. This is the origin of the term "Age of Aquarius" adopted by some people a few years ago to identify the present epoch.

STAR BRIGHTNESSES—REAL AND APPARENT

Continuing our study of the sky as seen with the unaided eye, we note that in some places stars seem to be located quite close together (for example, the stars Alcor and Mizar in the Big Dipper, Figure 2-11, which can be separated or *resolved* only by an observer with good vision). This suggests that even noting the positions of the stars might not make it simple to identify them individually. Clearly, if we record a star's brightness as well as its position, we make it easier for other astronomers to determine to which star we are referring. The brightnesses can also tell us much about the stars themselves, as we shall see in some later chapters.

Hipparchus (circa 150 B.C.) is credited with preparing the first star catalogue; he tabulated almost 850 stars and introduced a brightness classification or *magnitude* system. However, this work has not survived in its original form. A later catalogue, by Ptolemy (circa A.D. 150), contained positions and magnitudes for more than 1,000 stars and is believed to be an expanded version of Hipparchus' list. According to the modern definition of the term "magnitude" (which is in practice not very different from that used by Hipparchus and Ptolemy), the brighter stars are of the first mag-

Figure 2-11. Outline of the Big Dipper. Persons with good sight can resolve the two stars (inserted photograph) Alcor and Mizar; others see them as one.

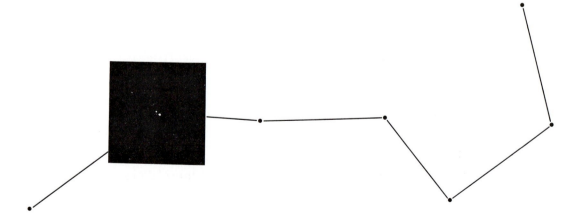

nitude, fainter stars are of the second, even fainter ones are of the third, and so on; the faintest stars visible to the naked eye are sixth magnitude. A few extremely bright stars have zero or even negative magnitude, as do bright planets like Venus and Jupiter. The faintest objects recorded with the largest telescopes are about + 25, whereas the sun is –26. A magnitude is defined such that a 1 magnitude difference between two stars means that one star is the fifth root of 100 (about 2.512) times brighter than the other star. Thus, if two stars differ by 5 magnitudes, one of them is 100 times brighter than the other. Since the brightest star seen at night (Sirius, also called alpha Canis Majoris) has a magnitude of –1, and the faintest stars easily seen by the naked eye are of magnitude +6, they differ by 7 magnitudes, or a factor of about 630.

We know from experience that a lamp looks brighter when it is nearby than when seen from a distance. Thus, if the stars are at various distances from the earth, the magnitudes that we note for them are not necessarily intrinsic properties. For this reason, magnitudes such as those listed by Ptolemy are called *apparent magnitudes*. When we know a star's distance, we can use its apparent magnitude to determine a true measure of the intrinsic brightness of the star, which is termed the *absolute magnitude*. It is obvious to even the most casual observer that there are more faint stars than bright stars; that is, the number of stars having a high apparent magnitude (faint apparent brightness) is much greater than the number of stars with low apparent magnitude.

ORIGINAL TASKS OF THE ASTRONOMER

Time and the calendar are of great significance to us today. If one is consistently late to a 9 o'clock class, one may flunk out. If we lose track of April 15 and thereby fail to file an income tax return on time, we are likewise in trouble. The ancients, who lived closer to the soil than we do, were equally dependent on the calendar. They had to know when to plant crops and when to pay tithe to the temple or the king. In Egypt they had to know when to anticipate the annual flooding of the Nile. The timing of religious occasions, as among the Hebrews, was especially important. The astronomer, who had the natural clocks provided by the rising and setting of the stars, the phases of the moon, and the annual apparent motion of the sun, was thus provided with a livelihood. By observing the moon, sun, and stars, the astronomer determined, corrected, and kept track of the calendar. In some cultures the importance of timing religious functions was paramount, and the capacities of priest and astronomer were combined, just as in later times (the Dark Ages in Europe) the all-important skill of writing was substantially confined to the clergy. Since the apparent positions of the stars depend not only on the time of day and time of year, but also on the *location* of the observer, the astronomer's ability to deduce his or her position on earth by observing celestial positions led to the art of navigation as an outgrowth of astronomy.

In America during the Revolutionary period, astronomers were still involved in these practical tasks. A famous example is self-taught Benjamin Banneker, the son of a slave. He built a clock, used astronomical techniques to help survey the present District of Columbia, and computed the positions of the sun, moon, and planets (Figure 2-12). Banneker is rarely mentioned in astronomy books, since he made no discoveries.

Figure 2-12. Title page of Banneker's almanac for 1792. (Courtesy of The National Museum of History and Technology of the Smithsonian Institution.)

However, it is an interesting footnote in the history of the United States that his expert knowledge of astronomy was used as an argument by Abolitionists to refute the proslavery claim that blacks were intellectually inferior. At least one of the original tasks of the astronomer, the determination of time, has remained to this day. Time signals are provided for the nation by the United States Naval Observatory in Washington, D.C.

HISTORY OF THE WORLD-VIEW

Curiosity as a human characteristic is perhaps best exemplified by the progressive development of our world-view. The experience of early humans was limited to a few square miles of foraging area in the vicinity of their lairs and to a certain consciousness of the horizon and the sky. Thousands of years later, scientific geography began to flower through the travels of the great medieval explorers, typified by Marco Polo (1254–1324), who brought back to the West an account of the vast extent of China. Aristotle had taught that the earth is round, and in the early sixteenth century, Ferdinand Magellan (1480–1521) organized, captained, and lost his life in the first successful circumnavigation of the globe. In the next few centuries, we discovered that the earth, long regarded as the center of the universe, in fact *moves*. It turns on an axis and also revolves around the sun. We gradually learned that there are other worlds in space, that the sun is just another star, and that the solar system is a tiny spot in an immense galaxy of some hundreds of billions of stars. There are countless other galaxies, mostly organized into clusters, that exist in space out to the limiting distances to which our largest telescopes can see. The history of astronomy documents the evolution of human thought concerning our place in the world, and the relation of the earth to the universe around us.

PLANETARY MOTIONS

Nearly every ancient culture possessed a world-view in which the earth was at rest in the center of the universe *(geocentric hypothesis)*. The sky was the arched body of a goddess (Egyptians, Figure 2-13) or rested on the tusks of an immense elephant (Hindus), was shaped like a bell jar (Babylonians), or, somewhat more recently, was regarded as an immense tent overhead (Arabs). But in any case, the remainder of the universe rested on, or revolved about, the earth. One gets the impression that the number of cosmologies that were proposed in Greece and Alexandria was about equal to the number of philosophers who lived there. Several of

Figure 2-13. An ancient Egyptian conception of the universe, showing the heavens represented by the goddess Nut, her starry body arched over the earth (figure adorned with leaves), and the ship of the sun in both its rising and setting positions. (From *Pioneers of Science* by O. Lodge. Dover Publications, Inc., New York, 1960. Reprinted through permission of the publisher.)

these sages recognized that the earth turns; one (Aristarchus, third century B.C.) also suggested that the earth and the other planets revolve about the sun. However, the views of the Egyptian astronomer Ptolemy of Alexandria, who compiled the star catalogue mentioned earlier, came to dominate Western thinking for centuries.

The fundamental problem that each cosmologist faced was to explain the motions of the planets. It was known that three planets (Mars, Jupiter, and Saturn) occasionally trace out loops (*retrograde* or backward motion) as they wander through the sky (see Figure 2-14). Furthermore, two planets (Mercury and Venus) never move very far from the sun in the sky (28 degrees for Mercury and 48 degrees for Venus). There was a great preference for assigning purely circular motions to the celestial bodies, since the circle was regarded as a pure or perfect geometric form. In Aristotle's system (Figure 2-15) the planets, sun, and moon followed circular paths about the earth. However, this failed to give satisfactory agreement with the observed motions of the planets. Ptolemy achieved reasonable agreement with his system, in which the orbits also were circles, but superimposed on these paths were smaller circles *(epicycles)*, each centered on a moving, imaginary point located on the larger circle (Figure 2-16). Considering the projected position of Mars against the starry background as seen from the earth (Figure 2-14), we can see how Ptolemy explained retrograde motion. By postulating that the centers of the Mercury and Venus epicycles always lay on the straight line that connects the center of the earth and the moving sun, Ptolemy accounted for the fact that these planets are always fairly close to the sun in the sky.

Thus Ptolemy employed a clever geometric arrangement to explain the observed motions of the planets within the context of a philosophy that required a central location for a stationary earth and circular motions for other heavenly objects. There is no particular physical reason why the planets should move according to these geometric patterns, but in the age of Ptolemy few physical laws were known or recognized as such. However, the need for some degree of sophistication was apparent, and a source of power to move the objects was required. In one reproduction of the Aristotelian system, four "elements" (earth, water, air, and fire) are

Figure 2-14. Retrograde or backward motion of Mars. The planet's apparent track through the sky is shown in (b), and the modern explanation is given in (a). For each of nine times the line of sight from earth to Mars is drawn to project the apparent position of Mars against the stars. (From *Structure and Change* by G. S. Christiansen and P. H. Garrett. W. H. Freeman and Company. Copyright © 1960.)

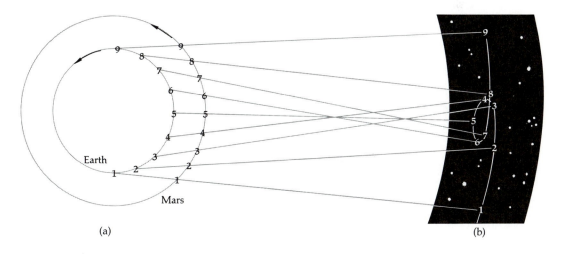

(a)   (b)

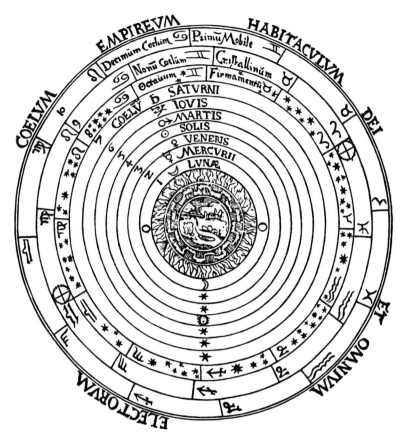

Figure 2-15. The Aristotelian universe as presented in Peter Apian's *Cosmographia* (1539). The earth is at the center. The moon, planets, sun, and stars occupy a series of concentric rings.

represented, as are the moving astronomical bodies (Figure 2-15). Beyond them is the crystalline firmament in which the stars were fixed (Aristotle and other Greeks taught that the stars were permanent, fixed, and immutable). Still beyond was the Primum Mobile (or Prime Mover) that powered all. Needless to say, the existence of the Primum Mobile would excite the curious (Figure 2-17)!

Some psychologists believe that a person must be at least mildly egocentric to function effectively in society. Perhaps a similar need explains the fervor with which the geocentric universe of Ptolemy was

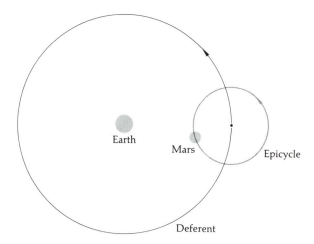

Figure 2-16. Epicycles and retrograde planetary motion. As explained by the Ptolemaic theory, Mars was thought to move around the epicycle as the center of the epicycle moved along the deferent, a larger circle centered on the earth. These two supposed motions combined to produce backward loops through the sky as seen from the earth.

adopted. It became the established and official world-view in Europe, despite its complexity. In fact, the Greek theories were so complicated that Alfonso the Learned, thirteenth-century astronomer and King of Castile, was moved to say that had he been present at the Creation he would have recommended a simpler plan!

THE HELIOCENTRIC WORLD-VIEW

The credit for the revolution in science that occurred in the Middle Ages, giving us the *heliocentric* (sun-centered) world-view, belongs to five men: Copernicus, Tycho, Kepler, Galileo, and Newton. Copernicus revived, extended, and discussed the theory of a planetary system centered on the sun. For 20 years, Tycho (without the aid of a telescope) made precise measurements of the position of Mars. Kepler showed that the motions of Mars, deduced from Tycho's observations, were not in good agreement with the Ptolemaic theory, but rather were in accord with certain general rules *(Kepler's Laws)* that put the sun about where Copernicus believed it must be. Among the many discoveries made by Galileo with his telescope, two in particular (the phases of Venus and the moons of Jupiter)

Figure 2-17. This fanciful drawing depicts a man who traveled to the end of the earth, putting his head through the sky in order to observe the machinery that moves the stars.

were strongly discordant with Ptolemaic philosophy. Galileo also began the scientific study of motion. Finally, Newton's development of the calculus, his formulation of the laws of motion, and especially his discovery of the law of gravity, enabled him to give a *physical* explanation for the accurate description of planetary motions that Kepler had derived from Tycho's measurements.

Nicolaus Copernicus (1473–1543), a Polish monk, studied his own and others' observations of the planets and proposed a heliocentric theory in which the earth revolved about the sun, as did the other planets, and in which the daily rising and setting of the stars was correctly ascribed to the rotation of the earth on its axis. Although the Copernican solar system (Figure 2-18) is usually remembered as a set of sun-centered circular planetary orbits, Copernicus in fact followed the practice of his predecessors in also including epicycles. Otherwise, the available observations of the planets would not have been completely accounted for by his theory. Thus, he was strongly influenced in formulating this model of the solar system by his continued belief in the circle as the ideal form for a planetary orbit.

Copernicus' ideas were published in 1543 as he lay on his death bed. Many of the questions that were asked about his heliocentric theory were of the same sort that children ask when they first become conscious of the

Figure 2-18. The sun-centered Copernican universe as drawn by Thomas Digges in 1576. The earth moves about the sun, but the moon travels around the earth. This diagram contains a feature unusual for the sixteenth century; the stars are not in a spherical shell at a certain distance, but extend to infinity. Most heliocentric models of the time placed the stars in a shell just beyond the orbit of Saturn, as in the pre-Copernican, earth-centered model shown in Figure 2-15.

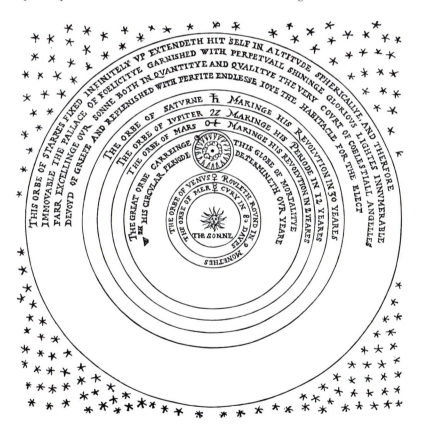

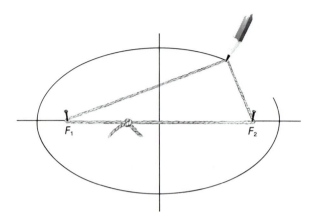

Figure 2-19. Drawing an ellipse. If a loop of string is pinned to a drawing board as shown, the figure traced out by a pencil which keeps the string taut is an ellipse with foci at F_1 and F_2, the locations of the pins. (From *New College Physics* by A. V. Baez. W. H. Freeman and Company. Copyright © 1967.)

world around them—for example, "If the earth moves, why don't we fall off?" But leaving aside simple-minded objections and details, his basic idea was not welcomed because established religious doctrine held that the earth did not move. About seventy years later, Copernicus' book was placed on the forbidden list, the *Index Librorum Prohibitorum*, by the Church in Rome. It was also reportedly denounced by Martin Luther, although some of his followers held it in great esteem.

Tycho Brahe (1546–1601) rejected the Copernican theory in favor of a geocentric cosmology of his own. According to this *Tychonic theory*, the sun revolved about the earth, but the other planets revolved about the sun. A more lasting contribution by Tycho was a series of very precise observations of the position of Mars, made with equipment of his own design at an observatory on the Danish island of Hven. The telescope did not yet exist, and a typical Brahe instrument consisted of little more than two pointers used to sight on a star and planet. The angle between the two pointers (and hence the angular separation of star and planet in the sky) was then read from a graduated arc.

One of the great heroes of astronomy, Tycho is also one of its more fascinating characters. At the time of his discovery of a bright new star[5] in

---

[5] Tycho's star was a supernova or exploding star, like the object that produced the Crab nebula (Chapter 16).

1572, an observation that was a direct challenge to the accepted dogma of the immutability of the stars, he was living in splendor at Hven, surrounded by servants and research assistants. Losing his nose in a duel, he replaced it with a silver one. When Tycho's patron, the King of Denmark, died, there was a falling out with the new authorities, so Tycho moved to Prague and composed a poem on Denmark's ingratitude to its greatest astronomer. In Prague he was continually plagued with such difficulties as collecting his salary. But the final indignity was his death, following a dinner attended by imperial representatives. No one could leave the table before they were finished, and Tycho suffered a burst bladder as a result. His dying wish (not granted) was that Kepler concentrate on proving the Tychonic theory.

Johannes Kepler (1571–1630), a German, worked as Tycho's assistant and inherited his observations. Centuries before the introduction of the electronic computer, he was a master of data analysis. He analyzed Tycho's long series of observations, together with some later measurements, and found that they fit three empirical rules. Kepler showed that Copernicus was correct—the sun is the center of the planetary system. And Kepler found that Copernicus was wrong—the planetary orbits are not circles.

KEPLER'S LAWS

1. Each planet travels in an *elliptical* orbit, with the sun at one focus of the ellipse.
2. The *radius vector* (that is, the imaginary line from sun to planet) of a planet sweeps out equal areas in equal periods of time (Figure 2-20).
3. The distance of a planet from the sun uniquely determines the length of time required for the planet to complete one orbital revolution (the *period*). (See Figure 2-21; the distances are conveniently measured in terms of the *astronomical unit* or average sun-earth distance.) In computational terms the cube of the average distance of a planet from the sun in astronomical units is equal to the period (in years) squared. We can check this for the planet Mars. The distance from the sun is 1.52 a.u.; cubing it we get $(1.52) \times (1.52) \times (1.52) = 3.54$. The period is 1.88 years, and $(1.88) \times (1.88) = 3.54$.

Kepler's Laws gave a good empirical description of planetary motions, but the physical explanation for them was still unclear. Kepler realized that some kind of force must keep the planets in orbit, but not until Newton formulated the laws of motion and gravitation was there a general scientific understanding of the motions of objects in space. As a result, Kepler's work did not meet with immediate favor. Kepler observed

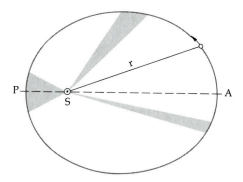

Figure 2-20. The law of areas. Imagine a line (marked r for radius) running from the sun (S) to a planet in its elliptical orbit around the sun. As the planet travels in its orbit, the radius sweeps out an area in such a way that equal areas are swept out in equal times. The area swept out (shaded) is equal in the three cases shown, and thus the planet must travel faster in its orbit at perihelion (P) than near aphelion (A).

another so-called new star, the supernova of 1604, although established authority (despite Tycho's supernova of 1572) still maintained the unchanging nature of the heavens. He also wrote a book on Copernican theory that was placed on the *Index*.

The philosopher Giordano Bruno fared even worse. He stated publicly not only that the earth moves, but also that the stars were other suns, surrounded by worlds similar to the earth. He was seized by the Inquisition, tried, and convicted of heresy because of his astronomical teachings and other radical ideas. A lengthy attempt to make him confess error failed, and he was burned alive in 1600.

In this setting lived Galileo Galilei (1564–1642), the greatest scientist of Renaissance Italy. He originated *dynamics* (the scientific study of motion), believed in the importance of experiments, and made fundamental advances in astronomy. Galileo's discoveries contradicted existing ideas and paved the way for Newton. For example, Aristotle taught that heavier objects fall much faster than light ones. Galileo tested this theory by dropping various objects from the Leaning Tower of Pisa (Figure 2-22). The heavy ones barely beat the light ones to the ground. Galileo theorized that the light objects would fall exactly as fast as the others if it were not for the resistance of the air. Since he had no vacuum chamber in

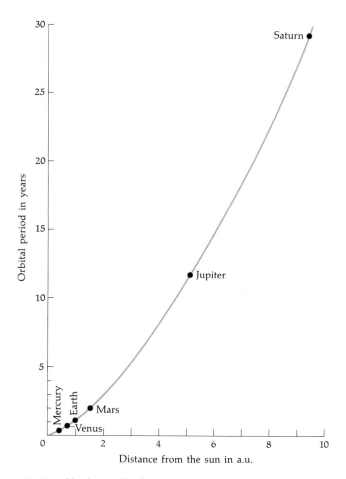

Figure 2-21. Kepler's Third Law states that the orbital period (in years) squared is equal to the heliocentric distance (in astronomical units) cubed. The planets out to Saturn are shown at the proper points on the graph. The curve can be used to find the period or heliocentric distance when only the other is known; for example, if an asteroid has a period of 4.6 years, its heliocentric distance is about 2.7 a.u.

Figure 2-22. Galileo's experiment at the Tower of Pisa. (After *Matter, Earth, and Sky*, 2d ed., by George Gamow. Prentice-Hall, Inc. Copyright © 1965. Used by permission.)

which to test this idea,[6] he performed an experiment in water. Water should have more resistance than air, so the difference in the drop times of dense and light objects would be more noticeable in water than in air. His experiment proved this point. To his students, if not to many of his fellow professors, Galileo established the superiority of experimentation to reasoning from logic alone.

Galileo applied the newly discovered telescope to astronomical research in 1609. Among his many great discoveries, several concern us here. Galileo found that the planet Venus exhibits phases (Figure 2-23) similar to those of the moon. This proved that Venus is not self-luminous but shines by the reflected light of the sun. The observed phases agree with a heliocentric model; the crescent is seen when the planet lies close to the line that joins sun and earth; full Venus occurs when it is on the opposite side of the sun. In the Ptolemaic model, however, Venus and earth are *never* on opposite sides of the sun (since the center of the Venus epicycle stays on the line from the earth to the sun), so Venus should never be full or nearly full.

Of lesser direct consequence to the geocentric versus heliocentric argument, but of perhaps even greater philosophical significance than the

---

[6]Apollo 15 astronaut David Scott, standing on the airless moon, dropped a hammer and a feather, beautifully demonstrating to the television audience that Galileo was right.

phases of Venus, was Galileo's discovery that Jupiter has four moons that orbit about that planet. Thus, the earth was clearly not the center of all motions. Later, Newton's discussion of the motions of celestial objects under the influence of gravity, a discussion that assumed no special properties for the earth or any other world, was shown to apply as well to Jupiter's satellite system as to the motions of the planets about the sun.

Figure 2-23.   Selected photographs of Venus, printed *to the same scale*. They show that Venus appears larger (that is, is closer to the earth) during the crescent phase, while it seems smaller (farther from the earth) when at full phase. (Lowell Observatory.)

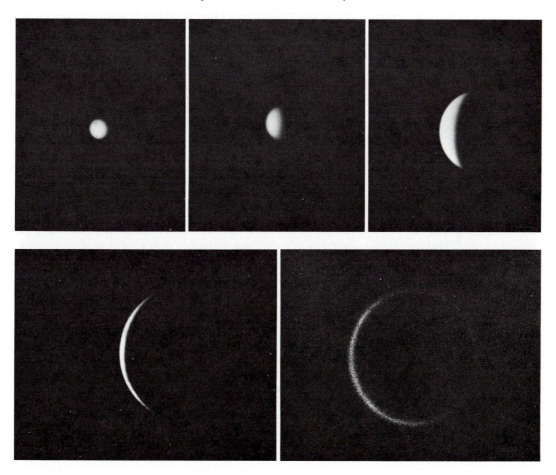

Galileo also observed the sunspots or "blemishes" on the sun and noted their motion (proving that the sun rotates), despite the fact that such transient markings were not permitted by the prevailing doctrine that maintained the unchanging nature of celestial bodies.

Galileo's adherence to the heliocentric world-view led to trouble with the authorities, and some of his doubters (including professors) even refused to look through his telescope. Eventually, he was summoned before the Inquisition in Rome and was asked to recant his views on the movement of the earth. After a time he did recant and was sentenced to house arrest for life.

Galileo's house arrest caused little reduction in his scientific activity, although he later became blind, perhaps because of the eye damage suffered when he first viewed the sun through a telescope. He invented a way of studying motion through the use of graphs, proposed the construction of pendulum clocks, and composed his famous dialogues on the science of motion. Galileo died in 1642 at the age of 78. He was denied a monument in the hope that he and his offending work would be forgotten, but in that year Isaac Newton was born, and Newton's work was to ensure the survival and triumph of the heliocentric theory.

Publish or Perish

At this point, if you were asked, "Who first applied the telescope to astronomy, studying such things as the moons of Jupiter and the properties of sunspots?" you would probably say, "Galileo." Indeed, he is the only person so credited in most textbooks. But in fact, Galileo had a contemporary in Elizabethan England who made similar observations. He was Thomas Hariot, science adviser to Sir Walter Raleigh. Hariot (who also brought tobacco back to England from the New World) independently used a telescope to investigate the moon, Halley's comet, sunspots, and the moons of Jupiter. He also studied the path of an object moving under the force of gravity long before Galileo did, and he made important contributions to the theory of how light is bent by a lens, and to algebra.

Why has no one heard of Thomas Hariot? Galileo corresponded with fellow scholars, wrote books, and lectured on his discoveries. Hariot did not publish or publicize his astronomical work, and thus it had no impact on other scientists. About 150 years after Hariot died, thousands of pages of his scientific notes were found and the story of his discoveries began to emerge—at a time when they had become of purely historical interest.

NEWTON AND THE LAWS OF MOTION

Like Galileo, Newton (1642–1727) was a man of many talents and interests. He studied light (Chapter 5), designed the first good reflecting telescope (Chapter 6), placed the rules of motion on a precise mathematical basis, and formulated the universal law of gravitation. Although he was not the sole discoverer of the calculus, he was the one who applied it to the study of moving objects, including the planets.

NEWTON'S LAWS OF MOTION

1. A moving object on which no forces act travels at a constant speed in a straight line. (It may appear that this law is contradicted by everyday experience; for example, a ball rolled along the ground comes to a stop, but the stopping is caused by friction, a type of force exerted on the ball by the surface along which it is rolling. Experiments show that if the friction is reduced by rolling the ball on a smoother surface, the ball will go farther.)

2. A force acting on an object produces a change in its motion in the direction in which the force acts, and the amount of this change *(acceleration)* is inversely proportional to the mass of the object.[7] (This law also agrees with the evidence of our senses—it takes much more force to get a massive bowling ball moving rapidly than to roll a marble along the same floor at the same speed.)[8]

3. For every action (force) there is an equal and opposite reaction (opposing force). (A familiar example: the angler rows his boat out to a good fishing spot, then throws—exerts a force on—the anchor overboard; the boat moves a slight way in the direction opposite to the throw, and this reveals the reaction—the force exerted by the anchor on the man as it is thrown.)

In stating Newton's Laws, we have used the term *mass*. Mass is defined in terms of Newton's Second Law as the property of an object that determines the amount of acceleration that the object undergoes when subjected to a given amount of force. In effect, an object's mass is a measurement of the total amount of matter it contains, in terms of individual atomic particles.

---

[7]For example, if object A (mass = 10 grams) and object B (mass = 20 grams) are both acted on by the same force, the acceleration of A will be twice that of B. A gram is equal to the mass of one cubic centimeter of water at 4° C.

[8]This example assumes that each force acts on its object for an equal length of time.

Given Newton's Laws, what about the motions of the planets? According to Kepler's results, they travel in elliptical orbits, but Newton's First Law says that a moving object will follow a straight line, unless a force is acting on the object. What force acts on the planets? According to a famous story, Newton, inspired by an apple falling from a tree, realized that the same force of the earth's gravity that acted on the falling apple might act on the moon, keeping it from pursuing a straight course through space. By analogy, a similar force might keep the planets in curved orbits around the sun. Other scientists were reasoning along similar lines at about the same time, but they were unable to test their theories. Newton, using the mathematical methods of the calculus that he himself derived, proved that his own hypothesis of *universal gravitation* could explain the planetary motions described by Kepler.

LAW OF UNIVERSAL GRAVITATION

Every object exerts a force of attraction on every other object, and the force between any two objects decreases as their separation increases, varying inversely as the square of the separation. (Thus, the gravitational force between two objects separated by 10 km is 10 squared or 100 times less than the force that would exist between the same objects if they were separated by 1 km.) Further, the force is proportional to the product of the masses of the two objects.

Orbital motion is achieved through a balance of the tendency for an object to keep moving in a straight line and the force of gravity.[9] If the gravitational attraction between earth and moon were miraculously shut off at some instant, the moon would fly off in a straight line tangent to its orbit at that point because no force would be pulling it toward the earth.[10] On the other hand, if the moon's orbital motion were suddenly stopped but gravity remained operative, the moon would fall directly toward the earth. The balance of these two opposite tendencies produces the actual orbit of the moon around the earth (see Figure 2-24). The moon is always falling toward the earth, but its own instantaneous motion (at nearly right angles to the direction toward the earth) produces an orbit. An analogous situation can be achieved by whirling a weight around in a circle on the end of a string. If you continue to whirl it and supply a force, the weight

---

[9]This tendency, a consequence of Newton's First Law, is commonly called *centrifugal force* when motion in a curved path is under discussion.

[10]We have ignored the effects of the gravitational force exerted on the moon by the sun for the purpose of this discussion.

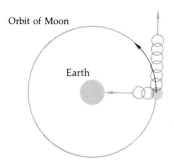

Figure 2-24. The balance responsible for lunar orbit around the earth. If the earth exerted no gravitational pull, the moon would proceed in a straight line into outer space. If the moon were held stationary and released, the earth's gravity would pull it to the earth. These two competing processes produce an orbit.

travels in a circle. If you stop exerting the force by releasing the string, the weight flies off; it continues moving, but not under your influence.

Newton showed that the laws of motion could explain planetary motion if the force acting on the planets were gravity. In fact, using Newton's Laws and the hypothesis of universal gravitation, it is fairly easy to derive mathematically the other laws Kepler had deduced from laborious examination of Tycho's measurements. Combining Kepler's Third Law and the Law of Universal Gravitation, we find that the precise relation between the period and the distance of a planet depends on the mass of the sun (about 2×10^{33} grams). This is the basis of all fundamental mass determinations in astronomy. The orbit of the earth around the sun tells us the mass of the sun; the orbit of the moon around the earth tells us the mass of the earth. The mass of any body with a satellite can be measured by studying the orbit of the satellite.

Newton's explanation of the motions of celestial bodies was not immediately accepted, because it supported and extended the already unpopular work of Copernicus, Galileo, and Kepler, all of whom maintained that the earth moves. Several mathematicians did make calculations that supported Newton's ideas, but the most striking confirmation came in 1758 with the return of a comet, as predicted by Edmund Halley, who based his computations on Newton's theory. In fact, a striking aspect of this successful proof of the theory of gravitation was that the comet did *not* return at the exact time predicted by Halley. Other astronomers who studied Halley's fairly simple calculations realized that, according to the law of gravity, not only would the comet be attracted to the sun, but significant forces would act on it because of Jupiter and Saturn. Taking these into account, in addition to the gravitational force of the sun, they correctly predicted the comet's return.

Halley's comet reappeared in 1758; the law of gravity dates from 1684. During the intervening period, there were nagging doubts about both gravitation and the motion of the earth. Newton could not explain *why* gravitational force exists. He hypothesized the properties of gravity and showed that they could account for the planetary motions, but he could not explain how objects exerted forces across a distance in space.

THE EARTH'S MOTION—SOME OBSERVABLE CONSEQUENCES

A crucial test of the earth's motion involves the *parallax* of the stars. Their directions should be slightly different as seen when the earth is on opposite sides of its orbit, because they are viewed from a different position (this does occur; see Chapter 10). Tycho had searched for it, but the effect is so small that it was not until the 1830s that it was detected. The difficulty lay in the lack of appreciation of the vast distances to even the closest stars.

The motion of the earth in orbit gives rise to the small difference between the solar and sidereal days, as described earlier in this chapter. This difference can be traced to the fact that during the passage of one

Figure 2-25. Solar versus sidereal days. At noon on Day 1 the sun and a hypothetical reference star are directly overhead. While the earth rotates on its axis, its orbital motion also carries it to the position marked Day 2. A sidereal day has elapsed when the reference star is overhead. The sun, however, is not in the same direction at that time, and the earth must rotate through the extra angle (as marked) to complete the solar day.

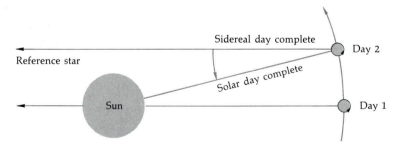

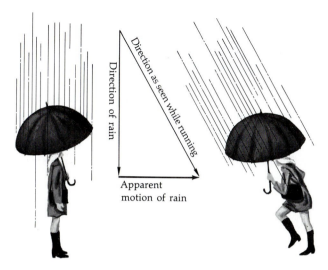

Figure 2-26. Aberration diagram showing that the direction from which raindrops appear to fall depends on the motion of the observer.

day (as determined by the earth's rotation), the earth has moved along in its orbit and, by consequence, the sun is seen at a slightly different position in the sky. This apparent motion of the sun accounts for the 4-minute difference in the lengths of the two types of day (see Figure 2-25). Note, however, that a real motion of the sun around a fixed, rotating earth would also explain it.

An additional fact easily explainable only if the earth moves is the phenomenon of *aberration*, which was discovered in 1728 by the English astronomer James Bradley. Bradley was trying to measure the parallax of stars, but his equipment was inadequate. Instead, he found an effect that had an opposite behavior to that expected for parallax. When the earth was situated favorably for observation of the shift due to parallax (see Figure 10-1), none was observed. When no shift due to parallax was expected, an easily measured shift of 20 seconds of arc on either side of the average position of a star was observed. The observations were somewhat perplexing until Bradley noticed one day that the direction in which the flag of a ship flew depended not only on the direction of the wind but also *on the motion of the ship*. This led him to understand that aberration was caused by the motion of the earth.

The effect of aberration can be visualized by considering the direction of falling raindrops. Suppose there is no wind; then the raindrops are falling vertically. However, if we run through the rain, we must tilt the umbrella forward to avoid getting wet because the running motion causes the rain to seem to fall at an angle (Figure 2-26). Thus, to have aberration, we must have both the motion of the objects to be observed and the motion of the observer. We see stars by means of their "light particles," or *photons*, which travel at the speed of light. The apparent directions from which the photons come are changed by the earth's motion, just as the raindrops' direction of origin seems to depend on the runner's motion. Thus, the aberration of starlight depends on the earth's speed and the speed of light. In fact, from the observed 20 seconds of arc displacement and the known speed of the earth in its orbit, it is possible to estimate the speed of light.

3

Our Planet Earth

Nine worlds, or *major planets*, exist in the vicinity of our sun. Among them, the earth, being nicely close at hand, offers unique opportunities for investigating the properties of a planet. The description and definition of our world has been a prime concern of science and philosophy, and the results of geological studies and exploration have given us a fair idea of its nature and the processes that determine its major surface features, including the oceans, continents, mountains, and volcanoes. In the present era of exploration of the moon by astronauts and investigation of the planets by unmanned spacecraft, the techniques and knowledge developed in studies of the earth are finding immediate applications. Similar characteristics of the earth and planets are providing clues to their common origin, while the differences among the worlds indicate the varied forces at work on their surfaces and interiors.

In this chapter we concentrate on the surface and interior of the earth. Its atmosphere is discussed in Chapter 8, together with those of the other planets.

The earth's shape and size once were matters of speculation. Aristotle (384 B.C.–322 B.C.) maintained that the earth was spherical. His opinion, although generally shared by scientists, was not universally accepted. In fact, popular sentiment for the flat earth idea existed until the actual circumnavigation of the globe in the sixteenth century dispelled any reasonable doubt that the earth was round.

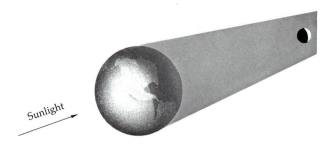

Sunlight

Figure 3-1. Sketch of a lunar eclipse. The earth's shadow is about three times the size of the moon at the moon's distance. If the shadow were centered on the moon, the entire earth-facing side would be dark. The circular shape of the earth's shadow is clearly visible, however, as it moves on and off the lunar disk.

MEASURING THE EARTH

Aristotle had philosophical reasons for arguing that the earth is round, but he also had evidence, in the form of the curved edge of the shadow cast by the earth on the moon (Figure 3-1) at the time of a lunar eclipse. The realization that the earth is a sphere led to the first accurate measurement of its size.

In Egypt, Eratosthenes (276 B.C.–194 B.C.) practiced geometry in its original sense—the measurement of the earth. Studying the shadow cast on a sundial at Alexandria on a certain day, he determined that the sun was not overhead at noontime but was one-fiftieth of a circle (slightly more than 7 degrees) south of the zenith.[1] He found that, on the same day, the sun was at the zenith at noon in the town of Syene (near modern Aswan). This was proven by the observation that the sun lit the water at the bottom of a deep well without shadowing any part of the well walls. Eratosthenes reckoned that if the earth were a sphere, the arc length between the two towns must be one-fiftieth of the circumference of the sphere. Syene was known to be a distance of 5,000 *stadia* south of Alexandria (surveyors had paced off the distance). Then the circumference of the earth must be 50 × 5,000, or 250,000 stadia.

How good was Eratosthenes' measurement? Certainly, the main achievement lay in his reasoning, that is, in originating the method to measure the earth, as illustrated by Figure 3-2. In fact, we don't really

---

[1]Zenith: the sky position directly above the observer.

know how accurate the result was, because we are not sure of the length of the *stadium* unit. According to one historian's estimate, it was equal to 157.5 meters (517 feet). If this is correct, then Eratosthenes' measurement corresponded to 157.5 × 250,000 meters, or about 39,400 km (24,500 miles) in modern terms. This agrees very favorably with the modern value of about 39,900 km (24,800 miles) for the circumference of the earth measured along a circle passing through the North and South Poles. However, the good agreement must be regarded as fortuitous, not only because of our uncertainty in the value of the stadium, but also because Syene was not exactly due south of Alexandria, and the sundial and surveyor pacing techniques were only rough methods.

It is important to recognize that Eratosthenes' method depended on the assumption that the sun is at a very great distance from the earth, so the sun's rays are virtually parallel as they fall on the earth. Such an assumption is valid when the distance to the light source is much greater than the separation of the two spots on earth (Figure 3-3a). This is certainly the case for Eratosthenes' work, because the distance from the earth to the sun is now known to be about 150 million km (93 million miles). We note for future reference that, as the stars are all much more distant from us than the sun, the light rays that strike the earth from a given star are virtually parallel.

Note that if the earth were flat and the sun were at a great distance, then when the sun is overhead at one place, it would be overhead everywhere (Figure 3-3b). This would contradict Eratosthenes' determination that the sun was at different positions in the sky as seen from the two towns at the same time. However, despite this result, despite the teachings of Aristotle, and despite such direct evidence of the senses as

Figure 3-2.   The geometry of the sun, Syene (now Aswan), and Alexandria on June 22 as used by Eratosthenes to determine the earth's circumference.

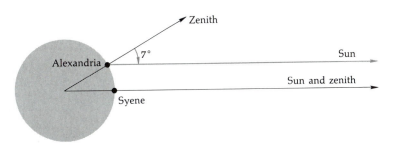

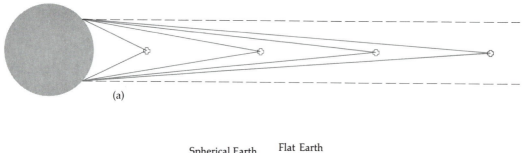

(a)

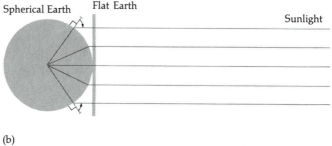

(b)

Figure 3-3. (a) Light rays become more nearly parallel for objects farther and farther away. The dashed lines are parallel, corresponding to a very distant object. (b) Side view of an imaginary flat earth and a spherical earth. The sun is shown when it is overhead on the flat earth; note that it will then be overhead *everywhere*. This is *not* the case on the spherical earth, and the angle between the overhead direction and sun's rays is indicated for two cases.

the sinking of a ship on the horizon as it sails out to sea, the idea of a flat earth persisted. Drawings warned the unwary navigator against sailing too close to the edge lest he fall off to his destruction! The many voyages of exploration in the sixteenth century finally put an end to this myth.

The curvature of the earth's surface is most easily seen on photographs taken from space, such as Figures 1-4 and 17-1. In fact, the earth is not exactly spherical but has a slightly larger radius at the equator (6,378 km) than at the poles (6,357 km). Compared to the radius at the pole, the equatorial radius is 21 km, or 13 miles, larger. This aspect of the shape of the earth is called the *equatorial bulge.* In addition, the earth and particularly the oceans are distorted by the gravitational pulls of the moon and the sun. These pulls cause *tidal bulges* and the associated phenomena of tides in the ocean.

THE AGE OF THE EARTH

Modern geological science began about a century ago, and the determination of the age of the earth was one of its fundamental problems from the outset. Besides the obvious geological interest, this datum is of great importance astronomically as a milestone in the history of the *solar system* (the sun and the planets, comets, and other objects that accompany it in space), and is of concern biologically for determining the maximum time that was available for evolution of life on earth.

The age of the earth has been estimated in a variety of ways at different times. Among the early calculations was that made by James Ussher of Armagh, Ireland, in the mid-seventeenth century. He simply added up the number of years given by a literal interpretation of the succession of human generations recorded in the Bible. In this way, he found that the Creation occurred in the year 4004 B.C. Soon this estimate was refined to 9 A.M., October 23, 4004 B.C.! Thus, the age of the earth could be reckoned to within a day, and according to this theory it is now approaching 6,000 years.

A greater age was derived on the basis of the earth's surface temperature by Lord Kelvin in the late nineteenth century. The temperature increases as one goes down into the earth. Hence, he considered that the deep interior could be molten, and that at one time in the past the entire earth was molten. From the known rate of heat loss as measured in deep mine shafts, he calculated the time required for the earth's crust to cool to its present temperature. (He assumed that there was no process that might have provided additional heat to the earth during the cooling period, and we will return to this assumption later in this chapter.) In this way, Kelvin estimated the age of the earth at 40 million years. This was a staggeringly large value as thinking went in his time. Kelvin himself considered the age a little too high because of simplifications adopted in making the computation. However, it was actually too low.

During the last half of the nineteenth century, geologists began to piece together an overall picture of the forces at work on the surface of the earth. There are, of course, many processes that are continually changing the surface. The earth is ground down by erosion; wind, water, and glaciers chip away at the mountains and highlands; and much of the resulting debris is ultimately carried into the oceans in the form of silt. On the other hand, there are mountain-building processes that cause the earth's crust to be uplifted or buckled into mountain ranges. On the North American continent the Rocky Mountains have been formed relatively recently (see Table 3-1), while the much older Appalachian Mountains have been extensively eroded. Mountain building also occurs through volcanism, as in the case of the Hawaiian Islands.

Table 3-1. Geologic Time Scale

Era	Period	Geological Event	Principal Fossils	Life Development	Years Ago
Cenozoic (recent life)	Quaternary (an addition to a three-part classification proposed in the 18th century)	Glacial activity in the Northern Hemisphere	Animals and plants similar to those living now	Humans	
					3 million
	Tertiary (third, from the 18th century classification)	Alps and Himalayas rise			
		Extensive erosion in Appalachians and Rockies	Mammals and flowering plants	Grasses become abundant	
		Climates warm, jungles widespread		Horses first appear	
		Appalachians reelevated			
					65 million
Mesozoic (middle life)	Cretaceous (from the Latin term for chalk)	Mountain building in Rockies		Extinction of dinosaurs	
		Seas invade parts of North America			
		Appalachians greatly reduced by erosion			
	Jurassic (Jura Mountains, Europe)		Conifers, dinosaurs, other reptiles	Birds and mammals first appear	
					150 million
	Triassic (from a time when this period was divided into three parts)	Extensive deserts in North America		Dinosaurs first appear	
					200 million

Table 3-1. (Continued)

Era	Period	Geological Event	Principal Fossils	Life Development	Years Ago
Paleozoic (ancient life)	Permian (after the Russian city Perm)	Appalachians complete their initial development			250 million
	Carboniferous (coal bearing); often split into Pennsylvanian and Mississippian periods	Ice age in Southern Hemisphere	Amphibians Spore-bearing land plants	Coal-forming swamps Reptiles appear	300 million
	Devonian (Devonshire, England)			Amphibians appear	350 million
	Silurian (ancient British tribe, the Silures)		Fishes		450 million
	Ordovician (ancient British tribe, the Ordovices)	More than 60% of North America covered by seas		First vertebrates (fishes)	500 million
	Cambrian (Cambria, the medieval name for Wales)	Seas invade North America	Marine plants and marine invertebrates (trilobites)	Marine invertebrates form first abundant fossil record	600 million
Precambrian		Mountain building in central North America		Primitive marine plants and invertebrates; one-celled organisms	3400 million
		Formation of the Earth			4500 million

As early as 1785, the Scottish geologist James Hutton had proposed the idea of *uniformitarianism* in natural processes on the earth. This theory was advocated and extended nearly half a century later by the English geologist Charles Lyell in a textbook, *Principles of Geology,* that had great influence on the subsequent development of the field. Simply stated, these men argued that the same processes of erosion and mountain building that we find now have been shaping and reshaping the earth for a very long time. This suggested that by carefully studying these processes we could infer much about the past history of the planet and the time required to bring its surface to its present state. The geological evidence indicated that most areas on the surface had been through the eroding and uplifting cycle several times. As far as the appearance of the geological features was concerned, there was no trace of a beginning, and the earth's surface might well be ageless. (Modern estimates based on silt flow in American rivers indicate that, on the average, erosion lowers a region by roughly a third of a meter in 5,000 years. Thus, a plateau 3,000 meters (or 10,000 feet) high will be worn away by the processes of erosion in 50 million years.)

Hutton and Lyell's uniformitarian view of the earth's evolution was not generally accepted in their time. The opponents of their view, called *neptunists,* held that the surface of the earth had remained essentially unchanged since Noah's flood. When investigation established the existence of distinct layers of rock *(strata),* the neptunists argued that such layers were sedimentary deposits from earlier catastrophic floods, of which the Noachian deluge was the last.

It was not geology, however, but the young science of nuclear physics that was to settle the question of the age of the earth. This science was born in 1896 when Pierre Becquerel discovered the radioactivity of uranium. Within a few years Lord Rutherford realized that uranium, through the process of radioactive decay (emission of particles from atomic nuclei), gradually changes into lead. The most common form of uranium decays at a rate such that one-half of the uranium in a given sample will have become lead after 4.5 billion years have elapsed. (This length of time is called the *half-life.* After 9 billion years, or two half-lives, have elapsed, only one-fourth of the uranium remains; three-quarters of it have become lead.) Thus, it is possible to determine the age of a mineral, defined as the time since it solidified, by measuring the relative amounts of lead and uranium that it contains. The oldest rocks found on earth have ages of about 3.7 billion years, as determined by the radioactive dating method. Other elements besides uranium can be used for radioactive dating. The half-lives of these substances range from the 4.5

billion years of U^{238}, the common form of uranium, to 5,600 years for C^{14}, a rare form of carbon.[2] The same technique has been applied to *meteorites* (rocks that were in orbit around the sun before striking the earth—see Chapter 7), and has revealed ages of about 4.5 billion years. Some of the soil and rocks brought back from the moon by Apollo astronauts are as old or even slightly older than this. In round figures, we can adopt 5×10^9 years as a good modern estimate for the age of the solar system.

The discovery of natural radioactivity not only led to the development of the best technique that we have for measuring the age of the earth, but it also explained why Lord Kelvin's estimate of 40 million years for this age fell woefully short of the mark. Kelvin did not include in his theory any *continuing* source of heat inside the earth. But the earth contains radioactive materials, and, as these materials decay, the particles and radiation that they release heat their surroundings. Because of the heat released by this process, the earth took a much longer time to cool than Kelvin suspected (the principal contributors to this heating are the elements uranium, thorium, and potassium).

THE GEOLOGICAL TIME SCALE

It is possible to deduce or reconstruct much of the earth's history by studying the traces (fossils) of ancient life forms that are found in rocks. These fossils are formed in several ways. Through the process of *permineralization*, the tiny voids in porous bone or other skeletal material are penetrated by mineral-bearing water; the minerals are deposited and gradually fill the voids. Other types of organic material, such as wood, may be slowly replaced by mineral deposits as the wood tissue dissolves in the water. A third type of fossil consists of an impression, such as an animal footprint or the outline of a leaf. Sample fossils are shown in Figures 3-4 and 3-5.

Rock layers that may be located far apart on earth often bear similar fossils, and thus their correlation in age can be established. For example, fossil trees are found in the coal-bearing strata that date from the so-called Carboniferous Period, whereas certain types of dinosaur fossils are found in layers of the Triassic Period (see Table 3-1).

The relative ages of rock layers are established through the principle of *superposition*, namely, that the lower layers must be older than the upper layers in any set of strata that has not been folded over by

---

[2]The superscripts in U^{238} and C^{14} refer to the isotope atomic weights; this subject is discussed in Chapter 5.

Figure 3-4. A fossil fern. (Courtesy of the American Museum of Natural History.)

Figure 3-5. Trilobites, fossil marine invertebrates. (Courtesy of the Smithsonian Institution.)

mountain-building processes. Not all strata in a time sequence are necessarily present at a given location (for example, a particular layer may have been eroded away in some regions). On the other hand, the layer that was destroyed at one place may have been preserved elsewhere. Through detailed comparison of rock strata and fossils found in various parts of the globe, it has been possible to establish the sequence in which major geologic events occurred during a substantial part of the earth's history, as summarized in Table 3-1.

The periods listed in Table 3-1 are further divided by geologists into epochs to indicate more precisely the times at which various prehistoric organisms existed and the various rock strata formed. Note that the Precambrian Era occupies seven-eighths of the total geological time scale. Unfortunately, fossils are hard to find in the Precambrian rocks; the primitive organisms of that era appear to have been mostly bacteria and algae.

The order of the periods in the geological time scale of Table 3-1 was determined from fossils and geological studies; the actual ages came from radioactive dating. Clues to the origin of the terms are given in parentheses; some of these are derived from the names of the places where particular rock layers were first studied.

The most familiar example of the physical evidence for the ancient history of the earth as portrayed by strata of different ages and origins is the Grand Canyon in Arizona. The Grand Canyon is the deepest portion of the longer and generally less spectacular canyon of the Colorado River, and it was cut by the river as geologic forces *raised* the terrain, a point that was first realized by the explorer John Wesley Powell a century ago.[3] Its numerous rock layers correspond to geological periods dating back at least 2 billion years. See Figure 3-6, where the numbers correspond to the layers labeled in Figure 3-7.

INTERIOR OF THE EARTH

There is actually very little of the earth that we can see or touch. Although its mean radius extends nearly 6,370 km (3,960 miles) from the center to the surface, our direct experience goes down only a few kilometers through caves, deep mines, and well drillings. Our knowledge of the interior is mostly based on inferences from earthquake measure-

---

[3]Powell's deductions can be summarized like this: since the walls of the Grand Canyon are higher than those of the adjoining portions of the canyon of the Colorado, the terrain of the Grand Canyon must have been uplifted gradually after the downcutting had begun, or else the river would have had to run uphill. One can think of other explanations for the Grand Canyon, but geological studies since Powell's time show that his idea was correct.

Figure 3-6. View of strata exposed east of Bright Angel Creek on the north side of the Grand Canyon. Numbers correspond to layers labeled in Figure 3-7. (From *Geology Illustrated* by John S. Shelton. W. H. Freeman and Company. Copyright © 1966.)

ments. The strength of a quake, the nature of the vibrations, and the times of arrival at each of many recording stations are analyzed to deduce the properties of the regions beneath the surface through which the vibrations or *seismic waves* must pass.

The earth's interior is studied by the techniques of the geophysicist, but the results are important to astronomers, because there may be some properties in common with those of other planets. In addition, the nature of the interior gives us clues about the manner in which the earth, and presumably other planets, was formed. In fact, the techniques by which earthquake phenomena are analyzed by the geophysicist have since been applied on the moon. Seismic detectors installed by the astronauts sent data back to earth via radio transmission and allowed us to deduce the

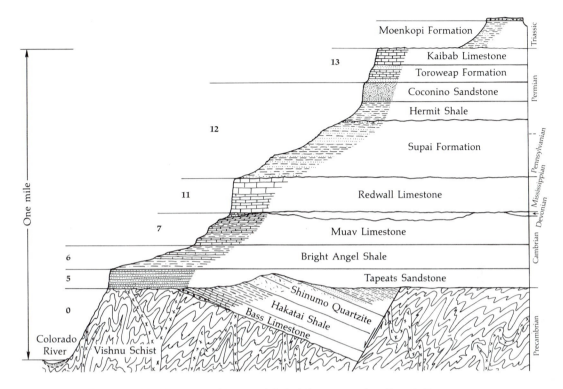

Figure 3-7. Diagram showing labeled layers in the Grand Canyon (see Figure 3-6). Names at the far right relate the ages of the rock layers to the geological time scale discussed in this chapter. (After *Geology Illustrated* by John S. Shelton. W. H. Freeman and Company. Copyright © 1966.)

properties of the lunar interior. Similar instruments were sent to Mars on the Viking spacecraft.

The disturbances produced by earthquakes travel across the globe in three basic types of waves: surface, shake, and push-pull. *Surface waves* are analogous to the ripples made by dropping a rock or marble into a pond or bathtub. In this case, the wave effect travels for long distances across the water's surface, but individual drops of water do not. Rather, the individual drops are vertically displaced as the wave passes by. (This agrees with our experience; a bather floating in the ocean bobs up and down, not moving toward shore, as a wave passes him en route to the shore. When the bather does drift, it is the result of a current, not the waves.) When the water is displaced above the average level, the *crest* of the wave is passing; when displaced below the average level, the water is in the wave *trough*. This type of wave phenomenon is called *transverse*, because the motion of the material is at right angles to the direction in which the wave is traveling. Surface waves produced by earthquakes are not important for probing the earth's interior, but their effects must be recognized and discarded from the earthquake data.

Shake waves are produced by a side-to-side (or up-and-down) shaking motion and thus are also transverse. This kind of wave can be simulated by taking the end of a clothesline and shaking it up and down. The individual parts of the rope move up and down, but the wave itself, produced by the motion of one's hand, moves horizontally along the rope. A transverse wave can also be generated by taking one end of a rope (other end tied down) lying on a smooth surface and shaking it from side to side, rather than up and down as in the clothesline example. Unlike surface waves, shake waves can travel inside the earth, but they do not pass through liquids.

Push-pull waves are common; sound waves are a familiar example. The production of push-pull waves can be visualized through the use of a tuning fork. When it is sounded, the prongs move back and forth, and the adjacent air is compressed once during each vibration of the tuning fork. This produces a series of regions where the density is alternately more or less than the original value. The motion of individual particles is back and forth along the direction of travel; hence, they are called *longitudinal* waves, and because the air is compressed, they are also often called compressional waves. As we know from experience with sound, this kind of wave can travel through solids, liquids, and air. The nature of shake and push-pull waves is illustrated in Figure 3-8.

The speed of a wave depends on the properties of the medium through which it travels. If the nature of the medium changes, both the wave speed and direction may be changed. Where there is an abrupt change (discontinuity) in the properties, the effect on the wave is more drastic, and it may, for example, be reflected.

Waves caused by earthquakes are routinely recorded by the seismic observatories located at many spots on the surface of the earth. Consider then the time of arrival of the P (push-pull) and S (shake) waves plotted against distance from the location (*epicenter*[4]) of an earthquake. This dis-

---

[4]The *epicenter* is defined as the point on the earth's surface directly above the actual rock movement that generates the earthquake (Figure 3-9). The measurements of seismic waves, as recorded at different locations, are used to locate the epicenter, as shown in Figure 3-10.

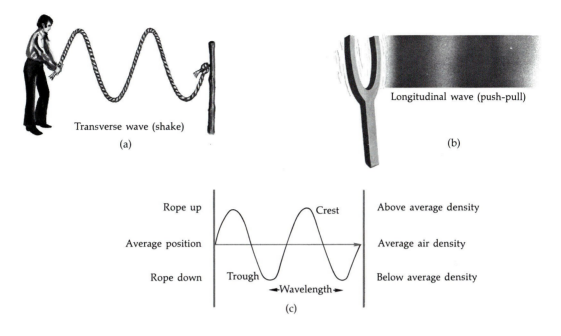

Transverse wave (shake)

(a)

Longitudinal wave (push-pull)

(b)

Rope up | Crest | Above average density

Average position | Average air density

Rope down | Trough | Below average density

◄Wavelength►

(c)

Figure 3-8. Schematic illustrations of (a) transverse waves (shake) and (b) longitudinal waves (push-pull) and (c) their graphical representation.

Figure 3-9. Sketch showing the location of the epicenter (on the earth's surface) above the focus of earthquake waves caused by slippage along a fault. (From *Principles of Geology,* 4th ed., by J. Gilluly, A. C. Waters, and A. O. Woodford. W. H. Freeman and Company. Copyright © 1975.)

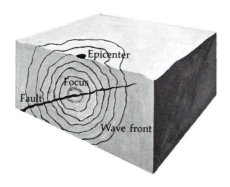

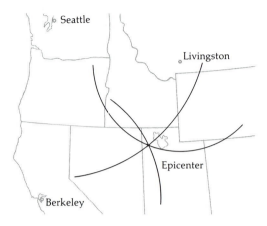

Figure 3-10. Sketch showing how seismic observations made at three distant locations (Seattle, Livingston, and Berkeley) are used to find an epicenter in Utah. Observations at a single station determine the distance to the epicenter and, hence, the knowledge that the earthquake occurred on the edge of a circle of known radius. Data from three or more stations determine an accurate location. (From *Principles of Geology,* 4th ed., by J. Gilluly, A. C. Waters, and A. O. Woodford. W. H. Freeman and Company. Copyright © 1975.)

tance is measured along the surface of the earth, but the P and S waves travel through the earth. Such a plot is shown in Figure 3-11.

Both the S and P waves can normally be detected as far away as 11,200 km (7,000 miles). The S wave disappears at this point, and as these waves cannot pass through liquid, we infer that a liquid region must exist inside the earth. Beyond 11,200 km, the P wave is still observed, but its arrival times at various locations indicate that it has passed through regions where the wave velocities are different. These are the kinds of seismic observations that reveal the presence and rough properties of distinct zones within the earth. The geometry used in this technique is shown in Figure 3-12.

The outermost layer of the earth is called the *crust;* it extends to depths of about 35 km (22 miles) beneath the continental surfaces. Under the oceans, however, the crust is only about 5 km (3 miles) thick. We have never sampled the rock from below the crust, but recently a deep-drilling experiment with this objective has been under way in the Soviet Union.

The existence of a distinct layer (the *mantle*) beneath the crust was proven by the work of a Yugoslavian geophysicist, A. Mohorovičić, in

1909. He studied the P and S waves from relatively nearby earthquakes and found that their speeds changed at a certain place beneath the surface. We now call this the Mohorovičić discontinuity, or *Moho* for short. It marks the boundary between the crust and the mantle, and lies at depths of from 5 km (3 miles) to more than 50 km (31 miles) at different points on earth.

The mantle (see Figure 3-12) extends down to a depth of almost 3,000 km (1,900 miles). The S waves do not penetrate below that depth, which indicates, as mentioned above, that a liquid zone must exist. The region below this discontinuity is called the *core*. Its outer part is certainly liquid, but a change in the traveling characteristics of P waves 2,100 km (1,300 miles) below the boundary of the core reveals the presence of another zone, the *inner core*, which extends to the center of the earth and is probably solid. A cut-away model with some representative seismic wave paths is shown in Figure 3-13.

The total mass of the earth is about 6×10^{27} grams (10^{21} tons). However, when we examine the rocks of the crust, we find that they are much

Figure 3-11. A plot of travel times for P and S seismic waves versus distance across the earth's surface. The S waves are not recorded farther than about 11,000 km from the epicenter. The P waves observed past this distance have penetrated the core. The split in the curve for the P wave beyond about 16,000 km is caused by the earth's solid inner core.

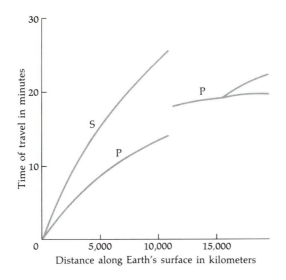

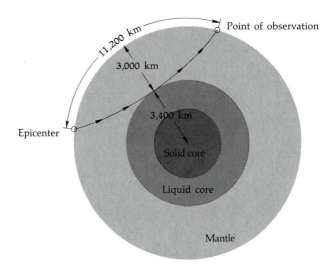

Figure 3-12. Schematic diagram of an S-type seismic wave, as produced and observed respectively at two widely separated locations on the surface of the earth. The example, which is drawn roughly to scale, shows the wave (curved line with arrowheads) that just grazes the liquid outer core. If an S wave were observed farther than 11,200 km from the epicenter of an earthquake, it would pass closer than 3,400 km to the earth's center. No such waves are observed (Figure 3-11). Since S waves cannot propagate through liquid, we conclude that the boundary of the liquid core is located as shown.

too light to make up the bulk of the earth and cannot account for its total mass. Therefore, much heavier material must exist at great depths, and in fact, the core is probably composed largely of iron. The discovery that the deepest part of the earth consists of heavier matter than we find in the outer regions is very important in our efforts to understand the origin of the earth. It suggests that all or nearly all of the earth was once hot and fluid, so that the heavier material could sink toward the center as the lighter substances floated upward.

THE CHANGING SURFACE OF THE EARTH

As long ago as 600 B.C., the Greek philosopher Xenophanes noted the presence of a kind of clam shell on mountain tops. The shells resembled those of living clams found along the shore, and so he concluded that the sea had once covered parts of the land. This view is now known to be entirely correct. It is clear that the sea position has changed in the past, and that many areas now far inland were once beneath the sea. It is only recently that we have been able to gain some insight into many of these changes and their causes. The earth's magnetism is fundamental to this discussion.

The practical value of the ordinary magnetic compass depends on the fact that the earth has a magnetic field. As we study the orientation of compass needles at different locations, we find that they behave as though they were a giant bar magnet inside the earth. The imaginary magnet would be tilted slightly with respect to the axis of rotation, as shown in Figure 3-14. (The basic field pattern of a bar magnet is shown in Figure 3-15.) It is well known that bar magnets and other similar devices lose their magnetism when heated. Further, the temperatures inside the earth are too high to permit a solid magnet to exist. In fact, the leading theory of the earth's magnetism asserts that it is generated by moving currents of liquid iron in the core.

Figure 3-13. (a) Cut-away model of the earth and (b) some sample seismic wave paths. (After *Structure and Change* by G. S. Christiansen and P. H. Garrett. W. H. Freeman and Company. Copyright © 1960.)

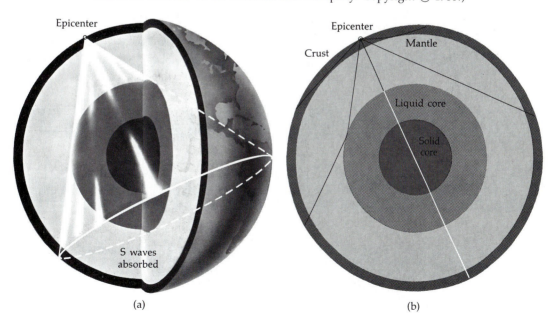

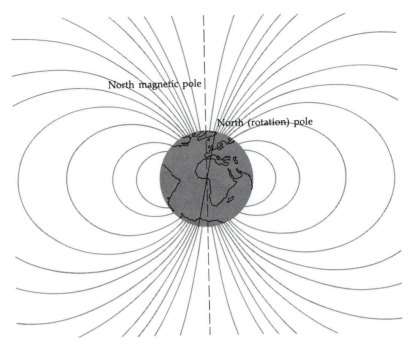

Figure 3-14. Sketch of the earth's magnetic field lines. The magnetic and rotational poles of the earth are separated by about 12° (1,330 km or 830 miles).

Figure 3-15. (a) Sketch of the magnetic field around a bar magnet as revealed by a compass needle. (After *Structure and Change* by G. S. Christiansen and P. H. Garrett. W. H. Freeman and Company. Copyright © 1960.) (b) The magnetic field of a bar magnet as revealed by iron filings. (After *The New College Physics* by A. V. Baez. W. H. Freeman and Company. Copyright © 1967.)

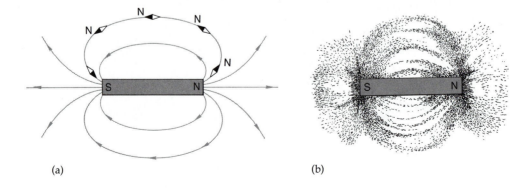

(a)   (b)

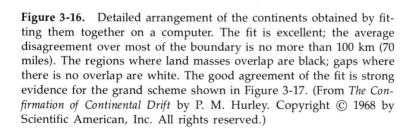

Figure 3-16. Detailed arrangement of the continents obtained by fitting them together on a computer. The fit is excellent; the average disagreement over most of the boundary is no more than 100 km (70 miles). The regions where land masses overlap are black; gaps where there is no overlap are white. The good agreement of the fit is strong evidence for the grand scheme shown in Figure 3-17. (From *The Confirmation of Continental Drift* by P. M. Hurley. Copyright © 1968 by Scientific American, Inc. All rights reserved.)

As rocks form, they are magnetized in the direction of the earth's magnetic field; it is as if many tiny bar magnets within the rock line up while it is forming, and are then locked in place when it has solidified. By careful measurements of this *fossil magnetization* in rocks, one can infer the positions of the earth's magnetic poles at the different times of rock formation.[5] The surprising result is that these poles seem to have moved around. However, an equivalent, very plausible interpretation asserts that they have remained in place, but the continents on which we find the rocks have drifted.

The idea of continental drift goes back at least to the time of the British philosopher Francis Bacon (1561–1626). It was largely stimulated by the remarkable fit between the coastlines of Africa and South America, as shown in Figure 3-16. A geologist once commented that "if the fit between South America and Africa is not genetic, surely it is a device of Satan for our frustration."

There are other kinds of supporting evidence. The presence of ancient glacial deposits suggests that South Africa once occupied a different position, and the locations of coral reefs also reveal the occurrence of major climatic changes that may have been due to continental drift. Coral grows in clear, warm, marine waters, but some fossil reefs are found in areas where the water is very cold. Likewise, fossils found in Antarctica show that it once had a much warmer climate. Finally, the close resemblance of certain fossils found on different continents indicates that the continents may once have been joined or at least were close together.

The evidence favoring continental drift is clearly widespread. Several efforts at reconstruction of the positions of the original land masses have been made. One such attempt (Figure 3-17) shows a set of hypothetical original positions.

Although the evidence for continental drift is reasonably strong, it poses a further, more difficult problem: What causes the drifts? Surely some great force must exist to move these huge land masses.

It was long thought that most of the important geological phenomena took place on the continents and the adjacent continental shelves, and that the floor of the deep oceans, beneath a thin layer of sediments, might be representative of the oldest part of the crust. In recent years, however, a variety of oceanographic observations (including the dating of sea-floor

---

[5]Rocks form in three basic ways: (1) by the cooling of liquids that come up from far beneath the surface (as in volcanic eruptions); (2) by the action of heat, pressure, and chemical effects at lesser depths beneath the surface; and (3) through the deposition of eroded material (sediments) on ocean and lake floors, where the material gradually becomes cemented together.

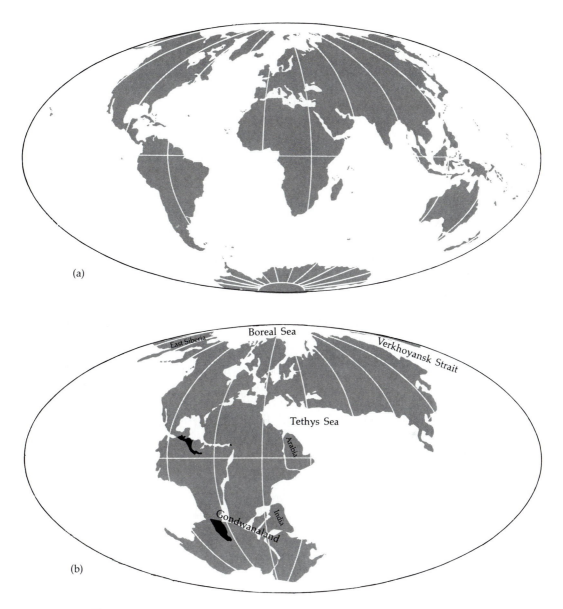

Figure 3-17. (a) The present-day distribution of continents as compared with (b) the single supercontinent that may have existed some 150 million years ago. Dark areas indicate overlaps. (From *Continental Drift* by J. T. Wilson. Copyright © 1963 by Scientific American, Inc. All rights reserved.)

rock samples) have profoundly changed this picture. Many geologists now conclude that quantities of fresh rock are slowly emerging from the mantle through rifts in the deep sea floor and are spreading out laterally, forming new crust that advances toward the continents at rates of up to several inches per year. It is believed that this *sea-floor spreading* is related to the drift, and that the continents consist of blocks of solid material (plates) which slowly "float" across the heavier rock of the deep mantle. The plates consist of material of the crust and upper mantle. The zone of solid material forming the plates is called the *lithosphere*. The region of the deep mantle beneath them, in which the rock is soft due to heat and pressure, is the *asthenosphere*.

The sea-floor spreading itself must originate in rock flow within the asthenosphere. The seismic wave studies show that the material there resembles a solid, and we usually think of solid rock as a rigid substance, more apt to break than to bend under force. However, rock will gradually soften and yield under the influence of a sufficient load or force, and the rock in the deep mantle, subject to immense forces, has a plastic, deformable character. The influence of rock currents in the asthenosphere provides a natural mechanism for mountain building. Where these currents meet or sink, rock can be piled up into mountains. Alternately, the collision of continental plates could produce mountains by squeezing and resultant buckling of the material. The Himalayas may have been formed in this way when the Indian plate merged into the main body of Asia. The folding and buckling forces that build mountains act at great depths, although it is at the surface that their results are so impressive (Figure 3-18).

Earthquakes and volcanoes commonly occur in the same areas, providing dramatic evidence of the forces involved in shaping the crust. These areas are often marked at the surface by *fault lines*. Perhaps the best known is the San Andreas fault, which passes very close to San Francisco. The plate to the west is slowly moving to the northwest with respect to the other plate. Figure 3-19 shows the effect of slippage along a related fault. Earthquakes occur when the deformation of the rocks exceeds their strength. The crust then fractures, and the plates shift so that the deformation decreases to an amount that the rock can tolerate. Farther to the north, along the Aleutian Island chain of Alaska, the plate boundary is marked by the presence of many volcanoes.

Volcanoes also represent a mechanism for mountain building and for altering the nature of the surface. An example is Parícutin (Figure 3-20), which was seen to form in 1943. This occurs through the build-up of cinder cones and the flow of lava. *Magma*, the molten rock within a volcano, originates in the mantle.

(a)

(b)

Figure 3-18. Two examples of folded strata. (a) Large-scale folds as seen looking northwest near Borah Peak, Idaho. Note the sizes of the folds as compared with the pine trees in the lower portion of the picture. (From *Geology Illustrated* by John S. Shelton. W. H. Freeman and Company. Copyright © 1966.) (b) Small-scale folds near Barranca de Tolimán, Hidalgo, Mexico. (Courtesy of Kenneth Segerstrom, U.S. Geological Survey.)

Figure 3-19. Displacements along a fault line are seen in an orange grove near Calexico, California. The view is toward the west and the displacement amounted to about 4.4 meters (14.5 feet) in a northwesterly direction. (From *Geology Illustrated* by John S. Shelton. W. H. Freeman and Company. Copyright © 1966.)

Most earthquakes and volcanic activity on earth occur along the boundaries between drifting plates. Thus, volcanoes are predominantly distributed in long arcs across the earth's surface. The earth's lithosphere extends to a depth of about 100 km (62 miles) below the continents, and to about 60 km beneath the sea. Mars has many fewer volcanoes than earth, but they are individually much larger than the average terrestrial volcano. It is theorized that the Martian lithosphere is much thicker than the earth's, and thus has not fractured into plates; without drifting plates, long chains of volcanoes don't form on Mars.

Finally, one other major agent in changing the surface of the earth should be discussed: *glaciation*, the effect of the motion of large masses of ice across the crust. During the ice ages, glaciers covered much of Europe and also the eastern North American continent down to approximately the Missouri and Ohio Rivers (Figure 3-21). The last great retreat of the ice began about 20,000 years ago and continued until about 6,000 years ago. Typically, glaciers move at about 50 meters (160 feet) per year. As they flow across the earth they alter the surface by abrasion, producing valleys, hills of debris, and other topographic features. Isolated glaciers occur today in major mountain ranges as well as the glacial areas of Antarctica and Greenland. Glaciers constitute a large reservoir of water, and as such, their growth or decline can cause dramatic changes in the sea level and thus the position of the seashore. We are currently in an age of relatively little glaciation, but if the remaining ice were melted, sea level would be raised by about 50 meters, enough to flood extensive coastal areas, including many major cities.

Figure 3-20. Parícutin erupting at dawn on February 20, 1944, one year after it first rose through the ground. (Courtesy of Tad Nichols.)

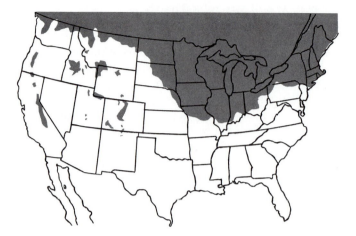

Figure 3-21.   Map showing the extent of glaciation in what is now the United States during the last 2 or 3 million years. (From *Principles of Geology,* 4th ed. by J. Gilluly, A. C. Walters, and A. O. Woodford. W. H. Freeman and Company. Copyright © 1975.)

Scientists have not reached a consensus on the cause of the ice ages, but it seems that rather small changes in the average temperature were involved. It was colder during the last glacial epoch than at present, but probably only by about 10° F. It may be important that the average temperature over the surface of the earth increased by 1° or 2° F in the 100 years following 1850, a date that corresponds roughly to the beginning of the industrial revolution. The gaseous waste (largely carbon dioxide) produced by man's activities might be involved in raising the average temperature of the earth (see the discussion of the "greenhouse effect" in Chapter 8). If so, increased pollution or even the maintenance of the current level of pollution could hasten the melting of the glaciers and thus cause a rise in the level of the sea. However, this effect of pollutant *gas* might be offset somewhat by the *dust* released into the air by industry and agriculture. The floating dust particles reflect some sunlight back into space, thereby reducing the amount of solar energy that heats the ground. Indeed, the average temperature of the earth may have begun to *decrease* in recent years as a result of this effect. Too much dust in the atmosphere could produce another ice age.

CLIMATIC CHANGE ON EARTH AND MARS

Although we do not know the cause of the ice ages, it is possible that these and other climatic changes that have occurred on the earth resulted

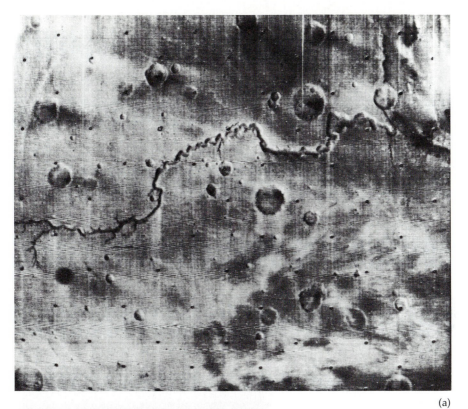

(a)

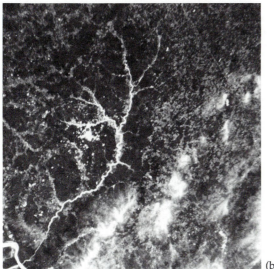

(b)

Figure 3-22. (a) A dry riverbed on Mars, photographed by Mariner 9. (b) A small river in Mississippi, photographed by an Apollo astronaut from space. (NASA)

from variations in the luminosity of the sun. If this is true, then climatic changes would also occur on other planets. Indeed, evidence of past climatic changes has been found on Mars. For example, photos from the Mars-orbiting spaceprobe Mariner 9 revealed that, although all the water on the surface of Mars is now in the frozen state, there once was liquid water there. Dry canyons on Mars, with the winding shapes and branched tributary patterns that characterize riverbeds on earth, are proof of this (Figure 3-22). Thus, at one time, at least briefly, it was much warmer on Mars, and water flowed there.

A series of layers in the Martian polar terrain, as photographed by Mariner 9, also suggests past eras of different climate. However, it remains to be proven that the terrestrial and Martian climatic changes occurred at the same times and arose from the same cause, and it likewise remains to be shown that significant variations in the light of the sun actually occur. An alternative theory ascribes climatic changes to very slow variations in the orbit of a planet and the tilt of the planet's axis with respect to the orbital plane.

4

Evolution and Life—
Terrestrial and
Extraterrestrial

In ancient times, Herodotus recognized certain fossils as the remains of living organisms, and Anaximander and other Greeks discussed some concepts of evolution. Later, the interpretation of fossils as evidence of ancient life forms was strongly opposed in Europe. The existence of fossils was indisputable, and the arguments used to dismiss their significance frequently appealed to the miraculous. Fossils were described variously as accidental "sports" of nature, tricks of the Devil, or as remains of life buried in the Biblical flood of a few thousand years ago. A few persons made significant contributions to evolutionary thinking during the seventeenth and eighteenth centuries, but although they recognized that evolution had occurred, they were unable to account for it or to back up their hypotheses with critical evidence. It remained for a careful observer of nature, Charles Darwin, to recognize in the properties of living organisms evidence for the nature of evolution, and thus to show (as did Hutton and Lyell in geology) that the present is the key to the past.

DARWIN AND NATURAL SELECTION

During the years 1832–1836, Darwin was an unpaid naturalist on H.M.S. *Beagle*, a ten-gun brig of the Royal Navy engaged in surveying Patagonia, the west coast of South America, and some Pacific islands. The voyage included explorations ashore in South America and in the Galapagos Islands off the coast of Ecuador. Darwin made extensive biological and geological observations, which he later analyzed and used as the basis for his theory of natural selection. In fact, he thought about the evidence and accumulated additional facts for more than twenty years, and only when he learned that another scientist had formulated a similar theory was he moved to announce his own conclusions. Darwin's classic work, *The Origin of Species*, was published in 1859.

In Chapter 3 we discussed the evidence for evolution that comes from the study of *fossil* remains of ancient life forms. Darwin's observations during the voyage of the *Beagle* led him to recognize the process of evolution on the basis of the nature of *living* organisms. The Galapagos finches are a famous example.

Thirteen species of finches have been recognized on the Galapagos, and some of them behave in a very unfinchlike manner. The "woodpecker-finch" has a long, pointed beak and preys on insects living in the bark of trees. Lacking the long tongue of a true woodpecker, it uses twigs to dislodge the insects (Figure 4-1). Other finches on the islands have short beaks, and they eat seeds that they find on the ground. Still another type has a beak suited to its habit of feeding on prickly pears.

Figure 4-1. The woodpecker-finch of the Galapagos Islands, where Darwin made a classic study. In this sketch it is using a twig to pry out insects from the bark of a tree. (From *Darwin's Finches* by David Lack. Copyright © 1953 by Scientific American, Inc. All rights reserved.)

In fact, all of the Galapagos finch species are roughly alike, their most significant differences being in the size and shape of the beaks, and the nature of their diets. Different islands among the Galapagos are inhabited by different groups of species. Darwin concluded that "from an original paucity of birds in this archipelago, one species had been taken and modified for different ends." At one time, a single finch species came to the relatively barren Galapagos from the mainland. Living in isolation, the birds on different islands evolved by adaptation and specialization into distinct species, each capable of most effectively harvesting a particular part of the food supply. Probably, no true woodpeckers were around to feed on bark-dwelling insects, and so a finch species evolved that could take advantage of this particular source of food. In the language of the biologist, they filled the *ecological niche* of a woodpecker on the Galapagos. The divergence of the original species of Darwin's finches into a variety of species, each able to survive in a different manner and thus not infringing very much on the resources available to the related species, is the classic example of the process of *adaptive radiation*.

The properties of the Galapagos that made them an excellent laboratory for evolution were isolation and a relative scarcity of animals that could effectively compete with the finches for food, or prey on the birds. But these same two characteristics made the Galapagos unrepresentative of the general situation on the continents. Darwin noted that populations tend to double at characteristic intervals. The human population doubles about every 35 years; a colony of rabbits allowed to reproduce freely in the laboratory will increase at a faster rate, and many insects even faster yet. The question arises: Why isn't the earth covered with rabbits or flies or whatever? This has not happened because each species is in a delicate *ecological balance* with its surroundings. Rabbits and flies have enemies that prey on them; predators, in turn, are limited by the available food supply.

Thus, we can envision a delicate balance with competition among all species. Any slight advantage or disadvantage may be crucial for survival. The climate may change or the food supply may run out; here we encounter the general problem of adaptability to a changing environment. Success for a species is essentially the ability to stay alive *and* to reproduce. A species that can survive the rigors of the environment and the attacks of predators is still doomed if its members cannot reproduce.

Darwin reasoned that reproduction was not perfect, but that changes occur, which we now call *mutations*. The mutations are essentially random; some of them produce species of more efficient competitors (beneficial mutations), but many do not. The *principle of natural selection* states that the organisms with a higher survival probability will eventually dominate the population, as they can best feed themselves and reproduce. Natural selection, or the "survival of the fittest," while not the fundamental cause of evolution, determines the direction of evolution when changes occur.

Despite the evidence of an enormous variety of now-extinct life forms revealed by the fossil record, and despite Darwin's brilliant explanation of evolution as a result of natural selection, his theory met enormous resistance. In large part this was based on the conflict with the Biblical account of the Creation. There was also the rather unpleasant feeling that arose from the idea that we were descended from the apes or related to them through common ancestors.

There were also quite reasonable scientific objections that faced Darwin and his followers. For example, evolution is clearly a slow process, and the critics doubted that enough time had been available. (Remember that it was only in the present century that we determined the true age of the earth.) Another objection was based on the fact that evolution is described as a gradual process, but the purported evolutionary lines of descent of many species lack intermediate forms. Darwin's answer was that the fossil record is incomplete. Fossils may be destroyed by heat, pressure, and chemical processes within the earth, and by erosion. In addition, the conditions necessary to form and preserve fossils may not always exist. Finally, the completeness of the known fossil record also depends on our ability to find and classify the fossils. In fact, some fossils of crucial intermediate stages in evolution have been found; a famous example is the *Archaeopteryx* (Figure 4-2). Birds are believed to have evolved from reptiles, and *Archaeopteryx* (found in Jurassic rocks) is a mixture of bird and reptilian characteristics; it has the tail, brain, and teeth of a reptile and the feathers and feet of a bird.

A key difficulty with the theory of evolution was its failure to explain how a mutation first occurred and how it was passed on to succeeding generations. Darwin could not answer this; the explanation was provided later by the new science of genetics. He also did not explain the origin of life, a problem that chemists and biologists only began to explore in recent decades.

MUTATION AND INHERITANCE

The basic rules of heredity that apply to living organisms were discovered by the Austrian monk Gregor Mendel through experiments with the reproduction of sweet peas. An organism has a large set of traits that are determined by genes in the cells. In a human being, a pair of genes for

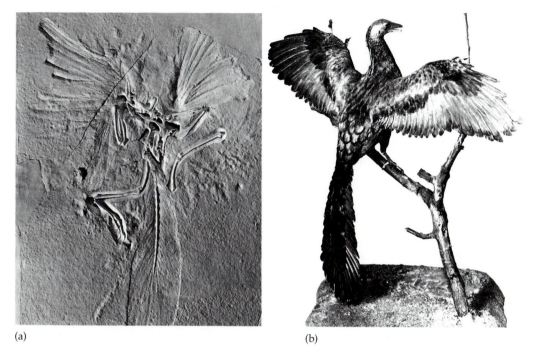

(a)   (b)

Figure 4-2.   (a) Cast of an *Archaeopteryx* fossil. Note the feather impressions. (Courtesy of the American Museum of Natural History.) (b) Model of how *Archaeopteryx* may have looked. (By permission of the Trustees of the British Museum—Natural History.)

each trait is inherited from the parents, one from the father and one from the mother. Genes may be *recessive* or *dominant*. The gene for brown eye color is dominant; blue is recessive. This means that if a person inherits a brown gene from one parent and a blue gene from the other, he will have brown eyes. Only when both genes are blue will the person exhibit the recessive characteristic of blue eyes. The situation is not quite as simple as we have indicated; for example, sometimes several genes will influence a particular trait, and some of these genes may also determine other traits. Mendel's work proved that genes must exist. Much of modern biochemistry has been devoted to determining what genes actually are and how they control inheritance. According to certain rules, genes pass on to succeeding generations the detailed specifications of an organism so that it will bear a clear resemblance to its parents. This detailed specification is called the *genetic code*.

A living cell is composed of a *nucleus*, a surrounding fluid of *cytoplasm*, and the cell walls (Figure 4-3). Biologists long ago determined that structures called *chromosomes*, which are found in the nucleus, must contain the genes. More recently, it has been shown that a substance known as *DNA* (deoxyribonucleic acid), which is contained in the chromosomes, is fundamental to the story of heredity.

The DNA molecule (Figure 4-4) is composed of subunits or building blocks called *nucleotides*. The molecule is very long (in man, DNA is composed of some 10 billion atoms) and consists of two long, corkscrew-shaped strands arranged in a double helix (Figure 4-5). Relatively short strings of the nucleotides in DNA determine inherited characteristics, and thus the genes, whose existence was first deduced from experiments in growing varieties of the sweet pea plant, can now be identified as chains of nucleotides in the DNA molecule.

The DNA molecule has another remarkable and vital property; that is, the ability to self-replicate. At the appropriate time the two sections of the double helix unzip or split down the middle and unwind. At this point the chemical bonds that formerly held the two parts together are unfilled. The two parts float in the cytoplasm, which also contains many nucleotides. Generally, only the proper kind of nucleotide can fill a particular broken bond. Thus, each of the coils of the DNA molecule acts as a mold or template and can copy itself (Figure 4-6) using building-block materials from its immediate surroundings. This process of DNA replica-

Figure 4-3.   View of human cheek cells (treated with silver nitrate) showing the nucleus, cytoplasm, and cell walls. (Courtesy of W. M. Copenhaver.)

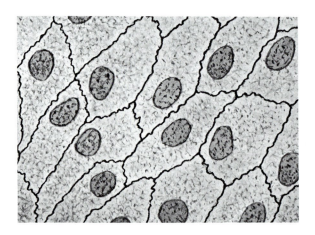

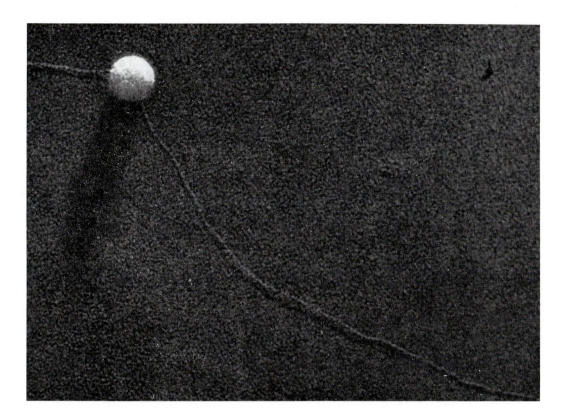

Figure 4-4. A small part of a DNA molecule, as recorded with an electron microscope. The white sphere, a tiny piece of plastic used as a measuring stick to indicate the size of the molecule, has a diameter of 9×10^{-6} centimeter. (Courtesy of Cecil E. Hall, Massachusetts Institute of Technology.)

tion allows a complete set of genetic instructions to be passed to each of the daughter cells.

If some event changes the arrangement of the nucleotides in a DNA molecule, and that molecule is then involved in reproduction, a mutation may result. Thus, changes in the genetic code are the raw material of evolution. The mutations have a variety of causes. First, simple copying errors may occur in the DNA self-replicating process; second, the genetic material may be changed by exposure to cosmic rays,[1] radioactivity (both

---

[1]Fast-moving atomic particles from space; see Chapter 15.

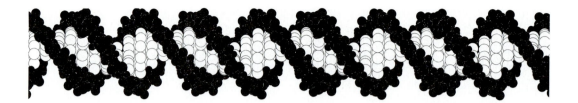

Figure 4-5. A model of the DNA molecule. The black spheres represent the repeating units of deoxyribose sugar and phosphate that give DNA its double-helix structure. The white spheres between the two helical strands represent the nucleotides that carry the genetic code. (From *Gene Structure and Protein Structure* by Charles Yanofsky. Copyright © 1967 by Scientific American, Inc. All rights reserved.)

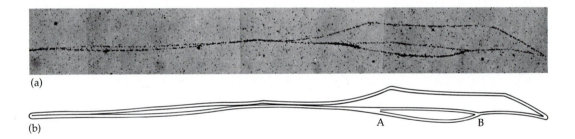

(a)

(b)   A   B

Figure 4-6. (a) Photograph and (b) sketch of a segment of DNA engaged in the replication process; thus far it has duplicated the portion of itself from A to B. (Courtesy of John Cairns.)

natural and artificial), and certain chemicals. The rate at which mutations occur due to the natural causes (copying errors, cosmic rays, mineral radioactivity) is estimated at 1 per 100 million reproductions. It appears that a small mutation rate is more advantageous from the standpoint of natural selection, because most mutations are probably undesirable. Recall that most species are in a state of rather delicate balance with their surroundings. The probability of a favorable mutation can be compared with the odds that your car will be improved if a mechanic removes one part at random from your car and replaces it with a part chosen blindly from the parts shelf. An improvement is not likely unless you have a real lemon!

In summary, Darwin showed that the natural selection of superior characteristics governed the evolution of life. Mendel's work was the basis of the genetic theory that established the rules for the inheritance of traits. Modern biochemistry, centered on the study of DNA, has revealed the detailed way in which inheritance of characteristics is actually controlled.

THE ORIGIN OF LIFE

Is there a scientific basis for the origin of life? Few questions have so great a philosophical importance or have so divided thinking persons. Several hypotheses concerning the origin of life on earth have been made. First, a supernatural event can be invoked. This possibility is unattractive to scientists because it seems more reasonable to accept a natural process if one can be found. In any case, the supernatural is by definition outside the realm of science.

Second, it has been postulated that life came to the earth in the form of spores or microorganisms that drifted through space from some other point in the universe. This *panspermia* hypothesis is not widely believed today. Although there is some dispute over whether or not such spores could survive the harsh environment of space, the theory in any case begs the question—it does not tell us how life began, but only claims that it began somewhere else. The same remark applies to what has been called the "garbage theory"—the idea that life began on earth at some time in the distant past when space travelers from another world visited our planet and contaminated it, whether accidentally or otherwise.

Third, life may have originated in common substances under physically reasonable circumstances. Modern research lends plausibility to this concept of *chemical evolution*.

The cells that make up all living organisms are distinguished by the presence of four major types of substance: *carbohydrates, fats, nucleic acids,* [2] and *proteins.* The modern approach to the problem of the origin of life has been to determine how these organic substances could have arisen on earth. Laboratory experiments have simulated conditions that probably existed during the early history of the earth, and chemicals have been produced that are among the subunits or building blocks of these four kinds of organic material.

The experiments on the origin of life are based on facts and deductions from several fields of science. For example, chemistry tells us that the proteins could not have formed in the presence of much oxygen, al-

---

[2]DNA and a similar molecule, RNA.

though there is a great deal of oxygen on the earth today. On the other hand, we do not find much oxygen when we study the atmospheres of the other planets (Chapter 8), and in fact the various planets of the solar system have different atmospheric compositions. Astronomers believe that the sun and all its planets were formed from the same cloud of matter, so at least neighboring planets, such as the earth and Venus, should have begun with fairly similar atmospheres. As the atmospheres are very different now, we conclude that they may have changed since the planets were formed. In the case of the earth, geology and biology provide the evidence for the occurrence of this *atmospheric evolution*. It appears that most of our oxygen has been produced since life began, and may even be a result of photosynthesis in plants.[3] Most of the other gases in the present atmosphere (and most of the water) have probably been released from the interior of the earth by volcanic activity.

The gas mixture that surrounded the earth during its early history is called the *primitive atmosphere* to distinguish it from the present atmosphere. According to one of the leading theories, the primitive atmosphere included gases left over from the formation of the solar system, notably methane, ammonia, hydrogen, and some water vapor. The fact that fairly similar mixtures exist in the atmospheres of some of the larger planets today lends credibility to the idea that the earth once had such an atmosphere.

Protein does not occur as a raw material in the crust of the earth. It is produced by living organisms and is the most complex kind of naturally occurring substance. Proteins play vital roles in the life processes of all organisms. For example, the enzymes that regulate so many of the internal chemical activities of our bodies are forms of protein. It is not surprising, therefore, that the investigations of the origin of life have emphasized the search for a way in which proteins could have been formed abiologically (without the presence of life).

The many proteins found in living organisms consist of distinct arrangements of only twenty subunits, or building blocks, called the *primary amino acids*. Laboratory experiments on the origin of life have not produced proteins, but they have yielded most of the primary amino acids. There have been a variety of experiments, but the basic methods are similar: a mixture of gases simulating the composition of the primitive atmosphere is exposed to an energy source, such as an electric spark, heat, ultraviolet light, or a beam of atomic particles (Figure 4-7). This causes

---

[3] Another possibility is that there once was much more water; according to this theory, some of the water vapor in the atmosphere was broken down by the sun's ultraviolet light into its constituents, hydrogen and oxygen. As hydrogen is the lightest gas, much of it escaped at the top of the atmosphere leaving an "excess" of oxygen.

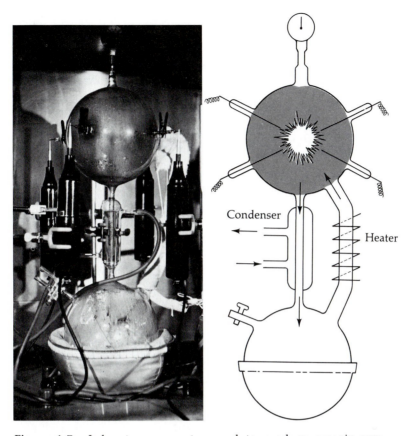

Figure 4-7. Laboratory apparatus used to produce organic compounds. The upper flask represents the atmosphere and contains methane and ammonia, while the lower flask represents the primordial seas and contains water. The side arm is kept hot to allow moisture to reach the upper flask, and the condenser in the central tube returns the moisture to the "ocean" below. The electrodes in the upper flask permit the simulation of lightning flashes. After typical experiments are run for 24 hours, there is a dark brown deposit of hydrocarbons on the sides of the upper flask, and organic compounds are found in the water below. (Courtesy of C. Ponnamperuma, University of Maryland.)

changes in the gas mixture, and the newly formed chemicals are analyzed; some of them have been identified as amino acids. These experiments have also produced subunits of the nucleic acids, and other constituents of living matter.

A great deal of further experimental work will be necessary to outline the methods by which life originated, but it does not seem beyond our reach to solve this problem. The present idea of the probable sequence for the development of life begins with the formation of the amino acids and other subunits of organic matter in the primitive atmosphere and oceans through the action of naturally occurring electrical discharges (lightning), heat, ultraviolet radiation (from the sun), and/or cosmic rays. The next important stage took place in water—perhaps the oceans, tidal pools, or even warm springs on land. Molecules of the amino acids came together by chance in various combinations. (Laboratory experiments show that amino acids in water will organize themselves into larger molecules.) Given millions of years, eventually the right combinations of amino acids were formed, and the first proteins were created. The same kind of process over a long span of time resulted in the formation of the nucleic acids from their subunits. Eventually, the chemical activity in this *primary broth* led to the development of *prebiodonts,* nonliving units with outer surfaces that separated them from the surrounding fluid. These objects, according to the Soviet chemist A. I. Oparin (who originated the theory of chemical evolution), constituted "open systems," capable of receiving and modifiying materials and energy from the environment as well as discharging material to the environment. These in turn gave rise to the first living cells, capable of reproduction and the transmission of genetic information. Oparin's laboratory experiments show that such prebiodonts can form by natural processes.

Obviously there are many gaps in this argument, but it seems to be a reasonable working hypothesis. It appears that, given the conditions astronomers and geologists believe to have prevailed on the primitive earth, life could have arisen through a long sequence of chemical processes. Once life existed, evolution by natural selection must have begun.[4]

THE FLOW OF EVOLUTION

The oldest known fossils, identified as bacteria and blue-green algae, have an age of about 3.4 billion years. A rough timetable of key events in the origin of life, based on the thinking of current researchers in this field, is given in Table 4-1.

The first evolutionary steps from the one-celled organisms to many-celled or relatively complex animals probably took place in the sea and

---

[4] Also see the discussions of amino acids in meteorites (Chapter 7) and of interstellar molecules in the galaxy (Chapter 12).

Table 4-1. Milestones in the Origin of Life on Earth

Years Ago	Event
4.5×10^9	Formation of the earth
4.2×10^9	Occurrence of the first self-replicating entity; beginning of evolution by natural selection
3.4×10^9	Appearance of bacteria and blue-green algae
2.1×10^9	Multicellular life

involved the specialization of activities among the different cells. Before such evolution could proceed very far, there would be a differentiation of life forms based on how food was obtained. Growth of an organism depends on the availability of nutrients; the supply produced by random chemical combinations is limited and unreliable.

For higher organisms to evolve, a dependable food supply was needed. It came in the form of plants that contain the chlorophyll molecule. Chlorophyll acts as a catalyst in the chemical reactions that produce carbohydrate nutrients (such as sugar) directly from the raw materials of sunlight, carbon dioxide, and water. This process, *photosynthesis*, releases oxygen as a by-product.

Plants provide the basic food for all animal life. In a sense, they are the only productive organisms, because only they assimilate raw inorganic materials into living matter. Animals are destructive in that they require nutrients in the form of plant tissues or the flesh of other animals to grow and survive. Thus, animal life could not have arisen until plant life developed.

Our knowledge of evolution is based on the fossil record, and the reader should consult Table 3-1 for a brief outline of the events recorded in the rocks. Multicellular life is believed to have started in the sea and may have been similar to the present-day *phytoplankton*, the one- or several-celled aquatic green plants. The phytoplankton are the basic food for the small water animals, such as the protozoa, and they may be the earth's single most important source of raw food and oxygen.

The principal mileposts of evolution are summarized in Figure 4-8, but some specific developments deserve brief mention. The fossil record for Precambrian times is scarce and consists primarily of aquatic plants. By about one billion years ago, animals had developed to such fairly complex forms as worms, and occasionally fossil worm holes or worm

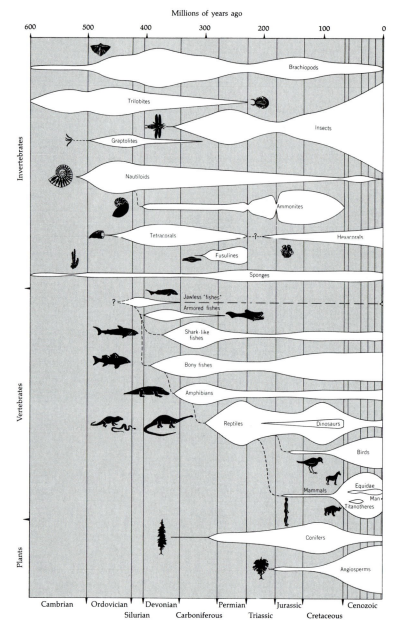

Figure 4-8. Historical evolution chart showing the geologic periods of principal life forms. The widths of the areas that represent various types of plants and animals show the extent of diversity of species. (From *Geology Illustrated* by John S. Shelton. W. H. Freeman Company. Copyright © 1966.)

Figure 4-9. Cambrian sandstone, showing the tracks of large worms that crawled across the wet sand, as well as an overall ripple pattern caused by the waves. (Courtesy of the Smithsonian Institution.)

tracks are found (Figure 4-9). The earliest really abundant fossils are the marine invertebrates, such as the trilobites shown in Figure 3-5. The Pennsylvanian part of the Carboniferous Period (280 to 310 million years ago) was characterized by a tremendous amount of plant growth, producing forests in swampy land. Geological processes turned the organic remains of the forests into coal over long periods of time (petroleum was probably also produced in this way); our coal supply dates largely from this period and is literally "fossil fuel."

The dinosaurs, perhaps best known of all the extinct life forms, appeared in the Triassic Period about 200 million years ago. They evolved into various species that were at home on the land, at sea, and in the air; they dominated the earth for some 100 million years. A famous example of evolution is the development of the horse over the 50 million years since its appearance as a dog-sized animal in the Eocene Epoch (56 to 37 million years ago). The modern animal is larger, faster, equipped with teeth suited to a wider variety of diet, and probably smarter than its ancient predecessor (Figure 4-10).

The Quaternary Period began 3 million years ago and includes the Pleistocene and Recent Epochs. During this period, extensive glaciation and the development of man have occurred. Many of the plant and animal fossils of the Quaternary resemble the living organisms that we find today.

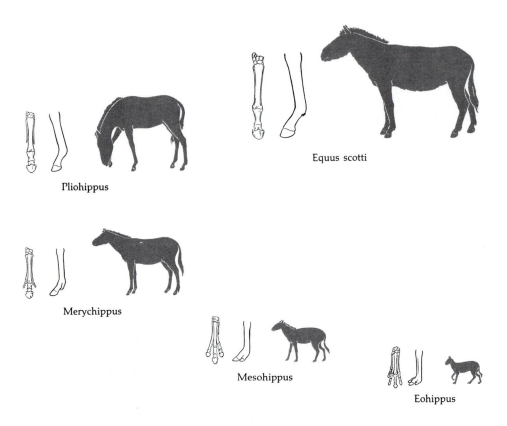

Figure 4-10. Evolution of the horse from the Eocene (bottom pictures) to the present (top) as reflected in the relative size and development of the bones in the foreleg. (From *Introduction to Biological Science* by C. W. Young, G. L. Stebbins, and F. G. Brooks. Harper and Row, 1938, 1951, 1956. Reprinted by permission of the publishers.)

MAN ON THE EARTH

In the preceding sections we have discussed the evidence for the evolution of species occurring on geological time scales. A similar case can be made for the evolution of man.

Fossils and contemporary anthropological evidence place the emergence of human beings in Africa some 3 to 4 million years ago. There appears to have been a sequence of ancient primate types that showed steady variation from ape-like properties to those found in modern man. The characteristics considered here are the shape of the hips and pelvis (to allow efficient erect walking, or *bipedalism*), the shape of the head and tooth structure, the size of the brain, and the use of tools and fire. The use of tools was made possible by the bipedalism, good binocular vision, and the grasping hand, a remnant of tree-dwelling ancestors.

The progression can be illustrated most simply in terms of brain size. The *Australopithecus* (3 to 4 million years ago) had a brain size of about 510 cubic centimeters, *Homo erectus* (0.5 million years ago) had a brain size of about 975 cubic centimeters, and *Neanderthal Man* (45 to 110 thousand years ago) had a brain size of 1,420 cubic centimeters. This latter figure is comparable to the brain size of modern man; by comparison, the brain size of a chimpanzee is about 395 cubic centimeters.

Among the earliest primate specimens in this progression was the *Proconsul* (15 million years ago), which is classified as an early ape and is perhaps the ancestor of the chimpanzee and the gorilla. The earliest direct ancestor of man appears to have been *Ramapithecus* (13 million years ago). We note that the ancestors of both man and apes were intermingled at the time in question, about 15 million years ago. Thus, man is not strictly descended from the apes. We are "cousins," descended from common ancestors of a long time ago.

This view of "African Genesis" is widely accepted today. Modern man, for better or worse, is the result of millions of years of human evolution. Many false starts are found in the fossil record; these were dead ends which became extinct. The thrust of the development was from fruit-eating tree dweller to ground-dwelling hunter-gatherer. Our fruit-eating days are mirrored in our teeth; they are not the teeth of a true carnivore. Stone tools were developed early (Figure 4-11), and enabled man to extend his own physical gifts in hunting and other activities. These tools are durable and are the chief evidence bearing on the life styles of early humans. Finally, social organization permitted men to hunt physically superior animals such as mammoths.

The climatic environment of our most immediate ancestors was very difficult indeed. The Pleistocene was an era in which the earth was subjected to repeated temperature changes and ice ages. This was a time of testing. Modern man, with a large, complex brain, the ability to formulate and communicate abstract thoughts, the use of tools, the ability to adapt to new situations (from a zoological point of view, to be a "specialist in non-specialization") is, to a certain extent, the product of those times;

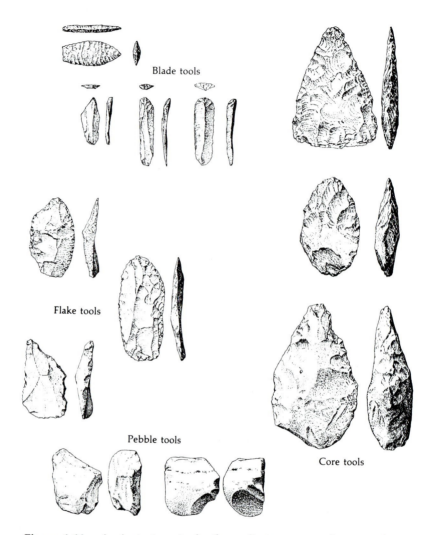

Blade tools

Flake tools

Pebble tools

Core tools

Figure 4-11. Ancient stone tools; the earliest ones are shown at the bottom, later ones at the top. *Core tools* were shaped by chipping off flakes; *pebble tools* were a crude form of these. The other tools shown were made from the flakes themselves; *blade tools* are flakes with nearly parallel sides. (From *Tools and Human Evolution* by S. L. Washburn. Copyright © 1960 by Scientific American, Inc. All rights reserved.)

even today human beings are often at their best under adverse conditions. *Neanderthal Man* flourished between 100,000 and 45,000 years ago. Toward the end, he gave way to *Cro-Magnon Man*.

Cro-Magnon Man and related species were widely distributed over the earth and developed rapidly. They made advances, including the conversion from hunting and gathering to agriculture and animal husbandry and the discovery and application of metals. The Cro-Magnons had the time or necessity for art (as preserved in the Lascaux cave paintings, Figure 4-12) and practiced ceremonial burial of the deceased. The first stirrings of religion and a curiosity about one's place in the scheme of things must have occurred by this time, if not earlier. The development of language, and thus the ability to pass knowledge on from one generation to the next, as inferred from their development, was clearly a phenomenon of immense importance.

Where did man's interest in astronomy begin? Surely at a very early time man was aware of the sun and the moon, of the basic division into

Figure 4-12. Prehistoric paintings in the Hall of Bulls, Lascaux Caves. (Courtesy of French Government Tourist Office.)

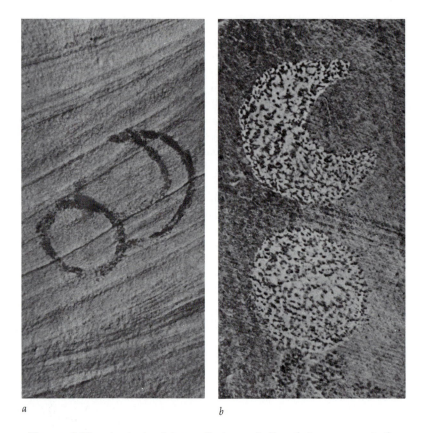

a   *b*

Figure 4-13. Ancient pictures that are believed to represent the supernova of A.D. 1054. Photographs taken (a) at White Mesa, Arizona, by William C. Miller, and (b) in Navajo Canyon, Arizona, by R. C. Euler. The crescent depicting the moon is reversed, but this common mistake is not considered to be significant. (Courtesy of William C. Miller, Hale Observatories.)

day and night. At some point in prehistory he began to speculate about the nature of the world and the origin of things. The obvious importance of the sun to all organisms and activities on earth was reflected in its incorporation as a motive power or deity in the earliest known myths and theologies of primitive civilizations. The deciphering of cuneiform inscriptions dating back to 3000 B.C.–2000 B.C. shows that the ancient Babylonians had an extensive knowledge of astronomical phenomena, and that even in those times astronomers were employed in observing the phases of the moon to establish, regulate, and periodically revise the calendar. It seems reasonable to conclude, therefore, that astronomy is among the oldest (if not in fact the very oldest) of the sciences. Observations of unusual astronomical events, such as eclipses and the appearances of comets in the sky, are found in the records of Oriental civilizations. Even primitive people who lacked a written language have left us what appear to be astronomical records in the form of rock paintings (Figure 4-13).

LIFE ON OTHER WORLDS

If life arose on the earth by natural means, might it not have appeared elsewhere as well? This might depend on the existence of similar, favorable conditions for chemical and biological evolution. Locales that deserve consideration include the other planets of our solar system and the hypothetical planets of other stars besides the sun. We say "hypothetical" because, despite tantalizing indications in one case (Barnard's star), we do not yet have proof of the existence of any planets besides the nine of our own solar system. At the present time, more than 30 nearby stars are under study by astronomers at the University of Pittsburgh for indications that they may have planets. However, since planets are much smaller and dimmer than stars, we do not yet have the technological capability to make a truly definitive search for them. In fact, if the existence of planets around a star is relatively rare, then the nearest star having planets might be quite far away and thus outside the scope of the Pittsburgh study.

In any case, as we will see in later chapters, we believe that the origin of the earth and other bodies of the solar system can be explained by known physical principles through a reasonable sequence of natural events. Therefore, it is only logical to assume that such events have occurred elsewhere in space, resulting in other planetary systems that might also harbor life.

We will also see in later chapters that organic molecules have been found to exist in the gas between the stars, in clouds of matter in space where new stars (and thus perhaps new solar systems) are forming, and also in meteorites. Whether this is directly relevant or not to the origin of life is unclear, since laboratory experiments mentioned in the preceding section on "The Origin of Life" suggest that such molecules may form independently on a planet with a primitive atmosphere. Exploration of our solar system by spacecraft has not turned up any evidence of life outside the earth so far, and the possibilities narrow as each additional planet is reached and studied by a spaceprobe. It appears that we must direct our attention to the planets of other stars in our search for extraterrestrial life.

Although very little is known about the conditions under which life may exist elsewhere in the universe—especially intelligent life with which we might communicate—the subject has attracted great scientific interest and has been featured at several international conferences and discussed in many books. Much of this discussion has been frankly devoted to what may turn out to be little more than informed speculation. Nevertheless, several basic principles have emerged. To begin with, if we assume that life must evolve on a planet with a nearby star as the energy source (just as the sun is the energy source for life on earth), this sets some important constraints on the kinds of stars that may be involved. As we proceed through this book, we will learn that stars are born, evolve and die, and that the lifetimes of different types of stars vary. In particular, stars that are much more massive than the sun have lifetimes that are shorter than the length of time that was required for the evolution of advanced life forms on earth. Thus, if terrestrial life is typical of life elsewhere in space, we do not expect to find advanced life on the planets (if any) of massive stars.

In the case of the sun, there is a certain "habitable zone," extending from just outside the orbit of Venus to roughly the orbit of Mars, in which the solar energy falling on a planet may be sufficient to keep water in the liquid state at least somewhere on the planet, while not evaporating it all to form steam. From our knowledge of biochemistry, it is thought unlikely that the proteins and other basic molecules of life could form and combine into more complex substances except in an aqueous environment. Assuming that terrestrial life is typical, we may ask the nature of the habitable zones of other stars.

For stars much more massive than the sun, the nature of the habitable zone is irrelevant because of the short stellar lifetime—the zone will not remain habitable long enough for life to develop and evolve. For stars like the sun, the habitable zone will be similar to that of the solar system; that is, there may be one or two planets in the zone. Stars that are less massive than the sun will have long lifetimes, adequate for the evolution of life. But since they are also cooler and dimmer than the sun, it turns out that their habitable zones are very small. The chances that a planet of such a star may orbit within the narrow habitable zone are very slight.

We conclude that the most likely prospects for the existence of life are planets of stars that are relatively similar to our sun. (Of course, this conclusion is based on the preceding arguments, which, almost inevitably, are biased by our knowledge of only one biology, the biology of earth.) Much thought has been given to the possibility of communicating with extraterrestrial civilizations, if they exist. Some experiments in listening for radio signals from other worlds have actually been conducted by radio

Table 4-2. Searches for Intelligent Life Elsewhere in the Universe

Project	Approximate Date	Nature of Search	Instrument	Description	Results
Ozma	1960	Passive	85-foot (26-meter) radio telescope	Searched for intelligent signals from the vicinities of two stars, located about 11 light years from the earth.	Negative
Gorkii Search	1968–69	Passive	50-ft (15-m) radio telescope	Searched for radio signals from 11 stars within 70 light-years and from the Andromeda galaxy.	Negative
Ozma II	1972–75	Passive	140-ft (43-m) and 300-ft (91-m) radio telescopes	Searched for signals from 659 stars within 60 light-years.	Negative
Copernicus Search	1974	Passive	80-cm (32-inch) ultraviolet space telescope	Searched for ultraviolet laser beam radiation from 3 nearby stars, using the Copernicus satellite.	Negative
Eurasian Network	Since 1974	Passive	Network of radio telescopes	Search under way for pulsed radio signals from any direction in space.	Pending
Arecibo Broadcast	1975	Active	1,000-ft (305-m) radio telescope	Three-minute message beamed toward M13, a globular star cluster 8,000 parsecs from the earth. In addition, a passive search for extraterrestrial communications is under way	Pending

astronomers in the U.S.A., Canada, and the Soviet Union, and a message has been beamed toward a distant star cluster with the huge radar telescope at Arecibo, Puerto Rico (see Figure 6-17). One scientist even searched for laser beam signals from nearby stars with the telescope of the Copernicus satellite. Table 4-2 summarizes some of these attempts.

An increasing number of scientists believe that intelligent life must exist elsewhere in the universe and that we may some day contact it. One proposed project, called Cyclops, would involve the construction of a huge array of radio telescopes for this purpose. However, it might cost 10 billion dollars or more. A Soviet astronomer, N. S. Kardashev, has considered the technological levels that such advanced civilizations may have reached. Kardashev believes that civilizations can be classified by their ability to generate or make use of energy. According to this scheme, we belong by definition to Type I, those civilizations with an energy consumption comparable to that of all of the nations of earth combined. Type II civilizations are those that can harness the total energy output of a star such as the sun; Type III civilizations would be capable of employing energy equal to that produced by a whole galaxy of a hundred billion stars. The discussion is now admittedly on the borderline of science fiction. However, if such advanced civilizations do exist, perhaps they are already in constant communication among themselves by means unknown to us. The exobiologist Carl Sagan wonders if we are like the people in "isolated valleys of New Guinea, who communicate with their neighbors by runner and drum, and who are completely unaware of a vast international radio and cable traffic over them, around them, and through them."

5

Matter, Heat and Light

We began our survey of the universe by examining the motions of the earth and planets and introducing some aspects of the physics of motion and gravitation. Next we described some relevant concepts of elementary geology and biology. Now we will present several basic ideas from physics and chemistry that are needed to understand the observations of stars, planets, nebulae, and galaxies.

HISTORY OF ATOMISM

We encounter matter in three forms in our everyday experience—as gases, liquids, and solids. Materials have different colors, different hardnesses, and so on. We sometimes see matter change its state—for example, when ice melts or water becomes steam. The complexity of even the most familiar things becomes apparent when we observe them carefully. Consider a campfire—flames and smoke rise and water sizzles from the burning wood. The smokes thins out with distance from the fire, eventually merging indistinguishably into the surrounding air. A small residue of solid matter is soon all that is left. Commonplace phenomena like these were of the greatest interest to the early natural philosophers. Among the Greeks, Empedocles (490–430 B.C.) suggested that all things are composed of four "elements": air, water, fire, and earth; a Chinese doctrine recognized five: earth, fire, water, wood, and gold.

The early attempts to describe the composition of matter led to the formulation of *atomism*, the theory that matter is not infinitely divisible but that discrete parts or building blocks that cannot be further divided are finally encountered at some level. The idea of ultimate building blocks, or *atoms,* was advocated by the Greek philosophers Leucippus and Democritus in the fifth and fourth centuries B.C. The atoms were thought of as endlessly drifting through a void; an occasional combination of them formed material objects. Later, Epicurus (341–270 B.C.) taught that atoms have weight.

The concept of atomism lay dormant through the Middle Ages, then reappeared in the seventeenth century. Newton discussed and favored the theory, as did the British chemist Robert Boyle (1627–1691), who conducted experiments showing that the four elements of the Greeks were not the primary components of matter, since they could not be extracted from other substances. Later, John Dalton (1766–1844), influenced by Newton's work, formulated an *Atomic Theory* and conducted relevant experiments. Considering the elements as primary materials that could not be broken down further by heat or chemical processes, but which did combine with each other to produce the various known substances, Dalton's Atomic Theory was based on three principles.

1. Elements consist of indivisible particles (atoms) whose properties are not affected by chemical reactions.
2. The atoms of a given element have equal weights and are also identical in their other properties, but atoms of different elements always have different weights.
3. A chemical compound is formed by the combination of specific numbers of atoms of different elements.

A familiar example of the third principle is the water compound (H_2O); each of its smallest units (molecules) consists of two hydrogen atoms and one oxygen atom. The fact that water can be broken down into two different substances (hydrogen and oxygen) establishes that it is a compound and not an element.

Dalton's theory proved capable of accounting for a wide variety of chemical experiments, and it became a keystone in the subsequent development of chemistry and physics. Today we know that although it can account for a wide variety of laboratory experiments, it is not strictly correct. The physicists have split atoms and found that they are built up from various combinations of three types of particles—the electrons, protons, and neutrons—and that a variety of other atomic particles exist as well. And it has been found that the atoms of a given element may exist in several forms *(isotopes)* that differ in weight. In fact, there exist isotopes of different elements that have virtually the same weights, but differ in their detailed makeup of electrons, protons, and neutrons. We now recognize that the atomic weights deduced for various elements on the basis of the Atomic Theory are really averages that depend on the individual weights and the relative *abundance* (frequency of occurrence) of the isotopes of each.

Certain elements have rather similar properties. For example, sodium, potassium, and lithium are all light-weight metals that readily react with other substances to form compounds. On the other hand, neon, argon, and xenon are gases that never combine with other elements under natural conditions.[1] Several scientists noticed that there appear to be numerical relationships between the atomic weights of elements that have similar properties. In 1863, the Englishman John Newlands suggested that (looking at the elements in order of increasing atomic weight) every eighth element was similar. The most important work on this subject was that of the Russian chemist Dmitri Mendeleev (1834–1907). He also tabulated the elements in the order of their known atomic weights. Believing in numerical relationships for similar elements, Mendeleev suggested that many elements were at the wrong places in the table, and therefore their measured atomic weights must be wrong. Furthermore, he pointed out that there were gaps in the table of atomic weights where no elements were known to exist. He deduced that six such elements should exist, and he predicted their physical properties on the basis of their locations in his Periodic Table.[2] Mendeleev's criticism of the old atomic weight measurements was confirmed by subsequent experiments; five of the new elements were later found in nature, and the sixth has been created artificially in the laboratory (Table 5-1).

Mendeleev's predictions of the properties of these new elements were also verified. For example, he suggested that "eka-silicon," the element of atomic weight about 72, would be a gray metal about 5.5 times denser than water, and that it would form a white compound when heated in oxygen. These predicted properties of eka-silicon were indeed found to characterize the element of atomic weight 72.6, which was discovered in 1886 and named germanium. This element has been of vital

---

[1] In 1962, chemists finally succeeded in creating compounds of xenon.

[2] As often happens in science, similar ideas were developed elsewhere at about the same time; in this case, the German chemist Lothar Meyer also formulated a periodic table of the elements.

Table 5-1.   Six Elements Predicted by Mendeleev in 1871

Mendeleev's Name	Year of Discovery	Present Name
eka-boron	1879	scandium
eka-aluminum	1875	gallium
eka-silicon	1886	germanium
eka-manganese	1937*	technetium
dvi-manganese	1925	rhenium
eka-tantalum	1917	protoactinium

*Produced artificially.

importance to twentieth-century electronics as a key ingredient of the transistor.

The brilliant success of Mendeleev's predictions showed the importance of investigating what it is about an atom that allows its physical properties to be determined by its position on the table of atomic weights. A modern version of the Periodic Table is given in Table 5-2. The elements are not arranged by atomic weight but by *atomic number,* the number of electrons or protons in the atom.

PARTICLES IN THE ATOM

The first component particle of the atom to be recognized was the *electron,* the negatively charged particle that carries the smallest (that is, indivisible) unit of electric charge. Its name is derived from the Greek word *elektron,* meaning amber, or fossilized pine resin. Amber figured in an electrical phenomenon that was known to the Greeks. They found that when a chunk of it is rubbed with a woolen cloth, the amber will attract light objects, such as feathers. Later experiments showed that other objects could be charged in this way, and that there are two kinds of electrical charge (called positive and negative) which can be distinguished because objects with like charge repel each other, whereas oppositely charged objects attract each other (Figure 5-1).

The interpretation of the experiment in which the initially neutral amber becomes attractive after being rubbed is that a certain amount of negative charge was literally rubbed off the amber, leaving it with a net positive charge. For a long time a controversy existed over whether electricity was continuous or consisted of discrete indivisible particles, like the

Table 5-2. The Periodic Table of the Elements

Hydrogen 1 H 1.008																	Helium 2 He 4.003
Lithium 3 Li 6.939	Beryllium 4 Be 9.012											Boron 5 B 10.81	Carbon 6 C 12.01	Nitrogen 7 N 14.01	Oxygen 8 O 16.00	Fluorine 9 F 18.99	Neon 10 Ne 20.18
Sodium 11 Na 22.99	Magnesium 12 Mg 24.31											Aluminum 13 Al 26.98	Silicon 14 Si 28.09	Phosphorus 15 P 30.97	Sulfur 16 S 32.06	Chlorine 17 Cl 35.45	Argon 18 Ar 39.95
Potassium 19 K 39.10	Calcium 20 Ca 40.08	Scandium 21 Sc 44.96	Titanium 22 Ti 47.90	Vanadium 23 V 50.94	Chromium 24 Cr 52.00	Manganese 25 Mn 54.94	Iron 26 Fe 55.85	Cobalt 27 Co 58.93	Nickel 28 Ni 58.71	Copper 29 Cu 63.55	Zinc 30 Zn 65.37	Gallium 31 Ga 69.72	Germanium 32 Ge 72.59	Arsenic 33 As 74.92	Selenium 34 Se 78.96	Bromine 35 Br 79.90	Krypton 36 Kr 83.80
Rubidium 37 Rb 85.47	Strontium 38 Sr 87.62	Yttrium 39 Y 88.91	Zirconium 40 Zr 91.22	Niobium 41 Nb 92.91	Molybdenum 42 Mo 95.94	Technetium 43 Tc (99)	Ruthenium 44 Ru 101.1	Rhodium 45 Rh 102.9	Palladium 46 Pd 106.4	Silver 47 Ag 107.9	Cadmium 48 Cd 112.4	Indium 49 In 114.8	Tin 50 Sn 118.7	Antimony 51 Sb 121.8	Tellurium 52 Te 127.6	Iodine 53 I 126.9	Xenon 54 Xe 131.3
Cesium 55 Cs 132.9	Barium 56 Ba 137.3	Lanthanide series 57–71	Hafnium 72 Hf 178.5	Tantalum 73 Ta 180.9	Tungsten 74 W 183.9	Rhenium 75 Re 186.2	Osmium 76 Os 190.2	Iridium 77 Ir 192.2	Platinum 78 Pt 195.1	Gold 79 Au 197.0	Mercury 80 Hg 200.6	Thallium 81 Tl 204.4	Lead 82 Pb 207.2	Bismuth 83 Bi 209.0	Polonium 84 Po (210)	Astatine 85 At (210)	Radon 86 Rn (222)
Francium 87 Fr (223)	Radium 88 Ra (226)	Actinide series 89–103	Rutherfordium 104 Rf (260)														

Lanthanide series

Lanthanum 57 La 138.9	Cerium 58 Ce 140.1	Praseodymium 59 Pr 140.9	Neodymium 60 Nd 144.2	Promethium 61 Pm (145)	Samarium 62 Sm 150.4	Europium 63 Eu 152.0	Gadolinium 64 Gd 157.3	Terbium 65 Tb 158.9	Dysprosium 66 Dy 162.5	Holmium 67 Ho 164.9	Erbium 68 Er 167.3	Thulium 69 Tm 168.9	Ytterbium 70 Yb 173.0	Lutetium 71 Lu 175.0

Actinide series

Actinium 89 Ac (227)	Thorium 90 Th (232.05)	Protoactinium 91 Pa (231)	Uranium 92 U 238.0	Neptunium 93 Np (237)	Plutonium 94 Pu (242)	Americium 95 Am (243)	Curium 96 Cm (247)	Berkelium 97 Bk (247)	Californium 98 Cf (249)	Einsteinium 99 Es (254)	Fermium 100 Fm (253)	Mendelevium 101 Md (256)	Nobelium 102 No (256)	Lawrencium 103 Lr (257)

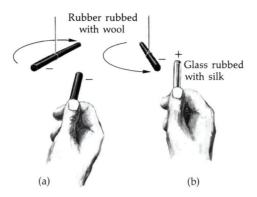

Figure 5-1.   Illustration of (a) the repulsion of like charges and (b) the attraction of unlike charges. (From *Structure and Change* by G. S. Christiansen and P. H. Garrett. W. H. Freeman and Company. Copyright © 1960.)

atoms of matter. During the second half of the nineteenth century, the discovery of *cathode rays* and subsequent experiments with them resolved this argument.

When each of the two terminals of a high-voltage battery is connected to a separate metal plate, and the two plates are housed inside a glass tube filled with air at low pressure, electricity flows through the gas, which shines like a neon sign. Pumping out the air, a point is reached where the pressure is so low that the glowing gas is no longer seen. However, a faint glow *in the glass itself* suggests that it is being struck by particles (Figure 5-2). Experiments showed that these particles emanated from the negatively charged plate (cathode); hence they were named cathode rays. Physicists disagreed about them; some believed that the rays were really particles of light *(photons),* whereas others suggested that they were the long-sought individual particles of electric charge. About 1897, J. J. Thomson at Cambridge University measured the velocity of the particles, showed that it was less than the velocity of light, and thus demolished the first theory. Thomson also demonstrated that the cathode rays behaved in the same way regardless of what metal was used for the negatively charged plate or what gas was placed in the tube. Thus, it appeared that they were neither particles of light nor atoms of a particular element. The evidence suggested that the cathode rays (now called *electrons*) were indeed particles of electric charge.

It was possible, from a variety of experiments, to estimate the value of the electron's charge. But in each case, this value was found as the average of a great many particles, and it was not known if *each* particle had an identical charge. This was resolved in a famous experiment by the American physicist R. A. Millikan (1868–1953), who developed an ingenious method to measure the electrical charges of tiny oil drops. He found that the charge values were all multiples of one negative number, and this largest common denominator was thus identified as the charge of the electron. It was also found that all electrons have the same mass, and that this is equal to $1/1837$ of the mass of a hydrogen atom.

At this point in historical development, scientists knew that electrons are a common constituent of matter, that they are charged, and that they have a specific mass. However, ordinary matter such as wood is not electrically charged. Hence, atoms must also contain an equal and opposite (positive) charge. Thomson proposed a "plum pudding" model, in which the atom consisted of a homogeneous sphere of positive charge with the electrons embedded like raisins (Figure 5-3).

The validity of the plum pudding model or any other atomic model could be checked if we had a way to probe the structure of the atom. A technique was in fact developed, making use of particles released in the radioactive decay of some atoms (recall the discussion of natural radioac-

Figure 5-2. The Crookes tube, with electrons moving from left to right. The glow is due to the electrons striking the glass. The presence of the shadow shows that the electrons move in straight lines. (After *Chemistry* by Linus Pauling and Peter Pauling. W. H. Freeman and Company. Copyright © 1975.)

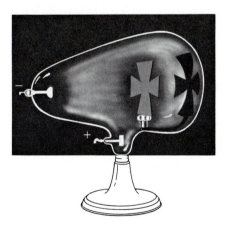

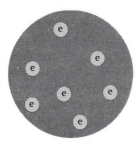

Figure 5-3. The Thomson plum pudding model of the atom. The shaded area represents the positive charge.

tivity in Chapter 3). Three kinds of rays were known to be emitted in the process of radioactive decay:

alpha rays: positively charged, also called alpha particles; now identified as helium nuclei, consisting of two protons and two neutrons.

beta rays: negatively charged; now identified as electrons.[3]

gamma rays: now identified as high energy photons.

The type of electrical charge of each kind of ray was determined by seeing how it behaved when passing through the field of a magnet (Figure 5-4). Positive and negative charges were known to be deflected in opposite directions by the magnetic field, whereas uncharged particles were not deflected.

In atom-probing experiments directed by Lord Rutherford in 1911 (Figure 5-5), a thin gold foil was bombarded by a stream of alpha particles. Most of these particles passed through the gold foil, but one in 8,000 bounced off at an angle. This suggested that most of the alpha particles hit nothing or, at most, hit the much less massive electrons in the gold. But every once in a while, an alpha particle did hit a relatively heavy, positively charged object. The fact that the positive charges were encountered so rarely meant that the plum pudding model was wrong, and the positive charge was probably confined to a minute fraction of the atom's volume. Thus, Rutherford was led to an atomic model in which the elec-

---

[3]Electrons were called cathode rays when they were found in the electrical experiments described earlier; they were called beta rays when they were found as a decay product of radioactive atoms; eventually it was realized that these two kinds of "ray" were identical.

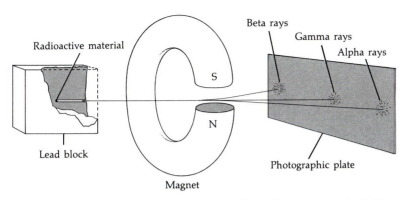

Figure 5-4. Alpha and beta rays are deflected by a magnetic field, but gamma rays are not. (From *Chemistry* by Linus Pauling and Peter Pauling. W. H. Freeman and Company. Copyright © 1975.)

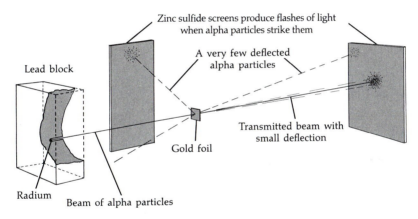

Figure 5-5. The Rutherford experiment. The fact that very few of the alpha particles were deflected by nuclei in the gold foil showed that the nuclei in the foil atoms were small compared to the atoms themselves. (From *Chemistry* by Linus Pauling and Peter Pauling. W. H. Freeman and Company. Copyright © 1975.)

trons circle about the positively charged nucleus in great orbits like planets around the sun (Figure 5-6). In the case of the simplest isotope of hydrogen (which is the simplest element), the atom consists of one electron in orbit around a particle of equal and opposite charge. This positively charged *proton* has a mass of 1,836 times the electron's mass.

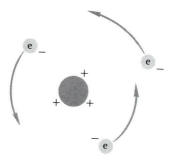

Figure 5-6. The Rutherford planetary model of an atom, for the case where the atomic number is 3. The orbits of the electrons are achieved in a manner analogous to the planetary orbits discussed in Chapter 2. In the atomic case, the effect of gravity is very weak, and the tendency for the electron to continue in a straight line is balanced by the electrical attraction of the nucleus. At the time of Rutherford's experiments, the neutron had not been discovered. In fact, the various isotopes of the element (lithium) with atomic number 3 have from two to six neutrons in their nucleii.

Rutherford's work suggested that the positive charge in the atom was concentrated in a small fraction of the atomic volume, but this was hard to understand. Protons have like charges and therefore repel each other. What process could confine them to the nucleus? This question is answered by the existence and properties of the neutron (discovered in 1932 by James Chadwick), a particle with approximately the same mass as the proton. As its name implies, the neutron has no charge. Free (that is, not in a nucleus) neutrons decay spontaneously with a half-life of 12 minutes, leaving a proton and an electron. Both protons and neutrons are found closely packed in the nucleus. They are attracted to each other by a *short-range force* (also called the strong nuclear force), which is much stronger than either gravitational or electric force over tiny distances like that in the nucleus, but which falls off in strength very rapidly with distance and so is insignificant on scales much larger than the nucleus.

Several terms are used in describing the properties of atoms in relation to their different numbers of particles. The *atomic number* is equal to the number of protons in the nucleus of an atom and is the same as the number of orbiting electrons when the atom is in its normal or electrically neutral state. The *atomic mass* is approximately equal to the total number of protons and neutrons in the nucleus. The *isotopes* of a given element are atoms with the same atomic number (equal numbers of protons) but with different atomic masses (different numbers of neutrons). For example, the common form of the hydrogen atom, which we described above as consisting of one proton and one electron, has an atomic mass[4] of about 1 and atomic number 1. On the other hand, *tritium*, a scarce isotope of hydrogen consisting of one proton, two neutrons, and an electron, has an atomic mass of about 3 and atomic number 1. There is an isotope of helium that also has an atomic mass of about 3: it has two protons, two electrons, and one neutron. Its atomic number is 2, indicating that it is an isotope of a different element than tritium.

The *atomic weight* of an element, as usually given in tables such as Table 5-2, is an average of the atomic masses of the isotopes of that element, calculated according to their relative frequency of occurrence in ordinary matter. For example, the tabulated atomic weight of helium is 4.003, due to the fact that the vast majority of naturally occurring helium on earth is in the form of the isotope helium-four, which has two protons, two neutrons, and two electrons, and an atomic mass of 4.003 daltons. The helium isotope with an atomic mass of about 3 amounts to only one ten-thousandth of one percent of the naturally occurring helium on earth.

Despite the existence of the short-range force, there appears to be a limit to the number of nucleons (protons and neutrons) in a naturally occurring atom. The heaviest atom found in substantial amounts in nature is the uranium isotope of atomic mass 238 (atomic number 92). Short-lived heavy atoms, notably the plutonium isotope of atomic mass 239 (atomic number 94), are found in trace amounts in uranium mines, where they are produced by reactions between the uranium atoms and neutrons. Heavier atoms have been made in the laboratory, but these *transuranium elements* all decay radioactively in times that are short compared to (for example) the age of the earth (indeed, some have half-lives of a fraction of a second). This has led some scientists to speculate that a number of the transuranium elements may have existed naturally in the past. There are also a few atoms of lesser atomic number than uranium, such as technetium (atomic number 43), that do not occur naturally on the earth. These elements also have been produced in the laboratory and they also do not have long-lived isotopes.

---

[4]More exactly, the atomic mass of the common form of hydrogen is 1.007825 daltons, where the *dalton* is defined as exactly one-twelfth of the mass of an atom of the carbon-12 isotope. Carbon-12 has six neutrons, six protons, and six electrons.

COMPOUNDS, GASES, LIQUIDS, AND SOLIDS

We will now discuss the nature of compounds and the chemical bond on the basis of Rutherford's model of the atom, in which the electrons are pictured in orbit around the nucleus. The basic properties of the model, in terms of the arrangement of the electrons in orbits and shells, were worked out by a number of scientists, notably the Danish physicist Niels Bohr. It has become known as the *Bohr atom*, or *Bohr model*, and is adequate for our purpose in explaining the emission and absorption of light by atoms in stars and other celestial objects, although some of the details have been superseded by more recent developments in physics.

The *noble gases* (including helium, neon, and argon) are so named because they are inert and almost never form compounds with other elements. This occurs because the electrons of atoms do not occupy randomly located orbits. There appear to be specific shells, each capable of containing some specific number of orbiting electrons. The innermost shell can accommodate up to two electrons. The atom with one proton, and therefore one electron, is hydrogen. That with two protons and two electrons is helium. Hydrogen reacts fairly readily with other elements, but helium, which has a complete (filled to capacity) electron shell does not. The other elements which, like helium, rarely form compounds (that is, the other noble gases) are precisely those elements that normally have complete electron shells and no electron outside. For example, the second shell can accommodate up to eight electrons, and the element that has ten electrons (two in the inner shell, eight in the second shell) is neon, a noble gas. On the other hand, we recall the metallic elements such as lithium, sodium, and potassium that react very rapidly with many common substances. These are exactly the elements in which the outermost electron is the sole electron in a shell, and the other electrons are in closed

Table 5-3.   Some Elements of Interest

Atomic Number	Noble Gases	Atomic Number	Alkali Metals
2	Helium	3	Lithium
10	Neon	11	Sodium
18	Argon	19	Potassium
36	Krypton	37	Rubidium
54	Xenon	55	Cesium
86	Radon	87	Francium

shells. For example, lithium has one more electron than the noble gas helium, and sodium has one more electron than the noble gas neon. Some noble gases and alkali metals are arranged in order of atomic number in Table 5-3.

The numbers of electrons that can exist in the various atomic shells are, respectively, 2, 8, 8, 18, 18, 32, and 32 (in order beginning with the innermost shell). If all seven shells of an atom were completed (that is, each contained the maximum possible number of electrons, as listed), then the atom would have atomic weight number 118, much higher than the heaviest known transuranic element (atomic weight 107).

The fact that the second and third shells each contain eight electrons, and the dependence of the chemical properties of an element on the arrangement of its electrons in shells, together explain why Newlands found some evidence for his idea that every eighth element was similar. The Periodic Table was designed so that the elements are listed along horizontal rows[5] in such a way that elements with similar properties occur in the same vertical columns (Table 5-2). The physical basis of the Periodic Table can now be understood from the atomic model discussed above. Each row in the table comprises those elements whose outermost electrons are normally confined to the same shell.

A given chemical compound, such as ordinary table salt (sodium chloride), consists of two or more elements that have combined to form a substance whose properties are different from those of the constituent elements. We have seen that the short-range force holds a nucleus together, and an electrical force (the attraction between unlike charges, usually called the *coulomb force*) holds the atom together in such a way that the electrons exist in certain discrete shells. The formation of compounds is related to the latter process. As the formula for table salt (NaCl) shows, the smallest unit or *molecule* of this compound consists of one sodium (Na) and one chlorine (Cl) atom. The tendency to complete an atomic shell of eight electrons is fulfilled in this case; sodium has one electron in its outer shell and chlorine has seven. The NaCl molecule can be visualized as forming when the electron in the outer shell of the sodium atom is transferred to the chlorine atom. However, the chlorine atom then has one extra electron and is a *chlorine ion* with a negative charge. The sodium atom now has one less electron than normally and is a *sodium ion* with a positive charge (see Figure 5-7). The unlike charges attract and an *electrovalent* bond is established. The electrovalent bond is not directional;

---

[5]The number of elements in a row was called the period.

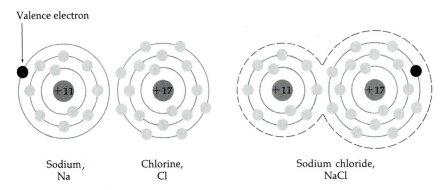

Valence electron

Sodium,
Na

Chlorine,
Cl

Sodium chloride,
NaCl

Figure 5-7. Schematic illustration of the formation of salt, NaCl, from sodium (Na) and chlorine (Cl) through the electrovalent bond.

that is, there is no preferred orientation for the two atoms in the NaCl molecule.

A complete outer shell of eight electrons can also be achieved by sharing some or all of the electrons between two or more nuclei. Thus, the chlorine gas molecule (Cl_2) is formed by sharing two electrons between the two chlorine nuclei to complete two shells of eight electrons. Water (H_2O) is formed by sharing the two electrons from the hydrogen atoms and the six electrons from the oxygen atom to form a common cloud of electrons. In these two examples, the electrons can be thought of as forming a common cloud around both nuclei. The bonds in these molecules are called *covalent*, and they have the remarkable property of being directional; that is, there are only certain possible directions of attachment. A diagram of the CH_4 molecule is shown in Figure 5-8. The large molecules encountered in biology (such as DNA) are the result of covalent bonds, as are many crystals (NaCl is an exception) with their regular arrangement and spacing. Carbon, with its ability to form four equally spaced covalent bonds (Figure 5-8), is well suited to the construction of large molecules. This property of the carbon atom, lending itself to the development of large, complex molecules, helps to explain why life developed from molecules based on carbon.

In a gas, the force that holds an individual molecule together is much stronger than any force that may attract one molecule to another, and so the individual molecules can move around freely. However, in a liquid the molecules are closer to one another, and some of the electrons of one molecule may be attracted by the positive ion of another molecule. This bond *(van der Waals force)* is much weaker than the electrovalent and covalent bonds, but it does hold the molecules loosely together, thereby distinguishing a liquid from a gas (Figure 5-9).

The bonds in a solid are much stronger than those in a liquid, and this allows a piece of solid matter to have a definite shape and size. For example, in a metal the matter is held together by a force *(metallic bond)* that results from the general attraction between all of the electrons that are not contained in closed atomic shells and a lattice or framework composed of the positive ions. An individual free electron is thus not bound to a particular ion and so it can move about, almost as a molecule in a gas. The presence of this *electron gas* causes metals to be good conductors of electricity and heat. Most solids are *crystalline;* that is, they have a regular internal structure consisting of a repeated three-dimensional pattern. Solids that lack this ordered structure are described as *amorphous*—charcoal is a good example.

Solid, liquid, and gas are the three most familiar states of matter; other states, such as plasma and degenerate matter, are discussed in later chapters.

Figure 5-8.   Structure of the methane molecule, CH_4, illustrating the directional property of the covalent bond. Methane is present in the atmosphere of Jupiter and may have been an important constituent of the primitive atmosphere of the earth. (After *College Chemistry* by Linus Pauling. W. H. Freeman and Company. Copyright © 1964.)

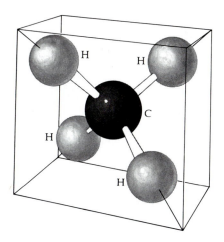

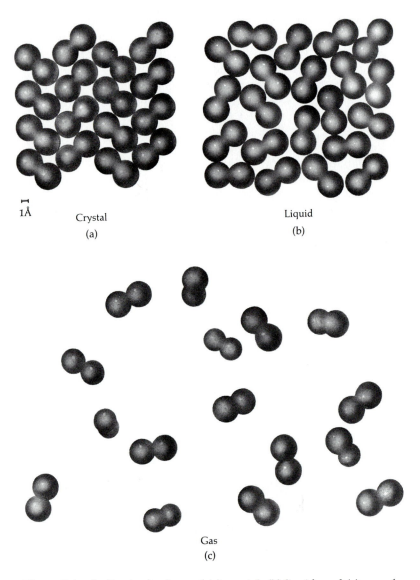

1Å   Crystal

(a)

Liquid

(b)

Gas

(c)

Figure 5-9.   Iodine in the form of (a) crystal, (b) liquid, and (c) gas of diatomic molecules (I_2). The scale is indicated by the short line (labeled 1Å) at the upper left, which represents a length of one Angstrom (10^{-8} cm). (From *General Chemistry*, 3d ed., by Linus Pauling. W. H. Freeman and Company. Copyright © 1970.)

HEAT, TEMPERATURE, AND ENERGY

Many of the basic facts about heat are apparent from personal experience. Heat can be transferred in several ways. A metal spoon placed in a hot cup of coffee soon becomes hot. In this case, the heat was transferred along the body of the spoon by *conduction*. On the other hand, when one boils water for the coffee, water at the bottom of the pot is heated and travels upward, transferring heat by *convection*. Finally, walking outdoors on a clear day we feel heat from the sun, transferred to us by *radiation*. Conduction is the transfer of heat by one individual particle (atom or molecule) to another; convection occurs through bulk motions of matter; radiation occurs through the transfer of photons from one object to another. In any case, the fact that heat flows from a hotter to a cooler object is clear from ordinary experience. Everyone knows that a block of ice placed in a bucket of hot water produces lukewarm water. One does not expect to place two buckets of lukewarm water together and somehow get a bucket of hot water and a bucket of ice.

The concept of heat is intimately related to the concept of temperature; things that are hotter than others are said to have a higher temperature. The temperature scales in common use have been set up in fairly arbitrary ways using identifiable points, notably the freezing and boiling temperatures of water (see Figure 5-10). The Fahrenheit scale (°F), named

Figure 5-10. Three commonly used temperature scales. (After *Standards of Measurement* by Allen V. Astin. Copyright © 1968 by Scientific American, Inc. All rights reserved.)

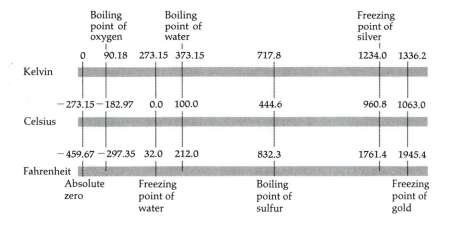

after the eighteenth-century German physicist who invented the mercury thermometer, defines the melting temperature of ice as 32° F and the boiling point of water as 212° F. Fahrenheit attempted to set up his scale so that the normal temperature of the human body would be 100° F, but he missed by slightly more than 1° F. Since this temperature is now generally taken as 98.6° F, it has been suggested somewhat facetiously that somebody had a fever when Fahrenheit calibrated his thermometer. Although this scale is the official system in a number of English-speaking countries, most nations have adopted the nearly identical Celsius and centigrade scales (°C). The centigrade scale is set up so that the interval between the temperatures at which water freezes and boils (again, under standard laboratory conditions) is divided into 100 degrees, and the two temperatures are defined as 0° C and 100° C, respectively. The fundamental unit of heat, the *calorie*, is related to the concept of temperature through its definition as the amount of heat required to raise the temperature of one gram of water (one cubic centimeter of water under specified laboratory conditions) by 1° C.

Heat is the energy of motion of the atoms and molecules that constitute matter. As the temperature in a room is increased, for example, the average speed of the air molecules increases. As the temperature decreases, the speed of the molecules decreases. We have seen that the commonly used Fahrenheit and centigrade scales were arbitrarily defined in terms of the properties of water, and thus they are not necessarily related to the physical nature of heat. To introduce the concept of a more physically meaningful temperature scale, we imagine an experiment in which the temperature of the air in a room is constantly decreased to such an extremely low value that the motion of the air molecules would essentially stop. This hypothetical temperature is called *absolute zero,* and it is used to define the zero point of the Kelvin temperature scale (K)[6] shown in Figure 5-10. The Kelvin degrees are the same size as the centigrade degrees, and the freezing and boiling points of water are found to be 273 K and 373 K, respectively. The Kelvin temperature scale is used in physics and astronomy because of its close connection with the energy of motion of atoms and molecules. Absolute zero has not been reached in laboratory experiments, but temperatures as low as 0.0001 K have been reached.

All atoms and molecules in the universe are in motion. Even the atoms in the crystal lattice of a solid move back and forth in a rather

---

[6]By international agreement, the symbol °K has been changed to simply K in scientific articles.

restricted range. If the temperature is increased, their speeds are increased, and eventually the bonds that hold the solid together are broken. This occurs at the melting point of a substance and usually produces a liquid in which the atoms or molecules can move around relatively freely, but are loosely bound together by the van der Waals force. If the temperature is increased still more, the speeds of the molecules in the liquid increase, the van der Waals force can be overcome, evaporation occurs, and we have a gas. Molecules in a gas are relatively far apart but do collide with each other.

The fact that the molecules of a fluid are in motion can be demonstrated in simple experiments. The most famous example is the *Brownian motion*, the zig-zagging of small dust particles floating in a gas or liquid. This effect was the subject of considerable study after it was noted by the Scottish botanist Robert Brown (1773–1856). If the zig-zag motions of the dust particles were due to currents in the gas, then the members of a

Figure 5-11. Brownian motion. The position of a 10^{-4} cm diameter smoke particle in air as recorded at one-minute intervals by the physicist Jean Perrin. (From *Matter, Earth, and Sky*, 2d ed., by George Gamow. Prentice-Hall. Copyright © 1965. Used by permission.)

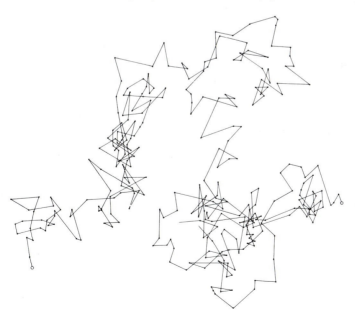

small group of adjacent particles would all appear to flow in the same direction. However, each particle takes its own randomly changing path, and this is understood as the effect of its being bombarded and knocked about by the molecules of the gas (Figure 5-11). Small particles are naturally more affected by collision with the molecules than are larger particles, and indeed they can be seen to undergo faster Brownian motion.

In gases, the molecular motions have effects important in our everyday lives through the phenomenon of *pressure*. The motions of gas molecules in a container cause the molecules to strike the wall of the container and be reflected. This constant drumming of gas molecules on the walls of a container produces an average force on the container wall, and pressure is defined as the force on a unit area. Often we are unaware of atmospheric pressure because it is the same inside our bodies as outside; thus, there is a balance. But, when we specify a tire pressure of, say, 20 pounds, we are actually specifying that the pressure inside the tire be 20 pounds per square inch higher than that in the atmosphere. Such a *pressure difference* can exert a large force; in the case of an automobile tire, the pressure difference supports the automobile's weight. Atmospheric pressure at the surface of the earth is sufficient to support the weight of a column of mercury about 30 inches high. This is the figure usually quoted in weather reports.

A famous demonstration of air pressure was carried out by Otto von Guericke, a seventeenth-century mayor of Magdeburg, Germany. He invented the vacuum pump and evacuated a pair of bronze hemispheres. The force holding them together was then so strong that two teams of eight horses each could not pull them apart. How did this force arise? As illustrated in Figure 5-12, the hemispheres initially were subjected to air pressure both on the inside and the outside, and hence no net force was exerted on them. When the hemispheres were brought together and evacuated, however, the force due to the difference between the outside atmospheric pressure and the negligible internal pressure held them together.

We have mentioned that heat is simply the energy of motion of atoms and molecules; macroscopic objects also have energy by virtue of their motions. Both the microscopic and macroscopic energies of motion are known as *kinetic energy*. We can think of several other kinds of energy, such as the *nuclear energy* that powers modern submarines, and the *chemical energy* that is released in the explosion of dynamite or the combustion of gasoline. Sometimes one form of energy is converted into another, as when the chemical energy of fuel is converted into the kinetic energy of an automobile.

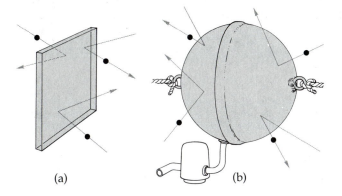

(a)   (b)

Figure 5-12. Gas pressure and the von Guericke demonstration. If air molecules are bouncing off both sides of a slab (a), the effects of pressure may be unimportant. However, if two hemispheres (b) are sealed together and evacuated with an air pump, the air molecules bombard *only* the outside of the hemispheres. This produces a strong force which holds the hemispheres together.

Related to the concept of energy is that of *work*. Work is done whenever an object is moved against a force; the work done on an object adds energy to the object. For example, if we move a book from the bottom to the top shelf of a bookcase, we increase its *potential energy* because we have moved it against the force of gravity. Giving the book a slight nudge so that it falls off the edge of the shelf and is accelerated downward by the force of gravity, we find that the potential energy is converted to kinetic energy as the book moves faster and faster toward the floor. The impact is greater for the same book when pushed from the top shelf than when pushed from the bottom shelf, because in the former case the book has more potential energy. Water stored in a reservoir behind a dam (in a hydroelectric power plant, for example) has potential energy. When the water is allowed to flow down through the dam, the potential energy is converted to kinetic energy. This energy of the moving water is in turn imparted to a water wheel or hydroturbine that drives a generator, which converts the kinetic energy to electrical energy.

CONSERVATION OF ENERGY

If we push a box along a rough floor, we are pushing against the force of friction. This work must add energy to the box; in fact, it appears as heat in the floor and the bottom of the box. Thus, kinetic energy of the moving box is changed to heat. A similar demonstration was cited by Count Rumford (circa 1798) on the basis of heat produced during the boring of a cannon. When work was done, heat was produced; these facts led to the idea that heat is a form of energy.

The principle of the *conservation of energy* states that the total energy of an isolated system is constant, although the form of the energy may change. This can be illustrated by a detailed look at a simple process such as picking up a rock and dropping it in a waste can. We can describe the energetics of this process as having started with the sun. The solar energy was transferred to the earth in the form of *radiant energy* or light. Photosynthesis in green plants converted this radiation into *chemical energy* stored in carbohydrates. A man ate some of the plants, and the calories of chemical energy stored in the carbohydrates were utilized by him to lift the rock. This added *potential energy* to the rock, which was converted into *kinetic energy* when the rock was dropped. As the rock hit the can a sound was heard, and the sound waves carried away *acoustical energy*. The waste can was also *heated* slightly. When the principle of conservation of energy is combined with the equivalence of mass and energy, as stated by Einstein, there are no known exceptions to this rule.

NUCLEAR ENERGY

Nuclear reactions take place in the interior of stars and in atomic bombs, achieving the goal of medieval alchemists—the transmutation of one element into another—and at the same time accomplishing the conversion of mass into energy.

The fundamental relation governing the conversion of matter into energy was determined by Albert Einstein (Figure 5-13):

$$E = mc^2$$

The energy, E, liberated by the annihilation of a certain amount of mass is equal to the mass annihilated, m, times c^2, the square of the velocity of light (all in the appropriate units).

This equation has an important application, as follows. A common sequence of nuclear reactions leads to the production of one helium atom from four hydrogen atoms; however, the total mass of four hydrogen atoms does not equal the mass of one helium atom, as can be seen from their atomic masses. Hydrogen has an atomic mass of 1.008, and the atomic mass of helium is 4.003. The difference in atomic masses between four hydrogen atoms and one helium atom is 4.032 minus 4.003, or 0.029.

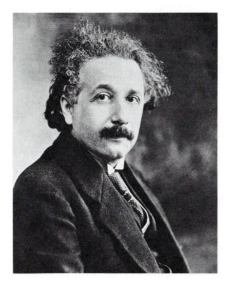

Figure 5-13.   Albert Einstein (1879–1955). (Yerkes Observatory photograph.)

Thus, in converting from hydrogen to helium, we lose slightly under one percent of the mass. This reappears as energy in the amount given by Einstein's equation. We will return to this subject in our discussion of the energy source of the sun (Chapter 9). Nuclear energy is released when the bonds due to the strong nuclear force are broken or otherwise rearranged. High temperatures (10 to 20 million degrees K) are required to do this in most cases (natural radioactivity is an exception); such temperatures occur in the central regions of stars. In nuclear reactions, energy appears to be created, but when we keep in mind that matter is at the same time destroyed, we find that a general law of *conservation of mass-energy* applies throughout the observable universe. No exceptions to this law have been shown to exist.

THE WAVE NATURE OF LIGHT

Astronomers receive the overwhelming majority of their information about the universe in the form of light or radiant energy emitted by the celestial objects. Astronomers cannot put their hands on a star or study it in the laboratory, so they must infer its properties by collecting and analyzing its light. Understanding the nature of light is therefore a matter of prime necessity to the astronomer.

Visible light is a specific example of what is more generally referred to as *electromagnetic radiation*. In fact, the *electromagnetic spectrum* (Figure 5-14) includes radio waves, infrared, visible light, ultraviolet, x-rays, and gamma rays. The detailed study of light has been one of the central problems of modern physics. In this book we limit ourselves to summarizing the conclusions of the physicists from an operational point of view as they apply to astronomy and the analysis of light received from the cosmos.

It is often difficult to describe light in simple terms; sometimes it acts like a wave and sometimes like a particle. In the first case, it appears to behave as though it were a transverse wave, that is, a moving disturbance, with the disturbance occurring at right angles to the direction of motion. Such a wave can be characterized by its amplitude and

Figure 5-14.   The electromagnetic spectrum. Note the very narrow range of visible light.

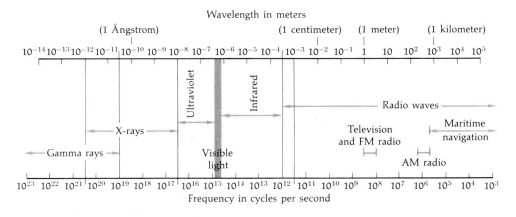

Figure 5-15.   Representation of light as a transverse wave; the electromagnetic disturbance associated with the wave is at right angles to the direction of motion.

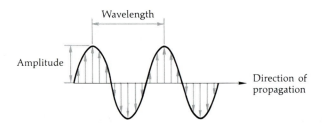

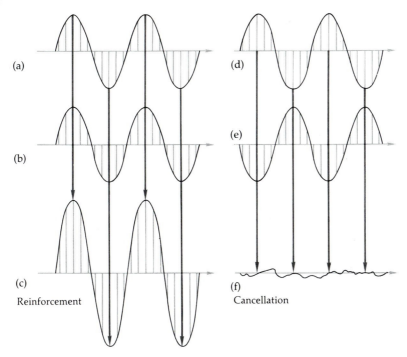

(a)

(b)

(c)

Reinforcement

(d)

(e)

(f)

Cancellation

Figure 5-16. Illustration of the phenomenon of interference through the examples of reinforcement—(a) and (b) are added crest to crest to form (c)—and cancellation—(d) and (e) are added crest to trough to form (f).

wavelength (Figure 5-15), which are found to correspond in familiar terms to the intensity and color of light, respectively.

Light waves are able to propagate through a vacuum. For a long time this point was disputed, and a hypothetical medium, the *ether*, was presumed to exist in space; the ether was supposedly disturbed by light waves just as the ocean water is disturbed by water waves.

Two important wave aspects of light are *interference* (Figure 5-16) and *linear polarization* (Figure 5-17). The interference effect is interpreted as the reinforcement or cancellation of light waves, depending on how their amplitude peaks and valleys coincide. For example, suppose that the crest of one wave at a certain point coincides with the crest of another. The two waves can be thought of as superposed, and the amplitude of the resultant wave at that point is greater than the amplitude of either of its com-

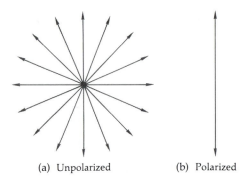

(a) Unpolarized   (b) Polarized

Figure 5-17. (a) Unpolarized and (b) polarized beams of light, sketched "end-on" as they travel at right angles out of the page. The lines represent the direction(s) of the electromagnetic disturbances in light waves that are moving toward the reader (that is, moving at right angles to the paper). In the polarized beam, all of the waves have the same direction of disturbance or oscillation. In the unpolarized beam, different light waves oscillate in different directions.

ponent waves. The result of this greater amplitude is a greater intensity of light at this position. On the other hand, suppose that the crest of a wave coincided with the trough of another wave. In this case, superposition results in a smaller amplitude. These effects are actually observed in experiments with light, and they find a natural interpretation if light is regarded as a wave phenomenon. Since the extent to which the crests and troughs of light waves coincide must depend critically on the respective wavelengths, we might expect interference phenomena to be important to the study of light of different wavelengths or colors. The way in which light interference caused by a *diffraction grating* separates light by color is shown in Plate 2.

Linear polarization is a concept rooted in the definition of a transverse wave as a disturbance perpendicular to the direction of motion. There are an infinite number of directions (represented by lines in a plane, Figure 5-17) that are perpendicular to the straight line that represents the direction of motion. When we are dealing with a great many light waves (say, the light received from a certain star), and we find that as many of them are disturbances oriented in one direction as in any other direction, then the light is *unpolarized*. However, if more of the waves are oriented in one direction than in any other, then the light is *partially polarized;* if all of the light is oriented in one direction, then it is *fully polarized.* A man floating on the ocean provides an excellent analogy—he only bobs up and down as a wave passes; he is not disturbed horizontally or at some intermediate angle because water waves are polarized in the vertical direction. There are other kinds of polarization, but this linear polarization is both the simplest and the most important type in basic astronomy. Like interference, polarization is a phenomenon that seems easily understood in terms of the wave nature of light. Some aspects of polarized light are illustrated by the familiar Polaroid sunglasses, which primarily pass light of a particular polarization. The sunglass makers have empirically determined the best orientation, so that little of the reflected, polarized sunlight (glare) comes through. If you look at the sky and rotate your Polaroid sunglasses slowly, it will be alternately bright and dim, depending on how the polarization of the light lines up with the preferred direction of the Polaroid material.

LIGHT AS PARTICLES

Sometimes, particularly when light interacts with matter, the light seems to behave as though it were composed of a stream of discrete particles, or photons, which have specific energies. A good example is the *photoelectric effect;* this term refers to the ejection of electrons from certain metal surfaces when light falls on them. Two basic aspects of this photoelectricity can be accounted for if we think of light in terms of particles. The first is an example familiar to astronomers, for the photoelectric effect is used to measure star brightnesses. A *photomultiplier* tube, placed at the focus of a telescope and illuminated by the light of a star, produces an electric current due to the electrons released by the photoelectric effect. When this technique is used to measure a very faint star, one finds that the current is intermittent and seems to be composed of individual pulses of electricity. These are hard to understand on the basis of the theory of light as a continuous wave, but are easily interpreted by the particle theory as being due to the arrival at the photomultiplier tube of individual photons from the star. (For a bright star, there are so many photons coming in that we don't notice gaps between their arrivals.) The second aspect is the dependence of the photoelectric effect on the color of light. After the discovery of photoelectricity by the German physicist Heinrich Hertz in 1887, experiments by other scientists revealed that the number of electrons emitted depended on the intensity of the light, but the velocities of the electrons depended only on the color of the light. Blue light releases more energetic (faster moving) electrons than does red light, and infrared light

Table 5-4. Correspondence of Colors and Wavelengths

Color	Wavelength Range (Å)
Violet	3,800–4,400
Blue	4,400–5,000
Green	5,000–5,600
Yellow	5,600–5,900
Orange	5,900–6,400
Red	6,400–7,500

may produce no electrons at all, depending on the particular metal involved. The facts are easily explained if light comes in discrete packets of energy. The energy of a blue photon exceeds that of a red photon, which exceeds that of an infrared photon, and so on. In terms of the wave theory, the blue, red, and infrared colors correspond to increasingly longer wavelengths (see Table 5-4). It takes a certain amount of energy to knock an electron out of a metal. Thus, Albert Einstein (who received the Nobel Prize in 1921 for this explanation of the photoelectric effect) reasoned that each photon of light that strikes the metal causes the release of either one electron or no electron at all, depending on the photon energy. If the individual photons do not have enough energy to knock the electrons out of the metal, no electrons are produced. When the photons have enough energy to produce electrons, part of their energy is used in removing the electron from the metal, and the remainder shows up in the speed (kinetic energy) of the electron. Blue light produces more energetic electrons than red light, and hence the shorter wavelengths correspond to more energetic photons. Observation shows that light also exerts pressure *(radiation pressure),* and this effect is also readily interpreted in terms of photons, just as atmospheric pressure is caused by gas particles or molecules.

How can light be both a particle and a wave? Perhaps we can turn the question around and ask why light *should* consist of a familiar entity like a wave or a set of particles? It is in fact a complex phenomenon that we can interpret or approximate in the familiar terms (wave, particle) that seem most helpful in particular cases.

ELEMENTARY SPECTROSCOPY

Spectroscopy, the analysis of the wavelength-dependent properties of light, was initiated by Sir Isaac Newton. While attempting to make some

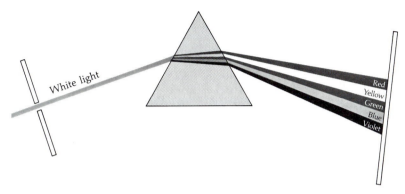

Figure 5-18. Sketch of Newton's spectral experiment. White light or sunlight passed through a prism breaks into a spectrum.

lenses for a telescope, he found that the images were always fuzzy. He could not correct the lenses, and so questioned whether or not the problem was with light itself. In a classic experiment, he passed a beam of sunlight through a prism and saw that it was broken up into colors of the rainbow, as shown in Figure 5-18. When the colors were focused on the same spot, the original character of the sunlight was restored. Newton immediately saw the reason for his fuzzy images. The bending of light by glass is used in a lens to produce a focus. His method of producing the spectrum showed that light of different colors is bent by different amounts by the glass of the prism. Thus, his lens had failed to focus sunlight sharply because the light of different colors was being focused at different distances from the lens.

The colors of light correspond to wavelengths or frequencies of oscillation (wave interpretation) or different energies (particle interpretation). Wavelengths of visible light are usually quoted in terms of the Angstrom unit (Å), which is 10^{-8} centimeters.[7] The approximate wavelength ranges for the major colors are given in Table 5-4.

A considerable advance in the development of spectroscopy was made in 1858 by the German physicist Gustav Kirchhoff. His experimental researches (in collaboration with the chemist Robert Bunsen) led to three important conclusions concerning the emission and absorption of light. These ideas, sometimes called *Kirchhoff's Laws of Spectral Analysis,* are:

---

[7]Named for A. J. Ångström, a nineteenth-century Swedish astronomer who studied the spectrum of the sun.

1. An incandescent solid or liquid, or a gas under high pressure, emits a *continuous spectrum* (that is, produces light at every wavelength over a wide range).[8]

2. Incandescent gases under low pressure give a spectrum composed of individual bright lines, and the number and wavelengths of these *emission lines* depend on the gas involved. (By line, we mean that the emission occurs at a particular wavelength or narrow range of wavelengths, as distinct from the continuous spectrum defined above.)

3. When the source of a continuous spectrum is viewed through a low pressure gas, dark lines appear in the continuous spectrum at exactly the same wavelengths as the bright lines emitted when the same low pressure gas is viewed in the incandescent state. (The dark features are called *absorption lines*.)

Kirchhoff's conclusions are illustrated by the experimental arrangement shown in Plate 6.

Understanding Kirchhoff's Laws requires an understanding of atomic structure. Specifically, we need to know why atoms absorb and emit light in certain characteristic wavelengths to produce lines. In addition, we need to know the physical conditions that lead to their production.

In the model of the atom as developed by Bohr, the electron orbits are not located randomly—only certain specific orbits are possible (Figure 5-19). Electrical force binds the electron to the positively charged nucleus. If an electron moves from one orbit to another, it must change in energy, as in the example of the book moved from one shelf to another. When an electron moves from an outer orbit to an inner orbit, the energy decreases. (In the terminology of atomic physics, we say that the atom changed to a lower *energy state.*) Since energy cannot be destroyed, the lost energy must appear in another form. In fact, it appears as a photon of light produced by the atom *(emission process).* Since the two atomic energy states corresponding to the outer and inner orbits have specific values, their difference is also a specific amount of energy. The energy of photons is related to the wavelength of light, and so when an electron moves from a particular outer orbit to a particular inner orbit, the atom in question always radiates light of the corresponding specific wavelength. This is the origin of the emission lines described above. In like manner, an electron

---

[8]Kirchhoff had visible light in mind. Today we know that any solid or liquid at a temperature above absolute zero emits a continuous spectrum in accordance with Planck's Law (see Figure 5-20). For cool objects, this continuum consists mostly of infrared light and is therefore invisible.

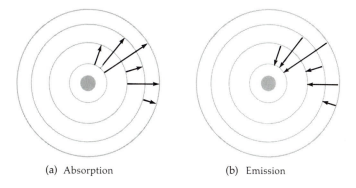

(a) Absorption   (b) Emission

Figure 5-19.   Allowed electron orbits and transitions. (a) Absorption.
(b) Emission. (Not drawn to scale.)

may move from an inner orbit to an outer one and, in this case, the atom
must have gained energy. Therefore, an energy source must exist, and it
may be a photon of light that the atom has encountered and absorbed
(absorption process). This photon has to have an energy just equal to the
difference of the two atomic energy states. Thus, if a hydrogen atom produces an emission line of red light at wavelength 6,563 Angstroms under
certain circumstances, it will also produce an absorption line at that
wavelength under other circumstances, as can be proved in the laboratory. An important corollary is that when we see absorption lines at the
wavelength of 6,563 Angstroms in the spectra of stars (as we frequently
do), we know that these stars must contain hydrogen atoms, and that the
circumstances in these stars are such that the atoms are able to exist in the
two energy states involved in the production of this line. A similar remark
applies to the quite common observation of an emission line at 6,563
Angstroms in the spectra of gaseous nebulae (Chapter 12); the nebulae
also must contain hydrogen atoms, and at least a crude idea of the nebular conditions can be gotten from the observation of this emission line.
For example, by Kirchhoff's second law, nebulae that produce this emission line must be composed of incandescent gas at low pressure.

(The existence of discrete energy levels in the atom is also expressed
by the statement that the atomic levels are *quantized*. For this reason,
photons are sometimes referred to as *quanta*. The scientific analysis of
quantized atomic energy states was largely begun by the German physicist Max Planck in the early twentieth century.)

It is found that when electrons are in the higher energy levels, they tend to drop into lower levels spontaneously, until they reach the lowest permissible level not already occupied by an electron. Thus, the process of absorption of a photon is eventually followed by the process of emission. A useful analogy is to think of the atom as a sort of quantum staircase in which the individual levels (steps) slope slightly forward and are not necessarily of equal heights. The steps are analogous to the energy levels in the atom and the slope is indicative of the tendency for *spontaneous emission.* Suppose we lift up a ball from the floor and place it on one of the higher steps. Due to the slight slope, the ball rolls off the step and the potential energy that we gave it by lifting it from the floor is converted into kinetic energy. In the atom, energy involved in raising an electron to a higher level reappears as one or more light quanta when the electron drops to a lower level or to a succession of lower levels.

Electrons in an atom can also be excited to a higher energy state by collisions with *free* electrons (those not bound to any atom), atoms, or molecules. The requisite amount of energy is taken from the kinetic energy of the colliding particle, so the latter particle moves at a slower speed after such a collision. This process occurs in a hot gas. The electrons in the atoms (*bound* electrons) then drop down spontaneously to produce the bright lines in the spectrum. Atoms in liquids, solids, and in gases under higher pressure have their energy levels smeared out by electric effects of their close neighbors. Since the allowable energy values are no longer highly specified, a great many energy differences correspond to the allowable changes in energy, and hence these atoms produce a continuous spectrum.

Two other phenomena associated with light quanta and atoms should be mentioned here. Often the absorption of a photon is followed immediately by the emission of a photon of the same energy in an arbitrary direction. To an external observer the photon appears to have simply changed direction. This process is called *scattering.* If, on the other hand, the re-emission does not occur immediately and the atom suffers a collision with some other particle, the energy gained by absorption of the photon may be transferred to the motion of the colliding particle. In this case a photon goes into the atom but does not come back out because its energy reappears as kinetic energy rather than radiation. This process is called *pure absorption.* Note that in all the phenomena described in this section, with electrons moving from one energy level to another, we refer only to the outermost electrons of an atom.

This brief review of atomic processes enables us to understand Kirchhoff's Laws as illustrated in Plate 6. Photons at certain specific wavelengths are absorbed by the sodium vapor to produce the absorption line spectrum in the light from the tungsten filament as viewed through the vapor. The light absorbed in the vapor is re-emitted at the same wavelength but in all directions (scattering), causing the vapor to become incandescent and show a bright line spectrum. In this case, the energy supply for the incandescent vapor is the light from the tungsten filament. In other circumstances, the energy supply could come from the heat of the vapor itself, with the bright lines excited by electron collisions.

The principles used to explain Kirchhoff's Laws enable us to identify elements in gases under observation, for example, in the sun's atmosphere. In a crude way we can think of continuous radiation from the sun's interior shining through its atmosphere, which produces absorption lines at specific wavelengths corresponding to the elements present. This is observed, and many elements have been identified (see Plate 7) from the presence of their characteristic lines. The strength of the absorption line (how much light is removed as judged by the strength of the light at adjacent wavelengths) depends in part on the amount of the element present. Obviously, the rarer elements are less likely to produce strong lines. Spectroscopic analysis of emission and absorption lines constitutes our primary evidence concerning the abundance of the elements in the sun and the universe beyond the solar system.

RADIATION LAWS

Two important laws describe the amount and wavelength distribution of light emitted from the surfaces of objects at different temperatures. Strictly, these laws apply to an ideal radiator, originally called a blackbody because it could be approximated by a surface covered with lampblack. In practice, many things, including astronomical bodies, are fairly good approximations to ideal radiators.

The correct wavelength distribution of light from an ideal radiator was first calculated by Planck in 1901. The intensity of emission as a function of wavelength for different temperatures is shown according to *Planck's Law* in Figure 5-20. The energy source for this radiation is the motions (kinetic energy) of the atoms or molecules in the emitting material. The motions increase with temperature, and such radiation is therefore called *thermal radiation*.

Inspection of Figure 5-20 shows two important features. As the temperature increases, so does the *total amount* of radiation emitted, and the wavelength of greatest emission occurs at *shorter wavelengths*. Thus, if the temperature of an emitting surface is slowly increased, the peak of the

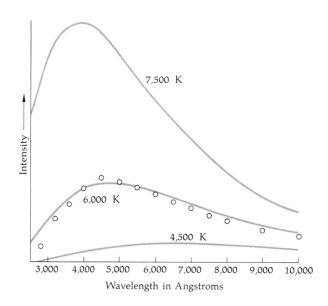

Figure 5-20. Planck's Law, the variation of intensity as a function of wavelength for an ideal radiator at different temperatures (solid curves). The circles are the observed intensities of solar radiation and show that the solar radiation can be approximated by an ideal radiator at about 6,000 K. The graph shows that a slightly lower temperature would be a better approximation, and the accurate value is 5,750 K.

emission will move in the direction red—orange—yellow—green—blue —violet, and so the color of an object, such as a star, depends on its temperature. (The color as perceived by our eyes also depends on the relative sensitivity of our eyes to light of different wavelengths.)

The variation of total energy output with temperature is governed by the *Stefan-Boltzmann Law*. The total energy outputs at different temperatures can be represented by the areas under the various curves in Figure 5-20. The total amount of energy radiated increases as the absolute temperature (K) raised to the fourth power. Thus, an ideal radiator with twice the absolute temperature of another one of equal size radiates (2) × (2) × (2) × (2), or sixteen times more energy.

These laws can be applied (for example) to the determination of the temperature of the sun's visible surface, or *photosphere*. Both Planck's Law and the Stefan-Boltzmann Law indicate a temperature of about 6,000 K for the photosphere. The agreement is not exact because the sun is not an ideal radiator.

DOPPLER EFFECT

An apparent change in the pitch (frequency) of sound waves, occurring when the sound-maker and the listener are in motion relative to one another, is a well-known phenomenon. For example, the whistle of an approaching train is higher pitched when the vehicle is approaching the listener than when it is receding. This Doppler effect (explained by the Austrian physicist Christian Doppler in 1842) occurs for both sound waves and light waves.

The crux of the matter is that it takes a certain amount of time for a cycle of a wave to be emitted. If a light source is moving, the beginning and end of the wave are emitted at different locations. This causes a lengthening or shortening of the wave (as perceived by a stationary observer), depending on the direction of motion of the source (see Figure 5-21). If the wavelength is shortened (approaching source) but the velocity of light remains the same, more waves must pass the observer per second. Thus, the light appears to have a smaller wavelength (bluer color). For a receding source, the wavelength is lengthened, fewer waves pass by an observer per second, and the light seems redder. Exactly the same effects occur in the case of a stationary source and a moving observer—approach causes an apparent wavelength decrease.

In astronomical spectra, absorption and emission lines are found shifted toward the red or toward the blue in comparison to laboratory standard spectra, due to the motions of the stars and other astronomical objects with respect to the observer on earth. Since a faster motion produces a greater Doppler shift, measurement of the amount of shift in Angstroms reveals the speed at which the object and the observer are receding from or approaching each other.

VELOCITY OF LIGHT

Light travels at a very large but finite velocity. Thus, light reaching an observer was generated at some time in the past. For example, the travel time for light from the sun to reach the earth is about 8 minutes. Radio waves are a form of light, and the time required for them to travel between the earth and moon caused the slight delays in communication with the astronauts on the moon.

After the basic ideas about light were established (such as traveling from the object to the eye and not vice versa as some early philosophers believed), the first serious attempt to measure the speed of light was carried out by Galileo. Assistants with lanterns were stationed on successive hills. The first man uncovered his lantern. As soon as the second man

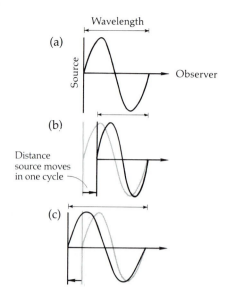

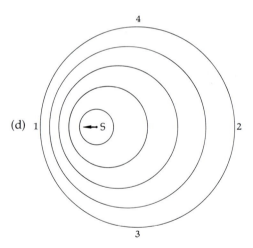

Figure 5-21. The Doppler effect. The wavelength (and frequency) of an emitting source *as perceived by an observer* is altered by motion with respect to the observer. (a) Stationary source; (b) moving toward observer; (c) away from observer. (d) Another way to understand the Doppler effect. Sketch of the positions of crests of waves emitted by a source, S, that is moving in the direction indicated by the arrow. The most recently produced crest is represented by the smallest circle. An observer at position 1 receives Doppler-shortened waves, whereas an observer at 2 receives Doppler-lengthened waves. At 3 and 4, essentially no Doppler shift is observed.

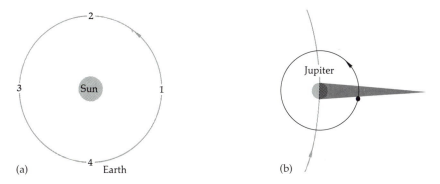

Figure 5-22. Timing of the eclipses of Jupiter's moons gave the first determination of the velocity of light (see text).

saw the light from the first lantern he was to uncover his lantern; the elapsed time required for light to travel twice the distance between the observers could be measured by the first man. Galileo and his associates soon realized, however, that what they were actually measuring was the reaction time of the observers and not the velocity of light. They could only conclude that the velocity of light was very great.

The first evidence for a finite velocity of light came from observations of the moons of Jupiter. These satellites pass behind Jupiter and are eclipsed, but a curious phenomenon was found by analyzing the times at which the eclipses occurred. Referring to Figure 5-22, the eclipses were on time when the earth was at positions 2 and 4, but early when the earth was at position 1, and late when the earth was at 3.

This phenomenon was discovered and explained in 1675 by the Danish astronomer Ole Roemer. He pointed out that the irregularities in the eclipse schedule could be understood if the light took more time to reach us when the earth was farther away from Jupiter; thus, its propagation could not be instantaneous. His crude data (especially concerning the scale of the solar system) led him to a determination of the velocity of light as about 210,000 km/sec. This value turned out to be low by about 30 percent, which is not bad for a determination made about 300 years ago. Most importantly, his method and reasoning were correct.

About 50 years later, Bradley's discovery of aberration (p. 45) confirmed Roemer's finding that light does not propagate instantaneously. The inferred value for the speed of light was approximately the same as

Roemer's. Many modern methods are available for determining the speed of light, and the accuracy has been constantly improved. The current value is 299,792.5 km/sec (about 186,000 miles/sec).

Telescopes and Observatories

With the exception of the occasional meteorite that reaches the earth, astronomy has been limited (until the recent advent of space travel) to *observational* studies. Unlike physicists and chemists, astronomers could not experiment with the objects under investigation. Whatever we learned about the stars came from analyzing their light from a distance. As most of the information in the remainder of this book is based on the results of studying light from the sky, it seems desirable at this point to review the instruments that are used for this purpose. In recent years increasingly complex instrumententation has become vital to advances in many areas of astronomy, but many observational results are obtained with relatively simple optical systems, and among them the human eye is a most important example.

THE HUMAN EYE

Light from a star or other source passes through the *cornea* and *pupil* of the eye. The size of the pupil is adjusted by muscles in the *iris* that widen it to accept more light from dim sources or contract it to exclude excessive light from bright sources. The light that enters is bent *(refracted)* by the materials that compose the various parts of the eye (Figure 6-1). Most of the refraction occurs in the cornea, but the *lens* does the fine focusing: the shape of the lens is adjusted by a muscle in order to form sharp images of objects at different distances. Thus, the iris acts like the diaphragm on a camera, which the photographer adjusts to various "f numbers," depend-

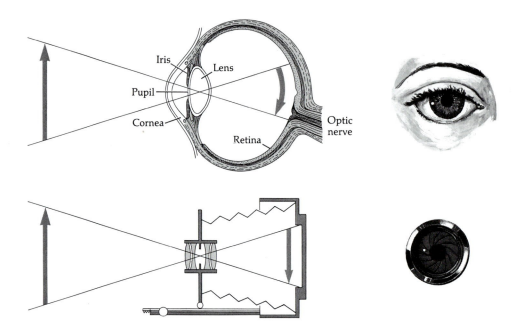

Figure 6-1. A comparison of the optical properties of the eye and the camera. (After *Eye and Camera* by George Wald. Copyright © 1950 by Scientific American, Inc. All rights reserved.)

ing on the brightness of the scene. The lens muscle is analogous to the focusing mechanism of the camera.

The image of the object is focused by the lens onto the *retina*. This surface contains a large number of small structures called *rods* and *cones*, which function, respectively, under conditions of dim and bright light. When light is absorbed by the molecules in a rod or cone, it produces an electrical effect that is sensed by a nerve, which transmits the information to the brain. Thus the retina corresponds to the light-sensitive surfaces in the astronomical detectors (such as photographic plates and photomultiplier tubes) discussed later in this chapter. Under ideal circumstances, the eye can sense as few as 2 photons, absorbed in adjacent rods within an interval of 0.1 second.

CAMERAS

The basic idea of a camera is simple. A lens (or lenses) at the front of the camera focuses an image on the film (Figure 6-1). A shutter opens briefly to allow enough light through to properly expose the film. The telescope used by an astronomer is not really very different; lenses or mirrors are used to form an image at a place called the *focal plane*. Depending on the purpose of the observation, the astronomer may place a photographic film or plate at the focal plane, or another instrument can be mounted there to analyze or record the light. Cameras are generally held in the hand, but when steadiness is important the photographer uses a tripod. Steadiness is almost always important in using a telescope, since many of the stars under observation are so faint that the astronomer must take a long exposure in order to record them properly. During this time, the stars move across the sky (due to the rotation of the earth), so the telescope must be appropriately mounted with a *clock drive* to move it at the correct rate to track the stars.

THE FIRST TELESCOPES

As far as can be determined, the first telescopes were constructed in Holland, perhaps as early as 1604. They were intended, at least in part, for military applications, but within a few years rumor of these devices reached Galileo in Italy. By 1609 he had built his own telescope and made the first astronomical observations with optical aid. An explanation of the principles by which such *refractors* (telescopes that use lenses) work was published by Kepler in 1611; by the mid-seventeenth century they were in use throughout Europe, although some astronomers continued to build new lensless instruments patterned after those of Tycho. In 1671, Newton built the first *reflector* (telescope that uses mirrors) in order to overcome some of the problems with glass lenses, including the poor focus of light of different colors, as noted in Chapter 5. (See Figure 6-2.)

GALILEO, THE TELESCOPE, AND THE SCIENTIFIC METHOD

For centuries the ideas of Aristotle, Ptolemy, and other ancient Greeks had dominated Western scientific thinking. As we have seen in Chapter 2, the work of Tycho and Galileo in obtaining new facts by observation and measurement led to the adoption of a revolution in thinking—the heliocentric world-view. In fact, Galileo, in his additional capacity as a laboratory physicist, conducted a variety of experiments that tested and sometimes revised previous ideas. Perhaps Galileo's greatest contribution was not a particular experiment or discovery, or even the sum of all his discoveries—it was his life-long fight against the dogmatic assertion (and meek acceptance) of scientific "truths" untested by experiment. His great

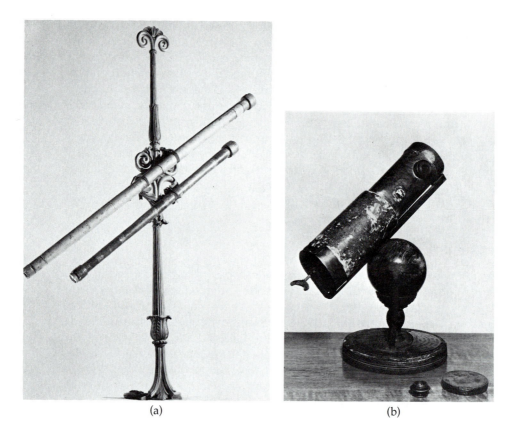

(a)   (b)

Figure 6-2. Early telescopes. (a) Galileo's refracting telescopes. (Courtesy of Prof.ssa Dott.ssa Maria Luisa Righini-Bonelli, Istituto e Museo di Storia della Scienza, Firenze, Italy.) (b) Newton's reflecting telescope. (Courtesy of the Royal Society.)

legacy is thus the active use of the *scientific method,* an approach to research that is now the basic procedure in all fields of science. Confronted with a set of "facts" consisting of data or measurements on some aspect of nature, a scientist proposes a theory or "model" to explain them; the characteristics of the model enable us to predict further properties of the subject being studied; new observations and measurements are then made to test these predictions. The last step in turn usually leads to a slightly (or greatly) improved theory from which new predictions are made, and the cycle continues. We generally find that the hard part in this scheme is to choose and perform a *critical* experiment, namely, one that really distinguishes between rival theories. An important adjunct to the scientific method is the test of *reproducibility;* that is, we are not prone to accept the results of one scientist if they cannot be duplicated by others who repeat the experiment under similar conditions.

Some of Galileo's astronomical discoveries were mentioned in Chapter 2. Several others are described here to illustrate the point that a new instrument capable of making observations or measurements not previously feasible will inevitably change our view of the nature of things.[1]

Galileo turned his telescope to the moon and found not the unblemished surface cherished by some earlier philosophers, but a *world* of mountains, craters, and vast dark plains. He looked at the sun, as did many of his contemporaries, and watched the sunspots move across the solar disk. Sunspots were known even in pretelescopic times, as some were occasionally big enough to be seen by the unaided eye. However, they were generally thought to be planets located between the earth and sun that were simply being seen in projection against the solar disk. Galileo correctly surmised that they were fixed to the solar surface, and he used them to measure the solar rotation rate, and thus to prove that the sun turns. He also found apparent appendages of Saturn that subsequent astronomers, with better telescopes, discovered to be rings. He looked at the smooth surface of the Milky Way with his telescope and realized that it was nothing less than the assemblage of a myriad of faint stars. As we mentioned in Chapter 2, he also made fundamental discoveries about Jupiter and Venus, which had extreme impact on the development of the heliocentric world-view. Some of these observations, as recorded in Galileo's notebooks, are shown in Figure 6-3.

REFRACTING AND REFLECTING TELESCOPES

The purpose of a telescope is to collect light and to focus it on a small area where it can be conveniently inspected, photographed, or measured. This purpose can be achieved through the use of lenses in a refracting telescope, or *refractor*, or it can be achieved by mirrors in a reflecting tele-

---

[1]Just as this was true for Galileo, it is true today. The first radio telescope (built by Karl Jansky at the Bell Telephone Laboratories in 1931) revealed unsuspected emission from the Milky Way galaxy (Chapter 12); the first attempt to detect x-rays from the moon (by Ricardo Giacconi of American Science and Engineering, Inc., in 1962) resulted in the discovery of the first example of a new class of objects, the galactic x-ray sources (Chapter 14); the 200-inch telescope at Mount Palomar was used in the discovery of the enormous red shifts of the quasars (Chapter 16).

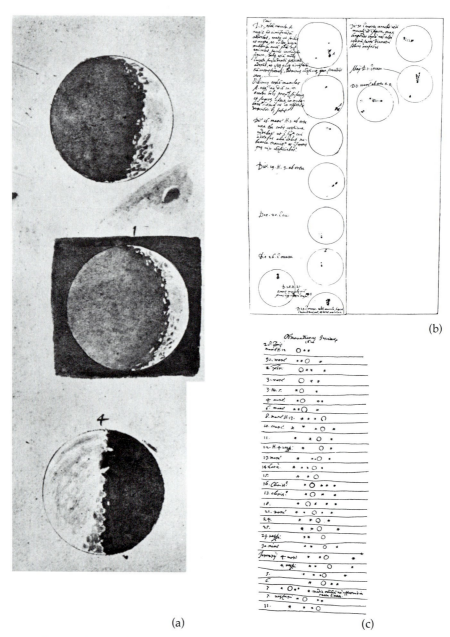

(a)   (c)

(b)

Figure 6-3. Galileo's observations. Drawings in his notebooks show (a) the surface of the moon, (b) sunspots, and (c) Jupiter and its four largest satellites at different times. (Yerkes Observatory photographs.)

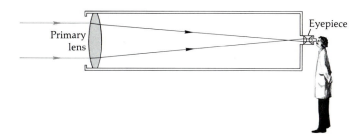

Figure 6-4. Sketch of a refracting telescope. In practice, the "primary lens" is usually composed of two closely spaced lenses.

scope, or *reflector,* as shown in Figures 6-4 and 6-5. The diameter of the light-collecting mirror or lens is called the *aperture* of a reflector or refractor. When we speak of a bigger telescope, we mean one with a larger aperture, not necessarily one that is longer.

Why is a big telescope better than a small one? Figure 3-3a shows that the light received from a distant point (such as a star) is essentially parallel, a fact Eratosthenes realized long ago. Now consider the parallel light from a star as shown in Figure 6-6; clearly a telescope with a larger light-collecting lens (or mirror) intercepts more of the starlight, and therefore it produces a brighter image of a given star. When photographing the stars, a larger telescope enables an observer to obtain an equivalent photograph in a shorter exposure time than if he had used a smaller telescope. Because fainter objects can be recorded with larger telescopes, and a distant star appears fainter than a similar star that happens to be close to us, larger telescopes allow us to see farther into space.

Many astronomical investigations require detailed analysis or measurement of an image. For example, when we observe the moon or Mars, we want to see the fine details of their surfaces, not merely measure the intensity of their light. Larger telescopes can provide sharper images, enabling us to *resolve* finer details in the images (Figure 6-7). Unfortunately, turbulent motions of the earth's atmosphere cause a blurring of telescopic images, so the full resolving power of the larger telescopes is rarely attained. This is one reason why astronomers are interested in using telescopes on spacecraft above the atmosphere.

HISTORICAL TELESCOPES

Historically, major new telescopes have led to significant advances in the astronomical knowledge of their times. Two examples are given here; in

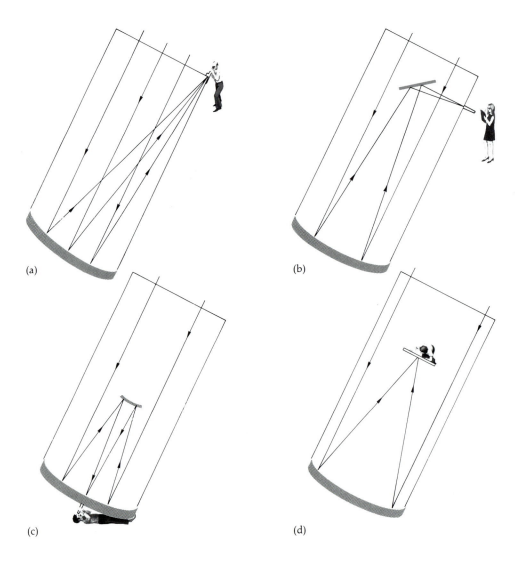

Figure 6-5. Schematics for reflecting telescopes showing four different ways of viewing the image: (a) method employed by Herschel; (b) *Newtonian* method; (c) *Cassegrain* method, the most common arrangement in modern telescopes: (d) *prime focus* method used with very large telescopes. Not shown is the Coudé arrangement, in which additional small mirrors are used to divert the light to a large stationary spectrograph. (a, b, c are after *Matter, Earth, and Sky*, 2d ed., by George Gamow. Prentice-Hall, Inc. 1965. Used by permission.)

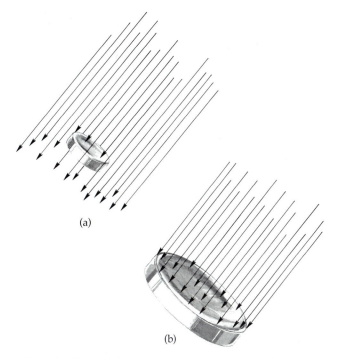

(a)

(b)

Figure 6-6.   (a) The smaller telescope intercepts (and therefore can collect) less light than (b) the larger telescope.

one case the size of the telescope was of prime importance; in the other case the quality of fabrication was paramount.

72-Inch Reflector in Ireland

In 1845, the third Earl of Rosse built by far the largest telescope of his time. The main mirror was fabricated from a metallic alloy called speculum; the great light-collecting and resolving powers of this 72-inch (183-cm) mirror[2] enabled Lord Rosse to discern the spiral nature of many galaxies and to discover the filaments in the Crab nebula. The origin of spiral structure remains as one of the great unsolved problems in the study of galaxies; observations of the Crab filaments later provided dramatic confirmation that the nebula is expanding into space as the result of an ancient explosion.

---

[2]In this chapter, we depart from our practice of giving precedence to metric units; many of the telescopes discussed here have been named in English units.

Figure 6-7. Resolution and the quality of an image. The spiral galaxy in Andromeda (M31) is shown as it would appear (a) as photographed with a large optical telescope, or with resolutions of (b) 1 minute of arc; (c) 3 minutes of arc; and (d) 12 minutes of arc. (Courtesy Leiden Observatory.)

Eighteenth-Century English Reflector

Sir William Herschel (1738–1822) attained great skill in shaping and polishing telescope mirrors. Although he began as an amateur astronomer, he was soon making telescopes that surpassed those of the Greenwich Observatory in optical performance. The sharp images obtained with one of his smallest telescopes enabled him to recognize the unusual nature of what might otherwise have seemed to be an ordinary star. Viewed through Herschel's telescope, this object had a slightly larger image than the other stars, and the image increased in size with greater magnification. Viewed on subsequent nights, it was found to have moved against the background of stars, and was soon shown to be a planet. Herschel had discovered Uranus, the seventh planet of the solar system and the first new world of the telescopic age. The importance of his telescope's quality in resolving the disk of the planet cannot be underestimated. Going back through the older records of other astronomers, it has been found that Uranus was observed many times before Herschel saw it, but his predecessors noted it as just another "star."

MAJOR TELESCOPES TODAY

The largest telescopes today are reflectors. The biggest one is the 6-meter (236-inch) reflector at Mount Pastukhov in the Caucasus mountains of southeastern Russia (Figure 6-8). It began operating in February, 1976, limited however by the somewhat unsatisfactory optical quality of the primary mirror. A new mirror is being made. Next largest is the 200-inch (5-meter) Hale reflector at Mount Palomar in Southern California (Figure 6-9). Ranking after this are three telescopes of about 4 meters (158 inches) in size at Kitt Peak in Arizona, Cerro Tololo in Chile (Figure 6-10), and Siding Spring, Australia. Many astronomers consider these 4-meter reflectors to be the finest telescopes in the world. They were completed in the 1970s by engineers who made full use of modern advances in electronics, optical materials, and optical design. This includes the use of new mirror materials that are less subject to the slight deformations caused by temperature change, the use of computers to control the telescopes and help process the measurement data (Figure 6-11), and prime foci with lower "f numbers" than heretofore attained in large telescopes. This last quality means that shorter exposure times are required to photograph faint galaxies and nebulae. In addition, the size of the field of view (in angular units) that can be photographed sharply at the prime focus is much larger, for example, than for the 200-inch telescope. Thus a smaller number of exposures (and therefore much less telescope time) is required to record an extended region, such as a cluster of galaxies.

Figure 6-8. The 6-meter (236-inch) telescope of the Special Astrophysical Observatory near Zelenchukskaya in the Soviet Union. (Courtesy of Dale P. Cruikshank, University of Hawaii.)

A special type of telescope, the *Schmidt camera,* uses both a primary mirror and a smaller lens, called the *correcting plate* (see Figure 6-12). It is used to photograph a large region of the sky at one time. The light-gathering power of this telescope is determined by the size of the correcting plate because, although the primary mirror is larger than the corrector, it is the latter object that intercepts the light from a star. Thus, the large Schmidt camera on Mount Palomar (Figure 6-13), which was used to make a photographic sky atlas (*Palomar Sky Survey*), is referred to as a 48-inch (122-cm) telescope, although the primary mirror has a diameter of 72 inches (183 cm).

The largest refractor, a 40-inch (102-cm) telescope at the Yerkes Observatory in Wisconsin, was built during the late nineteenth century. Nearly all large telescopes built since then have been reflectors (or Schmidt cameras, in which the largest optical component is also a mirror). There are several reasons for this. Among them is the fact that the operation of figuring (grinding and polishing) the surface of a large lens or mirror to the necessary optical precision can require a year or more of expert labor. This work must be done on both sides of a lens; however, it is required on only one side of a mirror because the light does not pass through the glass. By the same token, the difficult (and expensive—a modern 150-inch mirror costs about one million dollars) problem of casting a large piece of high quality, bubble-free glass is reduced for a mirror, compared to a lens, since the important region is limited to a thin layer near the reflecting surface. Also, the lens in a large refractor sags and changes shape slightly as the telescope is tilted at different angles to view stars in different parts of the sky, and this degrades the optical performance of the telescope. A mirror, on the other hand, can be mechanically supported not only around its rim (as is done for a lens) but also across its rear surface. Thus it is possible to mount even the largest mirrors so that

Figure 6-9. The 200-inch (508-cm) Hale reflector located on Mount Palomar. Many important discoveries, notably in the study of distant galaxies and quasars, have been made with this telescope. (Courtesy of the Hale Observatories.)

Figure 6-10. Cerro Tololo Inter-American Observatory near La Serena, Chile. The largest dome houses the 4-meter (158-inch) reflector, which is (by a few inches) the largest telescope in the southern hemisphere. Other domes contain telescopes ranging in size from 152 cm (60 inches) down to 41 cm (16 inches), including a Schmidt camera. The observatory location in the Andes, above the Atacama Desert, is considered one of the world's best. (The Cerro Tololo Inter-American Observatory.)

they produce sharp images regardless of the telescope orientation. Still another major consideration is Newton's discovery that a lens focuses light of different wavelengths (colors) at different distances. Due to this effect of *chromatic aberration,* while one color is in focus, all the other colors will be out of focus, and the recorded image will not be as sharp as one might expect from the aperture of a refractor. On the other hand, the focus of a mirror does not depend on the wavelength, and thus reflectors are not affected by chromatic aberration. Finally, new forms of glass and quartz developed in recent years make it possible to fabricate large mirrors that are lighter in weight and less sensitive to temperature changes. The primary mirror for a large modern telescope is shown in Figure 6-14.

Figure 6-11. The control console of the Cerro Tololo 4-meter telescope. (The Cerro Tololo Inter-American Observatory.)

Figure 6-12. Schematic diagram of a Schmidt camera.

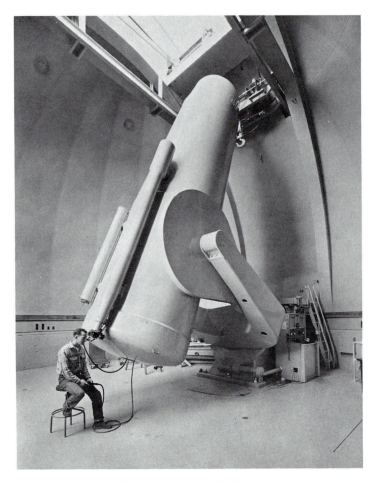

Figure 6-13. The 48-inch (122-cm) Schmidt camera at Mount Palomar. (Courtesy of the Hale Observatories.)

Telescopes of the future are unlikely to continue the trend toward larger and larger primary mirrors. For one thing, the brightness of the faintest objects that can be recorded with existing telescopes is severely limited by the existence of faint light in the night sky and by atmospheric effects that smear the images of stars. In fact, it has been shown that the faintest objects that could be detected with a 400-inch telescope would be only some 40 percent fainter than those recorded by the 200-inch telescope, although the money needed to build the larger telescope would pay for at least five 200-inch ones. For this and similar reasons there is considerable work underway in studying techniques to use telescopes of conventional size more effectively. Such methods include automation, combination of observations made simultaneously by several telescopes, and creation of more sensitive detectors for use with existing telescopes. In the last category are included the development of better photographic plates, and even the replacement of photography by electronic image recording techniques. A "Multi-Mirror Telescope," with six separate 72-inch

Figure 6-14.   The primary mirror of the 120-inch (305-cm) telescope at Lick Observatory. The development of a large telescope is a team effort requiring a variety of technical skills. Shown here are two astronomers, a mechanical engineer, and an optician. (Lick Observatory photograph.)

Figure 6-15. The 140-foot (43-meter) dish at NRAO. This telescope was used in Project Ozma, the first search for radio signals from extraterrestrial civilizations. (Courtesy of National Radio Astronomy Observatory, Green Bank, West Virginia.)

primary mirrors to simulate the performance of a much larger primary, has been installed at Mount Hopkins, near Amado, Arizona.

RADIO TELESCOPES

Light, as the term is used in astronomy, is not limited to the radiation that our eyes can perceive; it includes all the forms of electromagnetic radiation that we mentioned in Chapter 5. Each of these (radio, infrared, visible light, ultraviolet, x-rays, gamma rays) describes light within a certain range of wavelengths. In fact, among the most fundamental advances in twentieth-century astronomy has been the observation of the cosmos in radio waves, x-rays, and other wavelength regions not accessible with ordinary visible-light telescopes of the sort just described. The great importance of these observations has been that as each new wavelength region has been investigated, astronomers have discovered celestial sources of such radiation that have a completely different character than the familiar stars, planets, nebulae, and galaxies of visible-light or optical astronomy.

The most familiar radio telescopes are the "big dishes" that resemble radar antennae; they are simple reflecting telescopes for radio waves (Figure 6-15). However, the elaborate electronic techniques that have been developed for processing radio signals also make it possible to operate several dish antennae simultaneously as one instrument (*radio interferometer,* Figure 6-16). Furthermore, the *elements,* or individual antennae, of some radio telescopes are not reflectors, but remind us of those commonly used with short-wave and television receivers. In any case, the function of a radio telescope, like that of an optical telescope, is to collect electromagnetic radiation and feed it to devices that detect, analyze, and record it.

The laws of optics show that a desired telescope performance in terms of *angular resolution* (the ability to discriminate fine detail) depends on the ratio of telescope aperture (diameter) to radiation wavelength. Telescopes used for longer wavelength radiation must be larger than those used to observe short wavelengths in order to resolve equally fine detail. Since radio waves are enormously longer than optical light waves, radio telescopes are usually much larger than ordinary telescopes (they are also larger because the cosmic radio emissions are very weak, and because it is practical to build large radio telescopes). For example, a 10-inch (25-cm) optical telescope has a resolving power of about 0.5 seconds of arc at a wavelength of 5,000 Angstroms. In order to resolve equally fine detail at the wavelength of 21 cm, which is over 400,000 times longer than the 5,000 Angstrom wave, a simple dish-type radio telescope would have to be more than 60 miles (97 kilometers) in diameter. However, the radio telescope at Arecibo, Puerto Rico, which is the largest existing single-dish antenna, has a diameter of only 1,000 feet (305 meters; Figure 6-17).

To obtain higher resolution than possible with a single antenna, radio astronomers can combine measurements of the same object made simultaneously with the several widely spaced dishes of an interferometer (Figure 6-16). For ultrahigh resolution, beyond the limits of optical telescopes, they arrange for simultaneous measurements with separate radio telescopes located far apart on the earth (for example, at U.S. and Soviet observatories). In this technique of *very long baseline interferometry* (VLBI), the data from each telescope are recorded on magnetic tape along with high-precision clock signals, and are later combined in a computer to produce the final results.

The great sensitivity of radio telescopes means that interference from man-made signals, such as radio, radar, and television, is a serious problem. By international agreement, broadcasting on certain frequencies is prohibited to allow radio astronomers to receive cosmic signals with a

Figure 6-16. The computer-controlled, 5-km interferometer radio telescope, developed by Sir Martin Ryle at Cambridge University. Four of the eight dish-type antennas can be moved along a rail track. The telescope, completed in 1971, can measure positions of radio sources with accuracy equalling that of the best optical telescopes but produces maps that are not quite as sharply resolved as fine optical photos. (Courtesy of Sir Martin Ryle, Cambridge University.)

Figure 6-17. The 1,000-foot (305-meter) diameter telescope at Arecibo, Puerto Rico. This instrument has been used to map Venus by radar and for many studies of pulsars. (Courtesy of the Arecibo Observatory.)

minimum of interference. Our planet is so noisy due to radio transmissions that some people believe that life on other worlds may discover *us*.

OBSERVATORY SITES

Until fairly recently, observatories were located on convenient hills near their parent organizations, generally universities or government agencies. The objective was simply that the view of the sky be unobstructed by natural or artificial structures. Unfortunately, the development of modern civilization and its concomitant urban technologies has resulted in the many forms of environmental pollution that face us now. Among these, and perhaps least appreciated (except by the astronomers!), is *light pollution*.[3] The increasing brightness of the night sky, due to atmospheric scattering and reflection of artificial light, limits our ability to observe faint stars at night, just as the scattering of sunlight in daytime prevents us from seeing even the brightest stars. This is a prime reason for the location of all major new observatories at considerable distances from urban centers. At some locations in the United States, local governments actually have enacted protective ordinances limiting nighttime lighting in cities close to observatories.

Other factors besides sky brightness are important in the selection of an observatory site. Obviously, the frequency of clear skies is important, so one studies the weather and cloud cover records; a recent, very effective method for doing this is the analysis of weather photographs (Figure 6-18) obtained by meteorological satellites. Another criterion is the astronomical *seeing*, a term that describes the steadiness and size of star images as affected by turbulent motions in the atmosphere. (Seeing includes the twinkling effect familiar to anyone who has ever looked at the stars.) Clearly, the lens- and mirror-grinder's skill is of little avail if the telescope is mounted in a location where the atmospheric conditions ensure that the stars will usually be blurred by seeing or obscured by clouds. The seeing criterion is a principal reason for the location of observatories on mountain tops, where they are often above a good fraction of the turbulent air and the air-borne dust. In fact, the hypothetical 10-inch telescope mentioned earlier could rarely attain its theoretical resolution of 0.5 seconds of arc, because the atmosphere is seldom stable enough. The water vapor in the atmosphere is a very effective absorber of the infrared radiation that is of great interest to astronomers; this is still another reason for choosing a high and dry location.

In short, an observatory site should have dark and dust-free skies, low cloudiness, steady dry air, and unobstructed horizons. For radio astronomy observatories the dark sky criterion becomes one of low man-made interference. Also, low cloudiness and dry air are important at short radio wavelengths (millimeter and centimeter waves). Good optical

---

[3]Conservationists also recognize the problem of light pollution. In May 1971, the Sierra Club's Board of Directors adopted as a policy that the "Club opposes unnecessary night lighting in both urban and suburban areas. This practice is a waste of electrical energy, destroys the aesthetics of the nighttime sky, and seriously interferes with astronomical research."

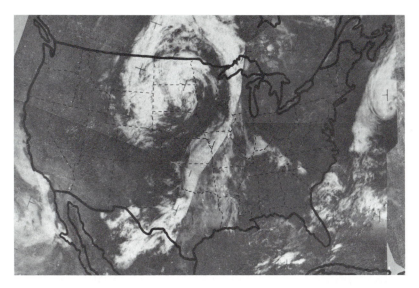

Figure 6-18. A composite satellite photograph showing cloud cover over the United States on September 14, 1967; also visible off the New England coast is Hurricane Doria. (National Environmental Satellite Service.)

observatory sites are generally on mountain tops; the better radio observatory sites are in valleys (when low interference is especially important—at the longer wavelengths) although a few (for which low water vapor is important—millimeter wavelengths) are on mountains.

Keeping in mind all the above criteria, the best American observatory locations appear to be in the southwestern states, including Southern California, and in Hawaii, but such sites are getting fewer every year due to pollution and urban expansion. The Kitt Peak National Observatory near Tucson, Arizona, provides a variety of telescopes and instruments at a good mountain-top location for use by astronomers and graduate students who lack access to comparable facilities at their own institutions. Potential users of the Kitt Peak telescopes send descriptions of their proposed research programs to the Observatory, which assigns telescope time on the basis of scientific merit. A similar facility for radio telescopes, the National Radio Astronomy Observatory, operates facilities in Green Bank, West Virginia, and in Arizona and New Mexico. In cooperation with the University of Chile, the Kitt Peak organization also maintains an Inter-American Observatory (Figure 6-10) at Cerro Tololo near La Serena, Chile, for studies of the southern stars. Several other observatories are located in the Chilean Andes, where the weather and seeing conditions are superb.

Some major observatories are listed in Tables 6-1 and 6-2.

DETECTION METHODS

The astronomer uses a telescope to select and track a target for observation and to collect light from the target. At the telescope focus he uses an instrument to record the light or analyze its properties. This may be a camera to photograph a portion of the sky, a *photometer* to measure light

Table 6-1. Some Major Optical Observatories

Name	Location	Chief Instruments
Hale Observatories	Mt. Palomar, Ca.	200'' (508-cm) reflector 48'' (122-cm) Schmidt
	Mt. Wilson, Ca.	100'' (254-cm) reflector
	Las Campanas, Chile	101'' (257-cm) reflector
Special Astrophysical Observatory	Zelenchukskaya, USSR	6-m reflector
Kitt Peak National Observatory	Tucson, Arizona	4-m reflector 84'' (213-cm) reflector 60''(152-cm) solar telescope
Cerro Tololo Inter-American Observatory	La Serena, Chile	4-m reflector 60'' telescope
Anglo-Australian Observatory	Siding Spring, Australia	4-m reflector
European Southern Observatory	La Silla, Chile	3.6-m reflector 1-m Schmidt
Lick Observatory	Mt. Hamilton, Ca.	120'' (305-cm) reflector
McDonald Observatory	Fort Davis, Texas	107'' (272-cm) reflector
Crimean Astrophysical Observatory	Nauchny, USSR	104'' (264-cm) reflector
Steward Observatory	Tucson, Arizona	90'' (229-cm) reflector
Sacramento Peak Observatory	Sunspot, New Mexico	30'' (76-cm) solar telescope

Table 6-2. Some Major Radio Astronomy Observatories

Name	Location	Chief Instruments
Arecibo Observatory	Puerto Rico	1,000-foot (305-m) fixed spherical antenna
Westerbork Radio Astronomy Observatory	The Netherlands	interferometer; 12 82-foot (25-m) dishes on 1-mile (1.6-km) baseline
National Radio Astronomy Observatory	West Virginia	140-foot (43-m) fully steerable reflector 300-foot (91-m) meridian transit reflector
	Arizona	36-foot (11-m) mm-wave reflector
	New Mexico	"Very Large Array" (under construction) consists of 27 dishes, each 82 feet in diameter in a Y-shaped array; arms of the Y are 11.8, 13, and 13 miles (19, 21, 21 km)
Max Planck Institute for Radio Astronomy	West Germany	328-foot (100-m) fully steerable reflector
Mullard Radio Astronomy Observatory	England	3-mile (5-km) baseline interferometer with 8 dishes
Goldstone Tracking Station	California	210-foot (64-m) fully steerable radar telescope (also used for tracking interplanetary spaceprobes)
Radio Astronomy Centre, Tata Institute	India	1,740 × 98-foot (530 × 30-m) reflector, also used as main element of 4-element interferometer with 2-mile (3.5-km) baseline
Radio Physics Laboratory, CSIRO	Australia	210-foot fully steerable reflector
Jodrell Bank Experimental Station	England	250-foot (76-m) fully steerable reflector
Algonquin Radio Observatory	Ontario, Canada	150-foot (46-m) fully steerable reflector

intensity, a *spectrograph* to study the distribution of light at the various wavelengths, or a *polarimeter* to determine the polarization properties of the light. Each instrument contains optical components (lenses, mirrors, filters, prisms, and the like) necessary for the measurements, but its key component is the light *detector* or sensor that records the light (as on a photographic plate) or produces a signal that measures the light (as, for example, a photomultiplier tube). A fundamental property of a sensor is its efficiency at producing a measurable signal or record from a given amount of light. For example, if a new kind of film requires less light to produce a recognizable image of a given star, then we can photograph stars that are fainter than those that could be recorded previously with the same telescope.

Photography

The photographic detector is usually a glass plate or a celluloid film coated with a light-sensitive emulsion. Photons of light that strike the emulsion produce chemical reactions in the grains[4] of emulsion material. On the developed film or plate (a negative), the areas that have been struck by light are dark, and the images of the brighter stars are larger and darker than those of the dimmer stars. The glass plates are usually preferable to film because they are less apt to shrink or bend, and thus measurements of the positions of stars are more reliable and a more permanent record is obtained. In addition to sensitivity differences, emulsions differ in such matters as grain size and spectral properties (response to light of different wavelengths). An emulsion with finer grains can yield sharper images, although it may not record stars that are as dim as those photographed with a more sensitive emulsion. A plate with high red sensitivity is better for photographing the red portion of stellar spectra, although it may be useless in observing blue light, and so on.

In 1839, Louis Daguerre announced the invention of photography in France, and within a year a daguerrotype of the moon was made by John W. Draper of New York.[5] The first great astronomical triumph of this invention came in 1860 when photographs of an eclipse of the sun proved that the red *prominences* at the edge of the sun were solar features and

---

[4]When viewed with a magnifying glass, the emulsion of a photographic plate or film is seen to consist of many tiny clumps of material. If you look at a photograph enlarged considerably from the original negative, you note that it seems less sharp. This happens because the enlargement allows us to notice the grainy pattern of the film.

[5]Daguerre actually perfected and popularized a process that was invented by J. N. Niepce in 1822.

(a)   (b)

Figure 6-19.   Photographs of the Pleiades star cluster taken with different exposures. (a) 600 minutes. (b) 6 minutes. Since these photos were made, the introduction of more sensitive photographic materials has made it possible to get similar results with much shorter exposures. (Yerkes Observatory photograph.)

not lunar phenomena as some astronomers had suggested. Early photography was crude, slow, and often quite awkward. Some early processes literally required that the photographic plate be wet during the exposure. But the techniques were improved and the development of emulsions with sufficient sensitivity to record dim stars produced a revolution in stellar map-making. In earlier times, astronomers charted the stars by making laborious measurements of their positions, one star at a time; now it became possible to record hundreds or thousands of stars on a single plate. Furthermore, information on both the positions and brightnesses of the stars is preserved. Comparison of several plates of the same sky region is a valuable tool for discovering the *variable stars* that change in brightness with time. Moving objects, such as asteroids, appear at different locations on plates obtained on different nights (or as streaks on individual plates). In addition, the photographic plate allows one to accumulate the effects of many photons during a long exposure and thus to produce images of fainter and fainter objects, as illustrated in Figure 6-19.

Often, the astronomer photographs the spectrum of a star rather than the star itself. These pictures, called *spectrograms*, are studied to determine the temperature, chemical composition, radial velocity and other properties of the star. Recall Figure 5-18; the instrument used to photograph spectra is similar in concept to Newton's experiment. A simplified diagram of such a *spectrograph* is given in Figure 6-20.

Figure 6-20. Schematic diagram of the prism spectrograph. A lens system takes light gathered by the telescope, passes it through a prism, and brings it to a focus. The fact (illustrated in Figure 5-18) that violet light is bent more than red light in passing through a prism produces separate images in the different colors on the photographic plate.

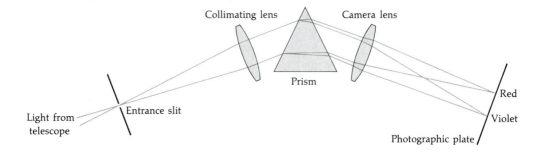

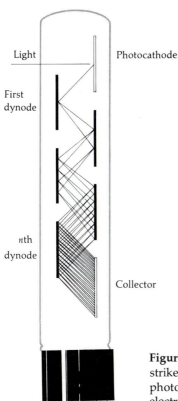

Light

Photocathode

First
dynode

*n*th
dynode

Collector

Figure 6-21. Schematic diagram of a photomultiplier tube. Light strikes the photocathode, and an electron is emitted through the photoelectric effect. Electric fields are used to accelerate and focus the electrons onto successive dynodes, where they knock out still more electrons at each step. In practice, the electric current at the collector is a million or more times the current in the first dynode.

Photoelectric Detectors

The photoelectric effect, in which light shining on a metal or other substance causes electrons to be ejected from the surface, was important in understanding the quantum nature of light. Because the flow of electrons from a metal surface constitutes an electric current, it is possible to use the photoelectric effect to determine the brightness of a star by focusing its light on a suitable metal surface and using electrical meters and other instruments to amplify and measure the resulting electrical current. Modern electronic technology has produced a variety of these sensors, known generally as photoelectric detectors. The type most widely used by astronomers is the *photomultiplier.* This is a vacuum tube in which, after light strikes the sensitive metal surface (photocathode), the ejected electrons

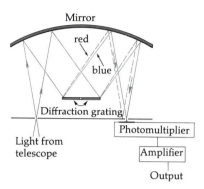

Figure 6-22. Schematic diagram of the grating spectrometer. The different wavelengths of light are supplied to the photomultiplier by tilting the diffraction grating to produce scans of a spectrum; an example is given in Figure 6-23.

strike another metallic surface (dynode), where each incident electron causes additional electrons to be released. These secondary electrons in turn strike another dynode, where again each incident electron produces many more particles. In this way, the number of electrons produced at the photocathode by a given amount of light is effectively multiplied (by a factor in the millions) so that a larger, more easily measured electric current is produced (Figure 6-21). In some applications, particularly when the incident light is very dim so that the number of electrons produced is rather small, the output of a photomultiplier is measured by counting the individual bunches of electrons, each of which corresponds to the arrival of one photon *(pulse counting technique)*.

Photomultiplers are being used in many modern spectroscopic instruments. These devices, called *spectrometers* or *photoelectric spectrum scanners*, may employ a prism to produce the spectrum, as does the spectrograph of Figure 6-20, or they may use a diffraction grating (recall Plate 2). The example sketched in Figure 6-22 was used in an artificial satellite to record spectra in the ultraviolet wavelength range. Since light of different wavelengths is reflected from a grating at different angles (Plate 2), it follows that only one narrow range of wavelengths falls on the photomultiplier in Figure 6-22 at a given time. However, rotating the grating (note curved arrow) causes the different wavelengths to sweep past the photomultiplier. When the signal from the photomultiplier is amplified and displayed on chart paper, the result is a graph of the intensity of the spectrum as a function of wavelength. An example of such a graph *(spectral scan)* is given in Figure 6-23. Before the intensities on a spectral scan are used in calculations or theories, they have to be adjusted for such effects as the absorption of light by the earth's atmosphere and the sensitivity of the photomultiplier, both of which depend on the wavelength.

Photomultiplier tubes are used to measure intensity of light. However, there are several types of imaging photoelectric detectors that use the photoelectric effect to produce pictures. In a common application of such *electronic image tubes* the image formed by a telescope is focused on the photocathode; at the opposite end of the tube a much brighter duplicate of the incident image is produced on a glowing phosphor screen (Figure 6-24). This brighter image is recorded on a photographic plate. The advantage of this method is that the output image of the tube is much brighter than the input image formed by the telescope. Therefore, using plates or film of a given sensitivity, it is possible to photograph much fainter stars than otherwise, or to reduce the exposure time required to record a star of a given brightness. Since astronomical photography frequently involves exposure times of several hours, the possibility of reducing the exposures through the application of new electronic techniques is very attractive. In another modern development, *vidicon* tubes, which apply the television process, have been used in conjunction with telescopes (Figure 6-25). The pictures taken with the vidicons can be preserved on magnetic recording tape and are relatively easy to feed into a computer for analysis.

Figure 6-23. Scan of the spectrum of Comet Honda (1968c) made with a grating spectrometer. The relative intensity shows the characteristic "spectral signature" of the C_2 molecule. (Courtesy of W. L. Gebel, State University of New York at Stony Brook.)

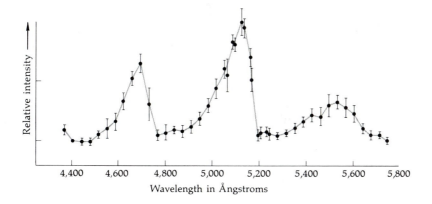

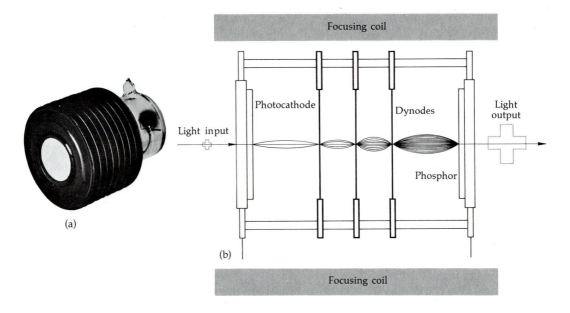

Figure 6-24. (a) Photograph and (b) sketch of an image converter tube. The basic principle of electron multiplication is the same as for photomultipliers (illustrated in Figure 6-21), but focusing properties are provided to produce an image, which is converted to light by the phosphor screen. (Courtesy of Westinghouse Electric Corporation.)

Other Detectors

The photographic plates and photomultiplier tubes described above are useful for detecting light in the visible wavelength region. Various types of devices are used in observing at other wavelengths. For example, *proportional counters* are flown on rockets to measure celestial x-rays, and *Germanium bolometers* are used to study infrared radiation. Although the operating principles of these other devices are different from those described above, in each case an electrical signal that corresponds to the light collected by the telescope is finally produced and measured.

SKY SURVEYS

Astronomy, like other sciences, seeks to describe and explain a certain class of natural phenomena. The first step in this process is to record and classify the subjects of investigation. From this, a collection of measurements and observations is obtained, which is useful in selecting particular targets for future investigations, and which provides a suitable basis for statistical analysis.

In studying the earth, the geologist uses maps of the topography, lists, descriptions, and locations of the different mineral and rock forms, and catalogues of the various types of fossils and their ages. Similarly, astronomers use maps of the sky and catalogues of the stars and their properties.

The construction of star maps and catalogues began in antiquity with the recording of the brighter stars visible to the unaided eye. In more recent times, sky surveys have been compiled by observing the positions of stars with the *meridian circle*, a special kind of telescope designed for precision measurements. Such a catalogue, or sky survey, is called a *Durchmusterung*, after the German term adopted by the creator of the first major survey, Friedrich Argelander, who worked at the Bonn Observatory

Figure 6-25. Astronomical television camera built by Hong-Yee Chiu and used to hunt for a pulsar in the constellation Vela with the 60-inch (152-cm) telescope at Cerro Tololo Inter-American Observatory, Chile.

in the mid-nineteenth century. Over a period of seven years, Argelander and his assistants, using the eye, the telescope, and the clock, mapped the northern half of the sky, plus a small zone just south of the Celestial Equator. Their maps and catalogue, which show the positions and approximate brightnesses of 324,000 stars, are known as the *Bonner Durchmusterung* (or B.D.) and are still used today. Later, other astronomers, working in such places as Argentina and South Africa, extended the surveys to cover the southern half of the sky and to include fainter stars.

By recording the stars according to position and brightness, and by giving each one a B.D. number, Argelander provided a convenient means for astronomers to locate individual stars and to identify those of special interest for further observation by their colleagues. The Durchmusterung is equivalent to a crude census that tells where people live. As the demands of the government statisticians have grown, so have the number of questions in the census, so that information is gradually accumulated on the personal characteristics, property, and so on, of the residents. Likewise, the next step after the Durchmustering was to catalogue the individual properties of the stars listed at the various positions.

Because the spectrum of a star gives valuable information on the chemical composition, surface temperature, and other properties, catalogues of the stars classified according to their spectra have been of fundamental importance to astronomers. Chief among these surveys has been the *Henry Draper Catalogue* (H.D.), which was compiled at the Harvard College Observatory around 1920 under the leadership of Annie Jump Cannon. The rudimentary state of astrophysics in those days did not yet allow reliable estimation of physical properties to be made from stellar spectra, so Cannon classified the stars according to the appearance of their spectra (H.D. types). Modern investigations have developed the physical interpretations of the various H.D. types, so that the nine volumes of the H.D., listing brightnesses and spectral types of 225,000 stars, are a major tool of the astronomical researcher.

With the advent of large telescopes, observations of stars and galaxies have been extended to such faint brightnesses that the job of cataloguing the millions of fainter stars by Argelander's visual techniques would exceed both the patience and the lifetime of any astronomer. The production of photographic sky maps was the only answer to this dilemma, and several star atlases have been compiled. These have culminated in the great survey made with the Schmidt camera on Mount Palomar, a survey used by all astronomers who work with the major telescopes.

In its original form, the *Palomar Sky Survey* comprises 1,870 photographic plates, each 14 inches (36 cm) square, arranged in pairs. Each pair consists of an exposure taken with a red filter and one taken on a blue-

Figure 6-26. Palomar Sky Survey plate (taken in red light) of the area marked by the square in Figure 6-27; the area covered by this plate is 6.°6 × 6.°6. An overexposed image of the Crab nebula (see also Plate 10) is indicated by the arrow at the lower, extreme right. Thousands of stars and some wisps of nebulae are also shown. This is a negative copy, so stars appear as black dots on the white background. (Courtesy of the Hale Observatories. Copyright © by the National Geographic Society—Palomar Observatory Sky Survey.)

sensitive plate. Examination of the plates reveals positions and brightnesses of stars, nebulae, galaxies, and so on, and, by comparing the two plates in each pair, one can determine the approximate colors of the objects.

The Palomar Survey covered the northern half of the sky plus a good portion of the Southern Hemisphere (down to Declination −33°). An enormous effort was required to obtain the 935 plate pairs: clouds, airplanes, poor telescope operation, bright moonlight, and the occasional heart-breaking dropped plate all interfered with the survey; over 3,200 plates were taken before the desired complete set of high quality photographs was obtained. The effort and expense involved explain why the original plates are stored in a vault in Pasadena with a high quality duplicate set kept on Mount Palomar. Additional copies, on glass and on photographic paper, are in use throughout the world. A sample of a Palomar Survey plate is shown in Figure 6-26. A survey of the southern sky is under way at the European Southern Observatory in Chile. Already, duplicates of many of these *ESO Sky Survey* photographs are available in astronomy libraries.

Radio astronomers have likewise surveyed the sky. Their early catalogues contained relatively few radio sources, each of which bore a name in the style Cygnus A, Cygnus B, and so on (first and second sources found in constellation Cygnus). Soon, larger antennae at such places as Cambridge University were in use, and individual research groups were listing as many as several hundred sources found in their observations. One of the most important of these studies was the *Third Cambridge Catalog of Radio Sources*. Many of the quasars and other radio sources are still known by their "3C" numbers as listed in this catalog. By 1964, a *General Catalogue of Discrete Radio Sources*, which consisted of a summary and cross-tabulation of the individual observers' lists, was published. It included nearly 1,300 radio sources both within and beyond our galaxy, and since then, more sensitive radio telescopes have detected tens of thousands more.

Figure 6-27. Radio map of a large portion of the sky which includes the area photographed in Figure 6-26 (indicated by the square). The contour numbers indicate the brightness of the galactic background radio radiation. The circles, triangles, and small squares represent individual sources such as the Crab nebula or Taurus A radio source. Note that the ecliptic (dashed curve) passes close by the Crab. As a result, the nebula is occasionally occulted by the sun and the moon. The Crab nebula is a strong radio source, but the bright star below and to the left of the Crab nebula on the Palomar plate is not. (Drawn after a map by W. E. Howard, S. von Hoerner, H. D. Aller, and L. R. Walker, National Radio Astronomy Observatory.)

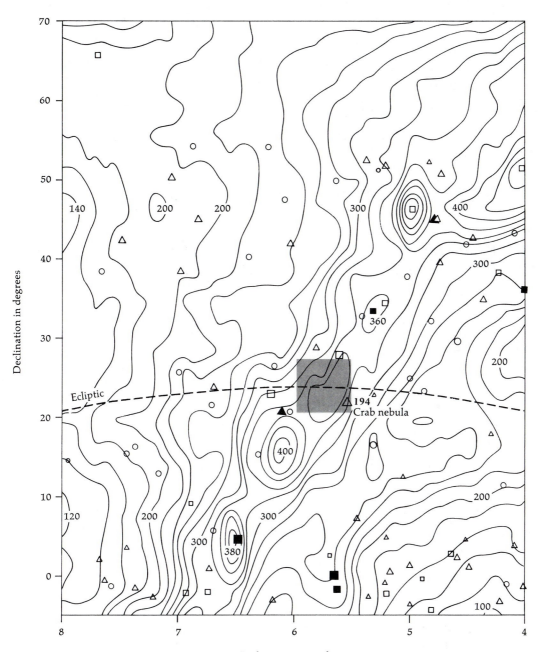

Figure 6-28. The constellation Orion as photographed from an Aerobee rocket in ultraviolet light with wavelengths from 1,230 to 2,000 Å (left) and in visible light with a ground-based telescope (right). Notice the different relative brightnesses of stars in the two photographs. The hot stars found in the Belt and Sword of Orion are much brighter in the ultraviolet than in the visual. Conversely, the numerous cool stars are not conspicuous in the ultraviolet photograph. Betelgeuse (upper left), the brightest star in the constellation as seen with the human eye, is not detected in the ultraviolet. (Visible photograph, Hale Observatories; ultraviolet photograph, Naval Research Laboratory.)

Figure 6-27 is a map which contains the same portion of the sky shown on the preceding sample Palomar Survey picture. This figure, however, shows the distribution of individual radio sources (represented by circles, triangles, and squares) in this portion of the sky, and the contour lines show the background radio emission of our galaxy.

CONTINUING EFFORTS

As modern technology advances and we observe with improved sensitivity, or especially as we look out in new regions of the electromagnetic spectrum (for example, ultraviolet observations of the stars from satellites), we find that the first step is generally the same as Argelander's—that is, we begin by mapping the skies. The first ultraviolet star atlas, recorded with a vidicon camera on the OAO-2 (second Orbiting Astronomical Observatory), does not provide full sky coverage, but does include pictures of many of the most interesting regions. A sample ultraviolet sky picture is shown in Figure 6-28. Other satellites have mapped the sky at the wavelengths of x-rays, gamma rays, and infrared radiation. As we shall see in later chapters, the observations in relatively unexplored regions of the spectrum are revealing dramatic new phenomena that enlarge and revise our concept of the universe.

7

The Solar System—
An Overview

The concept of the *solar system*, with the sun at the hub of an array of orbiting planets, comets, asteroids, and meteoroids, is so firmly engrained in our Space Age consciousness that the term has entered our language as a metaphor for a center of power attended by lesser luminaries. Nevertheless, as we have seen, the road to acceptance of a heliocentric system operating under the force of gravity was a long and hard one, and the recognition of the other planets as worlds was also slow in arriving.

The outermost of the nine known major planets, Pluto, was not discovered until 1930. Adding it to the list, the planets are (in order from the sun): Mercury, Venus, Earth, Mars, Jupiter, Saturn, Uranus, Neptune, and Pluto. Those who like to commit things to memory can use any one of several silly mnemonic devices to recall this list, such as: "Many Volcanoes Erupt Mulberry Jam Sandwiches Under Normal Pressure," or "My Very Eager Mother Just Served Us Nine Pizzas." Uranus and Neptune were discovered in 1781 and 1846, respectively. Searches for other major planets of the sun have not been successful. For a while it was thought that a planet, Vulcan, lay between Mercury and the sun, but the observations were apparently erroneous. Also, searches for trans-Plutonian planets have been fruitless.

BODE'S LAW

The orbital period of a planet, that is, the time required for one complete circuit about the sun, is easily established by observing the motion of the planet through the sky. In this way, for example, we find that the period of Mars is 687 days, or about 1.9 earth years. In the case of the earth the orbital period, one year, was determined by watching the apparent motion of the sun projected against the background of stars. Given the periods of 1.9 and 1.0 years, we can find relative sizes of the orbits of Mars and earth using Kepler's Third Law. The average distance of Mars from the sun is found in this way to be 1.5 times the average earth-sun distance of one *astronomical unit*. Table 7-1 (down to the double line) gives the approximate orbital periods and average planet-sun distances (from the Third Law) for the earth and the five other planets that are visible to the unaided eye and thus were known in antiquity. The telescopically discovered planets are listed below the double line.

In 1772, J. D. Titius found an intriguing numerical relationship, which he published in a little-noticed footnote. It later gained prominence through the work of J. E. Bode in Berlin and has become known as *Bode's Law*. The recipe is as follows: consider the simple series 0, 3, 6, 12, 24, 48, 96, 192, etc. Except for 0 and 3, the numbers in this sequence are obtained by doubling the preceding number. If we then add 4 to each number in the series, we obtain: 4, 7, 10, 16, 28, 52, 100, 196, etc. If we continue by dividing each of these numbers by 10, we now obtain the series of numbers in the last column of Table 7-1. Bode's Law certainly gave a reason-

Table 7-1. Approximate Orbital Periods and Average Planet-Sun Distances

Planet	Orbital Period (Earth Years)	Average Planet-Sun Distance (Astronomical Units)	Bode's Law Numbers
VISIBLE TO UNAIDED EYE			
Mercury	0.24	0.39	0.4
Venus	0.62	0.72	0.7
Earth	1.00	1.00	1.00
Mars	1.9	1.5	1.6
			2.8
Jupiter	11.9	5.2	5.2
Saturn	29.5	9.5	10.0
NOT KNOWN BEFORE INVENTION OF TELESCOPE			
Uranus	84	19.2	19.6
Neptune	165	30.1	38.8
Pluto	248	29.4	77.2

able representation for the distances of the six planets known in 1772, and it also predicted something beyond—at 19.6 a.u.

HERSCHEL AND URANUS

In our age of training and specialization, a Ph.D. degree is usually required for full membership in the American Astronomical Society and is, by itself, insufficient qualification for the International Astronomical Union. It is also widely held that a physicist must make his or her mark before the age of 35. The situation of the scientist was quite different in the eighteenth century, and it seems almost fantastic to us that one of the handful of astronomers throughout history who has been privileged to discover an unknown planet, a fundamental innovator in telescope design, a leading observer of nebulae, and the first man to apply statistics to the study of our galaxy was a professional musician who first took up astronomy as a hobby when he was approaching middle age! On March 13, 1781, William Herschel was looking through a reflecting telescope of his own manufacture and noted a faint "star" that differed from the others because it seemed to have a perceptible diameter, like the head of a faint comet, instead of appearing as just a point of light. When he increased the magnification, the stars still appeared as points, but the unusual image increased in size. It was being resolved. Repeating the observations on subsequent nights, he discerned a slow movement of this object across the background of distant stars. Mathematicians applied the rules of Newtonian mechanics to Herschel's observations and discovered that the newly found object was not moving in the elongated orbit typical of many comets. Situated at 19.2 a.u. from the sun, twice the distance of Saturn, and with an orbital period of 84 years, *Uranus* was moving in a nearly circular elliptical orbit. It was indeed a planet, the first new world discovered since antiquity. The occasion was fortunate for Herschel and for astronomy. In recognition of this accomplishment, King George III awarded him an observatory and a stipend that enabled him to concentrate on research.

The discovery of Uranus had come nine years after Titius published his original formulation of Bode's Law. Uranus' average solar distance of 19.2 a.u. was within 2 percent of the prediction of 19.6 a.u.

THE NEPTUNE EPISODE

After the discovery of Uranus, old records were examined for prediscovery observations of the planet. Sure enough, it had been recorded as a star on many occasions prior to its discovery. Tycho even had spotted it with the naked eye. Immediately, however, difficulties arose with computing its orbit. An ellipse that fit one set of observations did not seem to accurately predict the future motion of the planet. The discrepancy between the observed and predicted positions grew worse as time passed.

A first attempt at a solution was to discard the older observations on the grounds that they were more likely to be wrong. A new orbit for Uranus was calculated in 1821, but the problem had been merely postponed and not solved. The planet systematically departed from the improved orbit and the discrepancy continued to increase.

The problem could then be resolved in at least two ways. One approach questioned the validity of Newton's Law of Universal Gravitation under certain circumstances. Perhaps it must be modified at large distances from the sun. If so, understanding the universe beyond the solar system might prove to be extremely difficult. An alternative to this undesirable solution was to postulate that the discrepancies in the orbit of Uranus were due to the gravitational attraction of another, as yet undiscovered, planet beyond Uranus. The motion of Uranus would, of course, be controlled primarily by the sun, but the presence of another planet would *perturb* the orbit and cause the observed discrepancies.

This problem was tackled in 1841 by John Adams, a 22-year-old student at Cambridge University. His solution was, to a certain extent, one of trial and error. There were essentially two unknowns—the mass of the hypothetical planet and the size of its orbit. The procedure was then to try different values of the orbit size and planetary mass until a combination was found that could account for the observed perturbations of Uranus. Other information could be used as a guide. When the discrepancies between the predicted and observed positions of Uranus were largest, Uranus and the unknown planet were probably closest together, so that the gravitational perturbations were greatest. The relative directions of the two planets could be inferred by noting whether Uranus was being accelerated or retarded with respect to its nominal orbit.

After a great deal of study, Adams had a satisfactory solution. His new planet was at a distance of 36 a.u. from the sun and had a mass about 25 times that of the earth. This work was completed in October of 1845, and Adams took his computations to an astronomer at Cambridge University in the hope that the hypothetical planet could be observed. The Cambridge astronomer was "not interested" and passed the buck to the Astronomer Royal at Greenwich, who was "not interested" either. We have used quotation marks to point up a common situation. Any astronomer would be delighted to discover a new planet, but then, as now, scientists were besieged with new theories and revolutionary discoveries. Even when they originate with established authorities, a fair amount turn out to be wrong, and those that come from unknown persons are frequently just crank letters and bothersome inquiries. The astronomers who ignored Adams' requests are considered to be the villains of this story, but bear in mind the justification for at least initial skepticism.

The same theoretical problem was tackled by Urbain Leverrier in France, and by the summer of 1846 he had presented his results to the French Academy. By August, the news made its way back to England and a search was begun—and mishandled. Despite the close agreement between the predictions of Adams and Leverrier, the survey was performed in a leisurely fashion. The missing planet was actually seen twice by an astronomer in Cambridge during the month of August, 1846, but it was unrecognized! Meanwhile, Leverrier sent his predicted position to the Berlin Observatory because star charts for this region had just been prepared there. (Remember that neither the B.D. nor any other comprehensive star survey existed at that time.) The location was examined on September 23, and an object apparently having a disk was noticed. When the same field was examined on the next night, the object had moved. Thus Neptune was discovered. Detailed observations of this planet and its satellites show that Neptune is 30 a.u. from the sun and has a mass of about 18 times that of the earth (somewhat less than predicted).

The research by Adams had not been published when the discovery of Neptune, made on the basis of Leverrier's work, was announced. When Adams' claim was advanced, the French regarded it as an attempt to steal Leverrier's discovery. Bitter words were exchanged, and it was only after some years that calmer heads prevailed. The predictions of Adams and Leverrier were remarkably similar and were made entirely independently of each other. Both of these men are therefore usually regarded as the discoverers of Neptune.

Neptune is, in fact, closer to Uranus and to the sun than predicted. Its closeness is partially balanced by its smaller mass; this combination produces approximately the same gravitational perturbation. Although Adams and Leverrier had calculated the orbit and mass of a planet that would produce the observed perturbations on Uranus, there was no guarantee that their solution was the only one possible. The triumph was a substantial one; a new planet had been discovered using Newton's theory. Further probing of the universe using these laws was now regarded as very promising.

THE SEARCH FOR PLANET X

Following the discovery of Neptune, investigation of the outer solar system was continued because some small, but persistent, discrepancies existed in the orbits of both Uranus and Neptune. This suggested that another planet might lie beyond Neptune.

In 1906, Percival Lowell began a search for the trans-Neptunian planet at the observatory in Flagstaff, Arizona that now bears his name. In retrospect, it appears that the sought-for planet was slightly south of the region recorded on Lowell's closest photograph, and so it remained undiscovered for some time. Meanwhile, Lowell published his calculations of the likely properties of the hypothetical planet.

Although Lowell died in 1916, the search continued. The great difficulty was that thousands of stars appeared on each plate taken in search of the new planet. Plates of a given region taken at different times would show these numerous, essentially stationary star images, as well as (hopefully) the trans-Neptunian planet, which would show movement (Figure 7-1). No appreciable disk was expected in view of Planet X's great distance. To examine large numbers of plates without mechanical aid thus would be hopeless. A device called the *blink comparator* simplified the problem. Each of the two plates to be compared is mounted in the instrument, so that both can be examined through a common viewing device, but only one at a time. The instrument can be adjusted so that the images of the stars appear in the same locations on the two plates; then the view field is rapidly switched back and forth from one plate to the other. When this is done, any object that has shifted its position between the times when the two plates were taken appears to jump back and forth, while the stars stay fixed.

Figure 7-1. Photographs of Pluto taken one day apart with the 200-inch telescope, showing Pluto's motion with respect to the stars. (Courtesy of the Hale Observatories.)

By 1919, W. H. Pickering made independent calculations of the likely position and brightness of Planet X, and photographs were taken at Mount Wilson Observatory to check his prediction. However, only the parts of the plates that were most likely to contain Planet X (according to the calculations) were carefully examined. And so it was missed because, although actually photographed, it was not in any of the carefully searched areas. In 1929, 23-year-old Clyde Tombaugh joined the Lowell Observatory staff and went to work with a new 13-inch (33-cm) telescope, especially designed to record sharp images on large (14 × 17-inch = 36 × 43-cm) glass plates. Using one-hour exposures, he surveyed the predicted region in the constellation Gemini and other possible but less likely locations. Experience with the blink comparator showed that there were always some possible "planets" on a plate pair, usually due to defects in one plate or the other. So Tombaugh took *three* plates of each region. Then, finding a possible planet by blinking two of the plates, he could immediately check its reality on the third plate of the same area. Depending on how far the photographed region was from the Milky Way, a given plate contained from 50,000 to 300,000 star images, and it took Tombaugh several days to complete the blink comparison of a plate pair. Several thousand false alarms were eliminated by the third-plate technique. But finally, he was studying two plates of the region near the star delta Geminorum when,

> . . . one-fourth of the way through this pair of plates, on the afternoon of February 18, 1930, I suddenly came upon the image of Pluto! The experience was an intense thrill, because the nature of the object was apparent at first sight. The shift in position between January 23 and 29 was about right for an object a billion miles beyond Neptune's orbit. . . . In all the two million stars examined thus far, nothing has been found that was as promising as this object.

The planet was named Pluto for the ancient Greek god of the underworld (and also because the first two letters were Percival Lowell's initials).

Because the gravitational perturbations caused by Pluto are very small, its mass was known only very poorly until 1978. In that year, James W. Christy of the U.S. Naval Observatory discovered that Pluto has a moon. (By remarkable coincidence, the moon was found on photographs taken with a 61-inch (155-cm) telescope located near Flagstaff, in the vicinity of Lowell Observatory.) An analysis of the orbit of this moon, Charon,

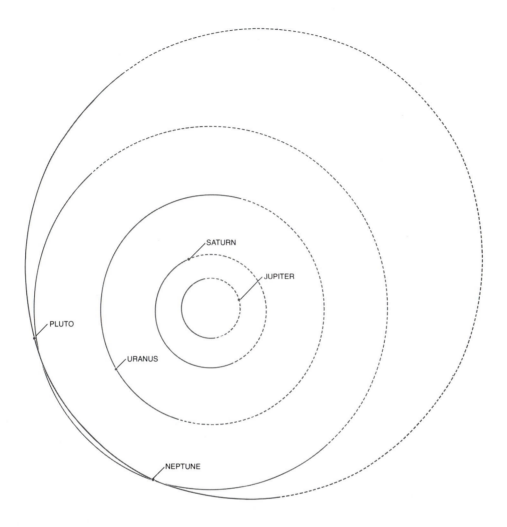

Figure 7-2. Orbits of the outer planets from Jupiter to Pluto are plotted on the plane of the ecliptic. Dots show where each planet was in September, 1975. Planets move counterclockwise in this diagram, as seen from "above" (from the north), with average speeds as listed in Appendix 5. Dashes show portions of orbits that, due to tilts of the various orbital planes, are below the ecliptic plane, while solid portions are above. In 1987, slow-moving Pluto will pass inside the orbit of Neptune. (After *The Solar System* by Carl Sagan. Copyright © 1975 by Scientific American, Inc. All rights reserved.)

revealed that Pluto's mass had been substantially overestimated, and that it is only about 1/640 the mass of the earth, or about one-eighth the mass of the earth's moon. It seems that so small a mass cannot fully account for the perturbations of Neptune's orbit, and hence they remain a mystery. Although Pluto is 39.5 a.u. from the sun on the average, it has a noticeably elliptical orbit, part of which lies within the orbit of Neptune (Figure 7-2). Their orbits are tilted with respect to each other, and the two planets can never collide.

ASTEROIDS

A review of the orbits of the nine major planets and their relation to Bode's Law (Table 7-1) can now be made. Except for the gap at 2.8 a.u., the law fits rather well in all but the outermost reaches of the solar system. But if Bode's Law is correct, then there ought to be a planet at 2.8 a.u. from the sun. In 1801, such a planet was discovered by the Italian monk Giuseppe Piazzi. Named *Ceres* after the guardian goddess of Sicily, the planet's mean solar distance was indeed 2.8 a.u. Things seemed to be falling into place, but then a year later *Pallas* was discovered, also near 2.8 a.u., and in 1804 *Juno* (2.7 a.u.) was found. *Vesta* (2.4 a.u.) was discovered in 1807. Now there seemed to be an embarrassment of riches—four planets where one was expected! In fact, all four are tiny; each is less than 1,000 km (800 miles) in diameter. Now referred to as *asteroids, planetoids* or minor planets, several thousand have been identified, most of them located in the *asteroid belt* between Mars and Jupiter, and most of them much smaller than the first four discovered. The motion of asteroids on a photograph is illustrated in Figure 7-3.

The success of Bode's Law in predicting the distances of Uranus and the asteroids from the sun raised the question of whether the Law was not just a numerical curiosity, but might have an actual physical explanation. Bode's Law predicted only one planet where the thousands of asteroids occur, but at one time it was believed that they might represent the debris of an ancient planet disrupted by collisions, or by some other cause. Today we know that the present total mass of the asteroids is no more than one-tenth of one percent of the earth's mass, and that the disruption of a planet requires more energy than is available from any known process in circum-solar space. However, some recent studies do suggest a sound scientific basis for Bode's Law and imply that the spacing of the planets is not accidental. We will return to this subject in Chapter 11, when we discuss the origin and evolution of the solar system.

Chiron, a distant object that is probably about as large as one of the "big four" (*Ceres, Pallas, Juno,* and *Vesta*), was discovered in 1977. It was

Figure 7-3. Two asteroids photographed in motion (short lines) against a background of stars. (Yerkes Observatory photograph.)

found by Charles Kowal on photographs made with the 48-inch Schmidt camera on Mt. Palomar. *Chiron* is remarkable for its location, far outside the asteroid belt; it has an elongated orbit, coming as near as 8.5 a.u. to the sun and going as far as 18.9 a.u. from the sun. Thus, it travels from within the orbit of Saturn to nearly the orbit of Uranus. Its orbital period is 50.7 years. Is *Chiron* a unique asteroid, a member of an outer asteroid belt, or something else?

THE ASTRONOMICAL UNIT

An important early use of asteroid observations was in the determination of the astronomical unit. The distance to an asteroid passing close to the earth could be determined by triangulation from different locations on the earth, and the distance so determined was used to scale the solar system.

The distances of the planets from the sun listed in Table 7-1 are all relative distances; that is, they are not given in miles or kilometers, but (from Kepler's Third Law) only in terms of the average sun-earth distance, the astronomical unit. Until the a.u. was determined, we were in the position of having an atlas of the solar system that lacks a scale. An analogous situation would be a state map with no mileages marked on it and the calibration statement of (say) "one inch equals ten miles" left off. In both cases, we can establish the scale by determining one distance on the map.

The basic idea is to determine the distance to an object passing close to the earth by direct triangulation from opposite sides of the earth. Photographs of objects close to the earth taken from different geographical locations at the same time should show the object in projection at different positions among the stars.

Mars was observed in 1862 and 1877, but it has a substantial disk, and the position shift was relatively hard to measure. A smaller, star-like image would be much better. The asteroid Eros filled the bill, coming to within 0.27 a.u. of the earth in 1900–1901 and to within 0.17 a.u. in 1930–1931. Many observations were made at those times to determine the a.u.

A new, more direct method for calibrating the distance scale of the solar system involves bouncing a radar pulse off a planet, such as Venus. The pulse travels at the speed of light, and the distance is determined with great accuracy simply by measuring the time required for the round trip. In 1975, when Eros came by at about 0.15 a.u. from the earth, it too was studied by radar. The modern value for the astronomical unit is 149,600,000 km (92,956,000 miles).

COMETS

Comets are somewhat unpredictable and sometimes spectacular members of the solar system. For centuries they were regarded with superstition and ignorance. Aristotle held that comets were fiery things in the upper layers of the earth's atmosphere. Their supposed proximity made them likely scapegoats for various calamities, including some of the plagues that swept medieval Europe. People sometimes prayed for deliverance from the evil of the comet.

If comets were, in fact, atmospheric phenomena, as was then believed, they would show a *parallax;* that is, they should have a different position with respect to the stars when viewed from various localities on the earth. The subject was investigated by Tycho Brahe, who observed the comet of 1577. Tycho could not find any evidence for parallax, and he concluded that the comet was quite distant, at least beyond the moon (60 earth-radii away). These observations should have established the comets as *bona fide* solar system objects, but, as usual, not everyone was convinced.

Observations of Halley's comet eventually led to the conclusion that comets were indeed solar system objects subject to the laws of motion, just like the planets. Newton's laws seem very reasonable to us today, but after all, when they were introduced they only accounted for *known* properties of planetary orbits, as summarized by Kepler. As mentioned in the discussion of the scientific method, the real test of a theory is not how well it fits the existing facts, but rather, how well it predicts as yet unobserved effects.

The Florentine astronomer Paolo Toscanelli, a contemporary and correspondent of Columbus, was foremost in scientific observations of the comet of 1456. This comet had a tail that extended across one-third of the sky, causing fear and consternation in Europe.[1] Centuries later, Toscanelli's estimates of the changing position of the comet among the stars were used to determine its orbit. Similar bright comets were observed in 1531, in 1607, and in 1682. In England, Edmund Halley studied the observations of the comet of 1682 to test one of Newton's conclusions. According to Newton's theory of gravitation, comets could travel in closed orbits (ellipses) or in open orbits (parabolae and hyperbolae). Halley's study showed that his comet must have an elliptical orbit. Based on the theory of gravitation, it should have an orbital period of about 75 years. Looking back through the old records of comets, he noted that:

$$1682 - 1607 = 75 \text{ years}$$
$$1607 - 1531 = 76 \text{ years}$$
$$1531 - 1456 = 75 \text{ years}$$

In Halley's time, it was not thought that a comet, once gone from sight, returns again. Although Newton's theory showed that the ellipse, a closed orbit, was a possible orbit for a comet, Newton's personal preference was for an open orbit, which provides only one trip past the sun. Thus, when Halley predicted that the comet of 1682 would return in 1758, he was really putting his reputation on the line. Well after his death, his hypothesis was vindicated when the comet indeed returned late in 1758.

---

[1] It was even blamed for the fall of Constantinople to the Turks, although that event had occurred in 1453!

In fact, the date when the comet reached *perihelion* (closest approach to the sun) was in accord with the more exact calculations of the French mathematician Alexis Clairaut, who improved Halley's work by allowing for gravitational perturbations of the comet by Jupiter and Saturn. It was a great triumph for the young science of celestial mechanics, which described motions in space as the result of gravitational forces. Any doubt that may have remained was dispelled by the discovery of Neptune. Halley's triumph also swept away much of the superstitious feeling about comets, as they were now seen to be predictable phenomena.

Later students, checking archives, have found extensive references to appearances of what we now call Halley's comet. In the Chinese records, in particular, there were reports of every one of its apparitions during almost thirteen centuries. However, lacking the necessary theory, and hampered by the fact that one man rarely lives to see two appearances of the comet, the Chinese astronomers had not realized that it was the same one that they continued to record. The records of other cultures also show traces of Halley's comet; for example, its apparition in 1066, on the eve of the Battle of Hastings, is depicted in the Bayeux tapestry (Figure 7-4).

The earliest known observation of a comet is, in fact, of Halley's comet in 467 B.C. It last appeared in 1910, and will be back in 1986, when hopefully we can all have a good look at the comet and raise a toast to the great mathematicians of the past!

Figure 7-4. Halley's comet as shown on the Bayeux tapestry. (Yerkes Observatory photograph.)

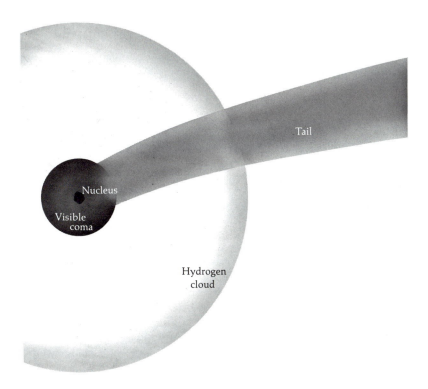

Figure 7-5.   The principal parts of a comet. (Not drawn to scale.)

The basic components of cometary forms are shown schematically in Figure 7-5. The central *nucleus* is believed to be solid and composed of "ices" (frozen gases, such as water, methane, carbon monoxide, and ammonia and dust particles. This structure has been likened to a large, dirty snowball. The *coma* is an essentially spherical cloud of gas and dust, centered on the nucleus. The radius of the coma can reach 100,000 kilometers (62,000 miles). The most spectacular part of the comet is the *tail*, which can be as long as 10^8 kilometers (6.2×10^7 miles). In 1970, observations of two bright comets from above the atmosphere disclosed a new feature—a cloud of hydrogen atoms about 10^7 kilometers (6.2×10^6 miles) in diameter. Radio observations of Comet Kohoutek in 1974 indicated that, as suggested by A. Delsemme, there may be a region of a few hundred kilometers around the nucleus populated by tiny ice grains a few millimeters in diameter. These grains are blown away from the nucleus by gas flowing out into the coma. This feature is called the *icy-grain halo*.

Figure 7-6. Comet Mrkos, photographed on August 22 (upper left), August 24 (upper right), August 26 (lower left), and August 27 (lower right), 1957, with the 48-inch Schmidt camera on Mount Palomar. (Courtesy of the Hale Observatories.)

The different types of tails are well illustrated in the photographs of Comet Mrkos (Figure 7-6). The long, straight tail with a great deal of filamentary structure is composed of ionized molecules (including carbon monoxide that has lost one electron, CO^+); this *ion tail* (by contrast, the coma is not ionized) is pushed away from the sun by a continuous out-flowing of gas from the solar corona, called the *solar wind* (Chapter 9). The shorter, smoother, curved tail is composed of dust particles with sizes of about 10^{-4} cm (3.9×10^{-5} inch); these particles are pushed away from the sun by the radiation pressure of sunlight. The ion tails are blue (because of the emission of light by the CO^+ molecules), whereas the *dust tails* are yellow (because of reflected sunlight). This effect is illustrated in a color picture of Comet Bennett (Plate 5). The two different types of tails can occur separately or together. Comet West developed a spectacular dust tail in 1976 (Figure 7-7b). A comet with an ionized tail only is shown in Figure 7-8. When comets pass close by the sun, the emission of light by atoms (notably by sodium) appears in addition to the molecular light mentioned above. Material in the tails is blown away and lost forever; a large comet, such as Halley's, probably has enough material for about 100 close approaches to the sun.

Little is known about the origin of comets. According to a leading theory, they were produced out beyond the present orbit of Pluto as a by-product of the process that formed the solar system, and a vast swarm (numbering 100 billion or more) of comets exists out there today. Their average distance from the sun is estimated by the Dutch astronomer J. Oort at 150,000 a.u. Since the nearest star is only about twice this far from the sun, it is clear that such remote comets, although orbiting the sun, would be perturbed by other stars. Sometimes these perturbations could enable a comet to escape from the solar system (that is, from a closed orbit around the sun), but occasionally a comet would be perturbed into the central part of the solar system, where it could be captured in a new, highly eccentric orbit that would make it periodically visible from the earth.[2]

---

[2]The point of closest approach to the sun is called *perihelion;* that of farthest removal is called *aphelion.* When the perihelion and aphelion distances are nearly equal, so that the elliptical orbit is nearly circular, it is called a low-eccentricity orbit; when the distances are very different, so the orbit is cigar-shaped, the eccentricity is high.

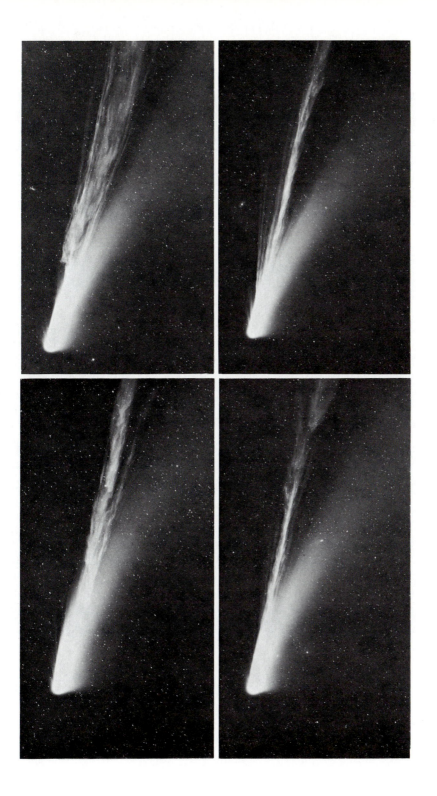

(a)

(b)

Figure 7-7. (a) Nuclear splitting in Comet West. The dates, left to right, are March 8, 12, 14, 18, and 24, 1976. On March 18, the projected average diameter of this cluster of nuclei was roughly 10,000 km. (New Mexico State University Observatory.) (b) Comet West on March 4, 1976, showing spectacular dust tail. (Courtesy of A. Stober, Goddard Space Flight Center.)

Figure 7-8. Comet Cunningham, photographed on December 21, 1940. (Courtesy of the Hale Observatories.)

As the cometary nucleus approaches the sun, the frozen material is heated by solar radiation and releases gases that form the coma and hydrogen cloud. If ionization (the stripping of outer electrons from atoms) occurs, an ion tail is formed and is pushed away from the sun by the solar wind. As the ices are vaporized, the dust embedded in them is also liberated and blown into the dust tail by the solar radiation pressure. Sometimes the nucleus splits, as that of Comet West did in 1976 (Figure 7-7a). As the comet moves away from the sun, the heating decreases, the liberation of ices and dust ceases, and the nucleus goes into cold storage.

When the comet is so far from the sun that its ices are not vaporized, it no longer has a tail. A new tail forms as the comet returns to the vicinity of the sun. The changing appearance of a comet as it moves in orbit is illustrated with photographs of Halley's comet taken in 1910 (Figure 7-9).

Comet orbits are mostly quite elongated, and their orientations in space are randomly directed, so they may cross the plane of the earth's orbit at quite steep angles. This is very different from the case of the planetary orbits, which are all tilted less than 8 degrees to the earth's orbit (except Pluto at 17 degrees). Occasionally, comets pass close to Jupiter and are perturbed into paths that are nearly circles. These comets have shorter periods of revolution and are observable through a larger fraction of each orbit, as compared with those in elongated orbits like Halley's comet (Figure 7-10).

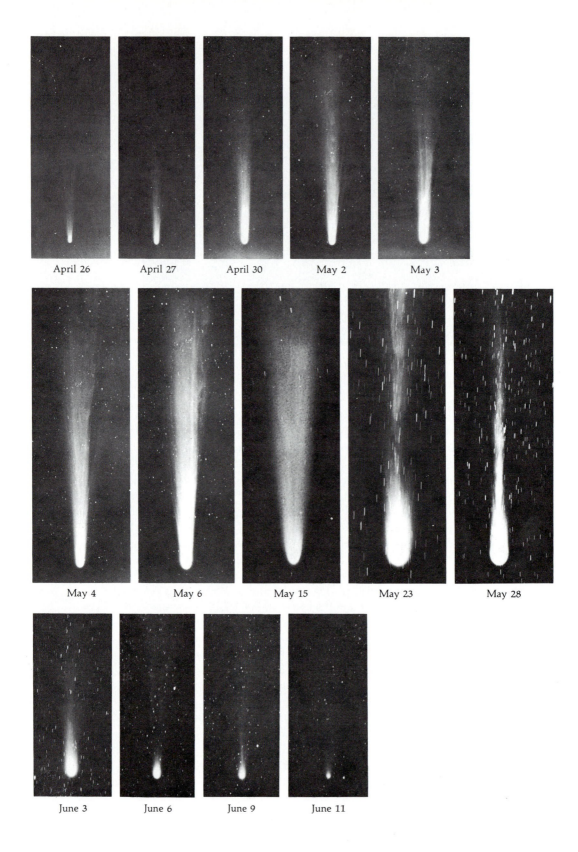

April 26 April 27 April 30 May 2 May 3

May 4 May 6 May 15 May 23 May 28

June 3 June 6 June 9 June 11

Figure 7-9. Views of Halley's comet in 1910. This comet was, of course, expected; but to the delight of many viewers, another, unexpected comet appeared in 1910 and stole the show! (Courtesy of the Hale Observatories.)

Figure 7-10. Orbit of Halley's comet compared with planetary orbits. The planetary orbits are nearly circular; the orbit of Halley's comet is a highly eccentric ellipse, and the comet travels around the sun in the opposite direction to that of the planets.

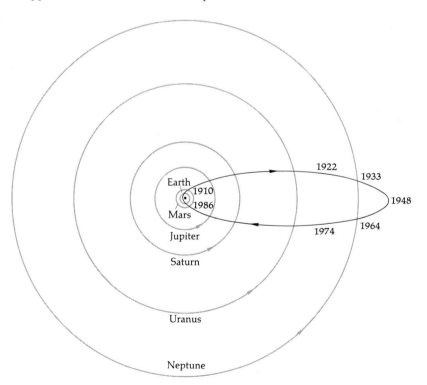

Bright comets that can easily be seen by the unaided eye are discovered from time to time, frequently by amateur astronomers and sometimes by airline pilots; three were discovered this way during the first half of 1970. Comets are usually named for their discoverers, although the most famous of all is an exception—Halley's comet was named after the man who first suggested that it had a closed orbit and predicted the date of its return.

METEORS AND METEORITES

Meteors are light-emission phenomena in the earth's atmosphere, once popularly called shooting stars. For the amateur they are great fun to observe from a comfortable horizontal position under dark country skies.

A photograph of a meteor is shown in Figure 7-11; it appears as a streak of light in the sky. The production of meteor light is caused by rocks *(meteoroids)* that were in orbit around the sun before they entered the earth's atmosphere. At the typical entry speeds of 30 km per second (19 miles per second), the rocks are heated by friction with the air molecules.[3] This heating can melt or actually vaporize the rock; the vast majority of shooting stars are caused by meteoroids no larger than gravel or grains of sand which are fully vaporized and never reach the ground. (Some meteoroids are so tiny that the amount of friction they cause is not enough to make them melt or vaporize. These *micrometeorites* drift down through the atmosphere like dust.) The collisions between a meteoroid and the air molecules also heat the air in the immediate surroundings. This hot gas emits the light we see as the meteor. If the initial meteoroid is sufficiently large, it can survive entry and reach the ground (Figure 7-12); such rocks of celestial origin found on the earth are called *meteorites*.

Photographs of a meteor from different locations on the earth's surface enable a determination to be made of the meteoroid's orbit in the solar system prior to encountering the earth. Sometimes many are found to have essentially the same orbit—they constitute a *meteoroid stream.* This can be explained by material distributed along an orbit as shown in Figure 7-13. In some cases, the stream orbit is known to resemble the orbit of a comet that was observed in the past but has faded out. Perhaps the stream is the debris remaining from the breakup of the cometary nucleus. When the earth passes through a meteoroid stream, a large number of meteors are seen for a few days *(meteor shower)*. The Perseid shower,

---

[3]This friction process was the basis of the reentry problem that had to be solved for manned spaceflight.

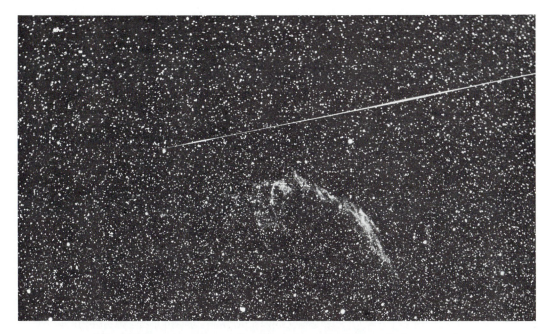

Figure 7-11. A meteor trail seen against a background of stars and a nebula in the constellation Cygnus. (Yerkes Observatory photograph.)

which occurs each year about August 12, is one of the most spectacular, with visible meteor rates often reaching 60 per hour. Other meteoroids, termed *sporadic*, appear to have unique orbits not associated with streams.

The frequency with which we see meteors varies throughout the night; more are seen after midnight. This effect has been known for centuries and has a simple explanation. Unlike the planets, which all go around the sun in the counterclockwise direction (as seen from a point above the earth's North Pole), some meteoroids are going in the same direction as the earth, and some are going in the opposite direction. On the side of earth where the time is before midnight, the meteoroids that enter the atmosphere are going in the same direction as the earth and must catch up to it. On the side where the time is after midnight, the earth encounters meteoroids that are moving in the opposite direction. They do not need to catch up and are, in fact, swept up by the earth. Many more meteors occur on the morning side because of this effect. The situation resembles that of the hapless motorist who enters a freeway via an exit ramp—he is going against the traffic and is more likely to meet a spectacular end than if he were moving with the traffic!

(a)

(b)

The rocks that actually strike the ground—the meteorites—are of great scientific interest as celestial material that can be analyzed in the laboratory. Some are stony and some are metallic (mostly iron). The age of 4.5 billion years (or slightly more) that we quoted for the oldest rocks in the solar system was first revealed by studies of radioactive decay in meteorites. Presumably, the environment of the meteorites in interplanetary space has been less destructive than the erosive processes and mountain-building processes on earth. More recently, lunar rocks and soil have been brought to earth by the crews of several Apollo missions and by two automated Soviet spacecraft. These samples also have been dated by radioactive means. The oldest soils and rocks from the moon are about as ancient as the meteorites.

There are several important reasons for studying meteorites in the laboratory. Historically, the radioactivity-dating experiments were of great interest, because they gave our first direct evidence for matter older than the oldest known rocks on earth. Secondly, chemists analyzed meteorites to determine the abundances of the elements in them. By combining information from the meteorites with data on elements detected in the spectrum of the sun and from studies of the material in the crust of the earth, a fair idea of the abundance of the elements in the solar system has been obtained. Clearly, theories of the origin of the solar system will have to explain why the elements are present in these amounts. Finally, just as we have recently been able to examine the lunar rock samples for possible biological material, meteorites have been studied for years in hopes of finding traces of organic matter from outside the earth. Occasionally, such traces seemed to have been discovered, but the evidence was not very convincing. The problem was that they might have been contaminated by organic matter *after* reaching the earth. For example, studies of some meteorites revealed tiny amounts of amino acids, but the particular assortment of amino acids that was found turned out to be the same as is present in fingerprints. The first convincing evidence came from several fragments that fell near Murchison, Australia, in September, 1969. Eight amino acids were detected in an assortment *not* resembling those found

Figure 7-12. (a) A crater, more than two meters in diameter, produced by a meteoroid fragment which fell to the ground in China's Jiling (Kirin) Province on March 8, 1976. (Wide World Photos.) (b) Chinese scientists examine a stony meteorite fragment weighing 123.5 kg from the same event. Another fragment recovered from this meteorite fall weighs 1,770 kg (3,900 pounds) and is the largest stony meteorite ever found. The meteorites are now in the Jiling City Museum. (Wide World Photos.)

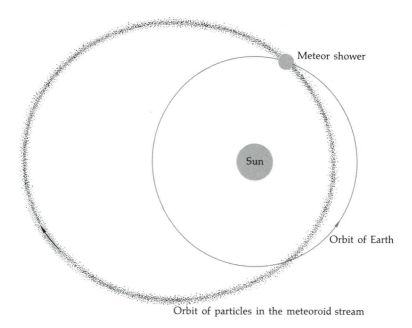

Figure 7-13. Sketch explaining the origin of annual meteor showers. Usually the orbit of the stream particles is tilted with respect to the earth's orbit, and there is only one intersection of the two orbits.

in fingerprints or other products of living matter on earth. The simplest conclusion was that these amino acids existed in the Murchison meteorite before it struck the earth. This does *not* prove that there is extraterrestrial life—amino acids also can form by nonbiological processes—but it does mean that some of the complex molecules that are needed for life as we know it are present in space.

Occasionally, very large meteoroids that may even have been asteroids have struck the earth, with spectacular effects. These meteoroids impact with great force, producing *meteor craters*. The most famous of these is Barringer Crater in Arizona (Figure 7-14), which was the first terrestrial carter to have its meteoritic origin proved. This crater is 1.2 km (4,000 feet) across and 170 meters (570 feet) deep, with a rim that rises about 50 meters (160 feet) above the surrounding countryside. Attempts have been made to mine the region beneath the crater floor for the primary body, which may have been a solid chunk of iron and nickel; these efforts have been unsuccessful. On the other hand, iron fragments are found throughout the countryside surrounding the crater and nowhere else in the region. Apparently, the meteoroid was disrupted by the impact, and these iron fragments are parts of it. Geological evidence indicates that the impact occurred about 20,000 years ago.

Proven impact craters, both singly and in groups, have been found at over 60 locations on earth, and many other round geological features are suspected to be of meteoritic origin, including craters as large as the 60-km (37 miles) Manicouagan Crater in Quebec. An object that probably was a small comet struck without warning and devastated the forest around the Tunguska River, Siberia, in 1908, but no identifiable fragments were found in the disaster area. In 1947, also in Siberia, the mid-air explosion of a meteorite produced about 100 craters, and a considerable number of iron fragments were recovered.

Figure 7-14.   Arizona Meteor Crater, located 18 miles west of Winslow, near Route 66 (view to the northwest). The crater is officially named for D. M. Barringer, who studied it for many years in the early twentieth century to prove his then-controversial theory that it had been caused by meteoroid impact. (From *Geology Illustrated* by John S. Shelton. W. H. Freeman and Company. Copyright © 1966.)

(a)

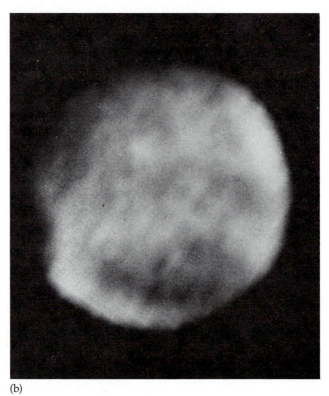

(b)

Figure 7-15. (a) Jupiter and the four Galilean moons. The image of Jupiter was overexposed in order to record the dimmer satellites. (Yerkes Observatory photograph.) (b) Close-up view of the largest Galilean satellite, Ganymede, obtained in December 1973 by the interplanetary spaceprobe Pioneer 10. Computer analysis of Pioneer 10 photos showed several dark areas similar to those found on our own moon as well as bright rings of an unexplained type on Ganymede's surface. With a diameter of 5,216 km (3,241 miles), Ganymede is larger than the planet Mercury. (NASA.)

The identification of certain meteor streams with comet orbits establishes a cometary origin for some meteoroids; others, the stone and iron meteoroids large enough to reach the earth's surface, may come from the asteroid belt between Mars and Jupiter. The relative numbers of meteoroids derived from these sources have not been definitely determined.

SATELLITES

The moon is the sole proven natural satellite of the earth, although tiny natural satellites have sometimes been suspected. A good idea of the nature of the lunar surface has developed in recent years through manned and automated spaceprobes. This is now a separate subject in itself, and we defer a description of the physical properties of the moon until we discuss space exploration in Chapter 15.

Studies of the satellites of other planets began in 1610 with Galileo's discovery of the four major moons of Jupiter (Figure 6-3). These are now called the Galilean satellites (Figure 7-15). In order of their distance from Jupiter they are Io, Europa, Ganymede, and Callisto. They are all roughly the size of our moon, but compared to Jupiter they are tiny, whereas our moon has about one-fourth the earth's diameter. Observations of the Galilean satellites of Jupiter led to the first successful measurement of the speed of light (Chapter 5). The discovery of satellites of other planets contributed greatly to the idea of a "plurality of worlds."

The planets Mercury and Venus have no known satellites, but Mars has two, Deimos and Phobos. They were discovered in 1877 by Asaph Hall of the U.S. Naval Observatory in Washington, D.C. They both are very small and very close to Mars. The Soviet astrophysicist I. S. Shklovskii once speculated that Phobos and Deimos were artificial satellites placed in orbit by an ancient Martian civilization. Photos obtained by Mariner spaceprobes show that both are oblong rather than spherical (Figure 7-16), and that each keeps one side always facing toward Mars. These photos also show that they resemble small asteroids and are not artificial satellites.

Jupiter has at least nine smaller moons in addition to the four large satellites observed by Galileo. Four of these were discovered by the American astronomer Seth Nicholson. Moon XIII was discovered in 1974 at the Hale Observatories, where a possible (but as yet unconfirmed) Moon XIV was observed in 1975. At least one of Jupiter's moons (Io) seems to be related to radio bursts observed from that planet (Chapter 8). Several of the small outer moons have retrograde orbits.

Saturn has ten moons, one of which (Janus) was discovered as recently as 1966 by the French astronomer A. Dollfus. The biggest, Titan, is larger than the planet Mercury and probably also larger than the Jovian

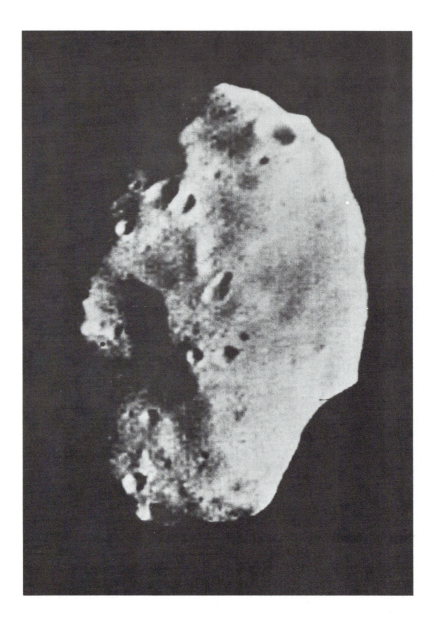

Figure 7-16. Phobos, one of the two Martian moons, as photographed on November 30, 1971 by the Mars-orbiting spacecraft Mariner 9. Irregularly shaped, with a largest diameter of 28 km (17 miles), Phobos has many impact craters. (NASA.)

satellite Ganymede. Molecular bands of the gas methane appear in its spectrum, indicating the presence of an atmosphere. The most distant moon of Saturn, Phoebe, has a retrograde orbit.

Saturn also has a large number of tiny satellites in the form of particles that make up the beautiful rings (Figure 7-17). Spectroscopic observations interpreted through the Doppler effect show that the particles in the rings obey Kepler's Laws for their various distances from Saturn. Thus, the rings are neither solid nor do they revolve as a rigid body; the particles are probably ices. It has been suggested that the rings are the remains of an ancient moon that disrupted. A more likely possibility is that they are composed of material that could not collect to form a moon because of the disruptive tidal force exerted by Saturn. The rings lie in the equatorial plane of Saturn and are exceptionally thin (less than 10 km, or 6 miles, thick); in fact, when viewed edge-on they are nearly invisible.

Uranus and Neptune have five and two moons, respectively. Uranus and some of its satellites are shown in Figure 7-18; the satellite orbits lie roughly in the plane of Uranus' equator, and hence are nearly perpen-

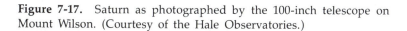

Figure 7-17. Saturn as photographed by the 100-inch telescope on Mount Wilson. (Courtesy of the Hale Observatories.)

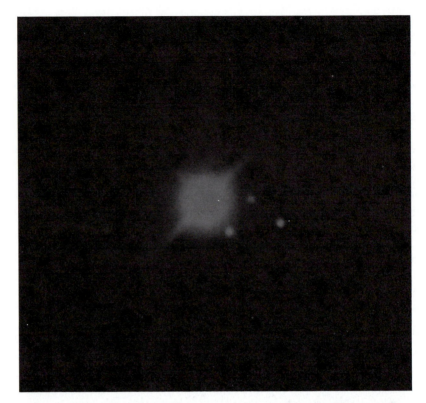

Figure 7-18. The planet Uranus and some of its satellites, as observed with a ground-based telescope. The rings of Uranus are too faint to appear in this photograph. Photos made in 1970 with the balloon telescope Stratoscope II showed that, unlike Jupiter and Saturn, distinctive cloud features are not visible on Uranus. (Lick Observatory photograph.)

dicular to its orbital plane. Their orbital motions take place in the same sense as the rotation of Uranus.

Uranus also has rings—at least five of them—but they were not discovered until 1977, when they were observed from a NASA airplane. The rings are in the plane of Uranus' equator and are very thin in the direction perpendicular to this plane. In this respect they resemble the rings of Saturn. However, each of Uranus' rings is also very thin along the radial direction (the direction from the ring toward the center of Uranus). In this respect they differ from Saturn's rings. The difference can be visualized in the following manner. Think of a long-playing record that contains several bands, each band corresponding to one song, or to a short movement of a symphony. Then one such band can be compared to a ring of Saturn; but the groove within such a band (traversed by the phonograph needle during *one turn* of the record on a turntable) corresponds to a ring of Uranus. Uranus' rings are so thin that they reflect very little sunlight, and thus they escaped observation for many years. It also appears that they have a relatively dark color.

Neptune's two satellites, Nereid and Triton, have normal and retrograde orbits, respectively.

Pluto has one moon, as mentioned earlier. Known as Charon, it is the largest satellite in the solar system, judged in proportion to its planet. According to the rough estimates available in 1978, Charon has a diameter of about 1,200 km (750 miles), fully 40 percent that of Pluto, and has a mass of about 5 to 10 percent that of Pluto. It appears to be revolving in the equatorial plane of Pluto in *Pluto-synchronous orbit,* meaning that Charon's orbital period of 6.4 days exactly equals the axial rotation period of Pluto, so that the moon is always located above the same spot on Pluto's surface. This means that, for hypothetical observers located on the surface of Pluto, those in one hemisphere would always see the moon, whereas those in the opposite hemisphere would never see it. In the case of earth's moon, the lunar rotation and revolution periods are equal, but neither is synchronized with the rotation of the earth. Note how this contrasts with the case of Charon.

SCALES AND REGULARITIES

The scale of the planetary orbits is shown in Figure 7-19. We have already noted the regularity of their spacing, as expressed by Bode's Law. Of great interest in studying the general properties of the solar system is the regularity in the planetary motions. All the planets revolve around the sun in the same direction, called *prograde,* that the sun rotates. (The sun rotates counterclockwise, as seen from a point north of the earth's orbit.)

Figure 7-19.   The planetary orbits to scale.

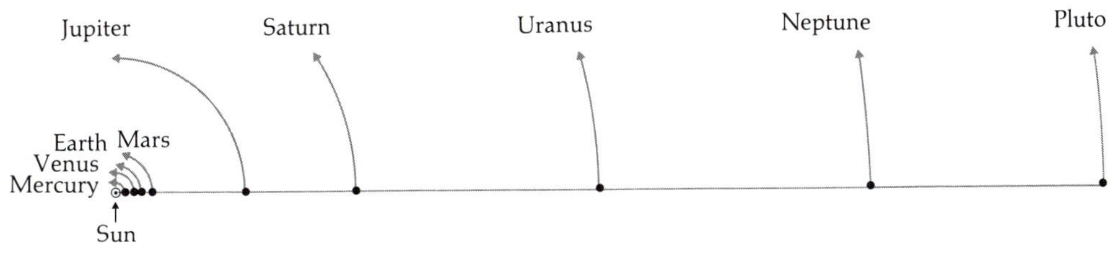

At least six of the planets also rotate on their axes in this same sense; Venus rotates the opposite way (retrograde direction) and Uranus rotates about an axis in the plane of its orbit. Twice every Uranian year, the axis is at right angles to the planet's motion, so it appears to roll or slide along the orbit. The rotational properties of Pluto are probably similar to those of Uranus, with the axis of rotation nearly in the plane of orbit. Also, as mentioned before, the orbits of the planets are nearly in the same plane. The chief exceptions are Pluto and Mercury, with orbits tilted at about 17 degrees and 7 degrees, respectively, to that of the earth. Thus, the planetary regularities are: (a) regularities of motion; (b) regularities of spacing; (c) orbits nearly in the same plane; and (d) most orbits nearly circular.

Some other fundamental properties of the solar system appear to be: (1) cometary orbits occur at all orientations to the plane of the earth's orbit; (2) there is probably a large swarm of comets beyond the orbit of Pluto; (3) asteroids lie chiefly between the orbits of Mars and Jupiter; (4) satellite orbits lie roughly in the equatorial planes of their respective planets, and most of the satellites move along their orbits in the same sense as the rotation of their planets. As discussed in the next chapter, the planets show a progression in chemical characteristics with distance from the sun.

These general properties of the solar system provide much food for thought. A comprehensive theory of the origin of the solar system would have to account for each of them and would indeed be an impressive achievement.

8

The Planets and Their Atmospheres

Once it was recognized that the other planets in the solar system are worlds in the same sense as the earth, many questions arose. How similar are they? Could life exist on Mars or some other planet? Investigation has revealed the striking fact that no other planet has an atmosphere like the earth's. On the other hand, it is likely that at an early stage in its history, the earth was surrounded by a primitive atmosphere of very different composition than that of the air we breathe today. How did the earth's atmosphere evolve to its present state? Why did other planetary atmospheres evolve so differently? The planets are not only of interest in their own right, they are also of concern because they may bear clues to the history of the earth. In recent years, our interest in the planets has been renewed by close-up investigations with space probes. Probes have already sent back valuable data from Mercury, Venus, Mars, and Jupiter.

In this chapter, we deal with the general properties of the planets as they relate to phenomena of their atmospheres. In Chapter 15, we shall return to the detailed study of surface features—the geology or *planetology* of other worlds.

The planets are shown to scale in Figure 8-1. For Mercury, Venus, Earth, Mars, and Pluto, the sizes indicated in the diagram are those of the solid planetary bodies. However, for Jupiter, Saturn, Uranus, and Neptune, the size of the solid interior bodies (if any) is unknown, and their nature is uncertain. These four planets are permanently shrouded by thick cloud layers in their massive atmospheres. Therefore, their sizes as presented in Figure 8-1 refer to the level of the visible cloud tops.

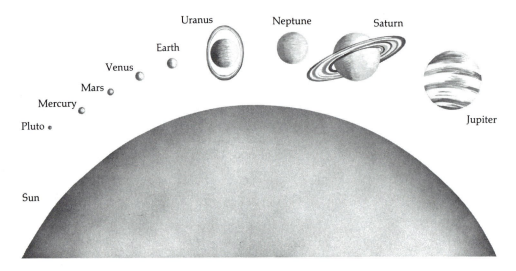

Figure 8-1. Planetary sizes to scale. Note the recently discovered, but poorly understood, rings of Uranus.

The masses of the planets are determined from the gravitational effects of each planet on the motion of other objects, such as its own moons, adjacent planets, or interplanetary space probes. The least accurate method is measuring the gravitational perturbation of one planet by another. However, every planet either has one or more known moons, or has been visited by spacecraft, or both. (Data on the planets are summarized in Appendix 5.)

INTERIOR AND EXTERIOR PLANETS

The four planets nearest the sun are relatively small, as shown in Figure 8-1. The vast bulk of their matter is concentrated in solid form, as indicated by Table 8-1. Their densities range from 3.9 to 5.5 times that of water. (For purposes of comparison, a typical earth rock has a density of 2.7, while iron has a density of 7.8. Deep inside the earth, a substance may have higher density than similar material near the surface, because the deep rock is compressed by the weight of the layers of material above it.) Thus, Mercury, Venus, Earth, and Mars are high-density objects, as contrasted with the four large outer planets, Jupiter, Saturn, Uranus, and Neptune. These four large planets, made up predominantly of the light elements hydrogen and helium, have densities ranging from only 0.7 to 1.7 times that of water. This is sometimes illustrated by the statement that Saturn, the planet with a density of 0.7, would "float in water." This is not strictly correct, since there is presumably a denser central core that would not float, but it gets the point across. Pluto, the outermost known planet, probably also has a density of about 0.7 but apparently no massive atmosphere.

Broadly speaking, the planets can be divided into two groups: the interior or *terrestrial* (earth-like) planets that have dense, solid bodies of rock and iron with relatively low-mass atmospheres (Figure 8-2), and the exterior, low-density planets with massive atmospheres. These have sometimes been called the *Jovian* (Jupiter-like) planets. Pluto is an exception, since it has a low density but no massive atmosphere. (Of course, its "atmosphere" may simply be frozen; spectroscopy reveals that methane, present as a gas in the Jovian planets, has the form of an ice on Pluto.) Pluto is also peculiar in having an orbit that is noticeably inclined with respect to the orbital planes of the other planets. Its orbit is so eccentric in shape that it actually passes inside a portion of the orbit of the next planet closer to the sun, Neptune. These facts suggest that Pluto may have a different origin from that of the other planets—speculation has centered on the possibility that it is an escaped moon of Neptune.

To speculate on the origin of the outer planets, we must anticipate some ideas (Chapter 11) concerning the birth of stars and the origin of the solar system. According to current thinking, the sun and the rest of the solar system formed from the *solar nebula* (Figure 8-3), a contracting gas cloud, flattened and rotating, with the same chemical composition found in the surface layers of the sun today: mostly hydrogen and helium with trace amounts of the heavier elements, including carbon, nitrogen, oxy-

Table 8-1.   Masses of the Interior Planets

Planet	Atmosphere Mass (grams)	Ocean Mass (grams)	Solid Mass (grams)
Mercury	$< 4 \times 10^9$	0	3×10^{26}
Venus	4×10^{23}	0	5×10^{27}
Earth	5×10^{21}	1×10^{24}	6×10^{27}
Mars	2×10^{19}	0	6×10^{26}

The symbol $<$ means "less than." The powers of ten notation in which the tabulated numbers are expressed is explained in Appendix 1.

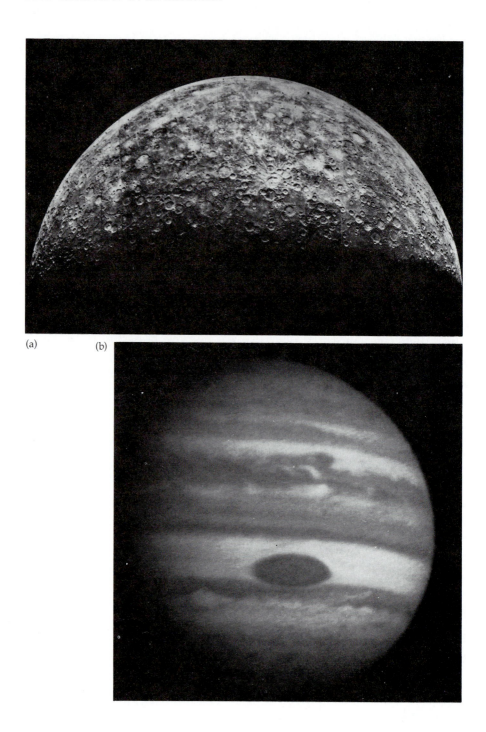

(a)   (b)

gen, and iron. The precise way in which the sun condensed at the center of this cloud, leaving the planets, satellites, and asteroids distributed through circumsolar space in a disk of dusty matter, perhaps surrounded at a very great distance by an enormous cloud of comets, is far from certain and indeed is a subject of considerable debate. If the composition of the solar nebula was uniform, one might think that the original planetary bodies would have begun with the same chemical composition as the sun. However, apparently this did not happen. The variation among the planets, especially the distinct difference in density of the terrestrial and outer planets, suggests that they formed from different substances, presumably under different circumstances. The key circumstance perhaps was the temperature in the region of formation.

It is thought that the planets formed at a time when the protoplanetary disk in the solar nebula was significantly heated, at least in its inner parts, by the newly formed sun. According to this theory, the central portion of the disk was so hot that only the most refractory materials, such as iron, could condense to the solid state and thus accumulate into planets. Hence the interior planets are largely composed of these relatively dense substances. Specific calculations based on one particular theory of this process indicate that Mercury formed at a temperature of about 1,500 K, Venus at 1,000 K, and the earth at 550 K. Farther from the sun, the Jovian planets may have formed at temperatures near 200 K, while on the extreme outskirts of the nebula, the temperature may have been as low as 20 K. Thus, the Jovian planets formed under conditions in which hydrogen, the most abundant element of the solar nebula, was present in considerable amount as a constituent of such frozen gases as water (H_2O), methane (CH_4), and ammonia (NH_3). They formed as such large objects that their gravity was sufficient to retain the thick nebular gases (mostly hydrogen) in their immediate vicinity. This would explain the difference

Figure 8-2. (a) *Terrestrial planets,* such as Mercury (shown here in a mosaic of photos by the Mariner 10 spaceprobe), are dense objects, largely made of rock and iron. Note the resemblance to the moon. Many planetologists consider the moon to be a terrestrial planet, even though it is a satellite of the earth. (b) *Jovian planets,* named for Jupiter, the nearest and largest of them, are mostly composed of the light elements hydrogen and helium, as is the sun. Whether they possess rocky cores is not definitely known. Substances such as frozen ammonia (NH_3) make up the bulk of the clouds, but do not account for the hues seen on color photos; the coloring may be due to trace constituents. This photograph, showing cloud bands, the great red spot, and smaller features of the atmosphere, was made by Pioneer 10. (NASA.)

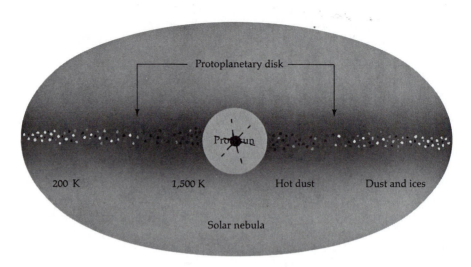

Figure 8-3. Schematic diagram, *not to scale*, showing a possible configuration in the solar nebula during an early stage in the history of the solar system. A thick disk of solid particles, with both mineral ("dust") and frozen gas ("ices") constituents, formed in the central plane. According to one theory, the terrestrial planets formed from the hot dust in the inner region of the disk, while the Jovian planets amassed much of the ice in the outer zone. (Sketch adapted from a theoretical model described by the astrophysicist Hubert Reeves.)

in density of the two groups of planets, and suggests that the makeup of the Jovian planets (especially the two largest ones, Jupiter and Saturn) more closely resembles the chemical composition of the original solar nebula. According to one *very* speculative theory, Pluto and its moon are "chips" broken off Neptune by a close encounter with another planet during the formative stage of the solar system. If so, there is a slight chance that a tenth and more distant planet may yet be discovered.

In addition to differences stemming from the conditions of formation, the planets have a variety of striking characteristics, which relate to the different ways they have evolved since formation.

RADAR ASTRONOMY AND PLANETARY ROTATION

The period of rotation of a planet on its axis is a key property. It may determine or affect the length of the day, the surface temperature, the presence of a magnetic field, or even the shape of the planet. For some planets, such as Jupiter and Mars, this period can be measured by observing distinct features in the clouds (Jupiter) or on the surface (Mars) as the planet turns. Another method (used, for example, to measure the rotation of Uranus) consists of studying the Doppler shift of a line in the spectrum of the planet. In the case of Mercury and Venus, however, neither of these methods was successful in the past, and in fact the rotation periods that were calculated in these ways turned out to be completely wrong.

Mercury is difficult to view from earth because of its small size; nevertheless, its rotation was determined by the Italian astronomer Giovanni Schiaparelli in the late nineteenth century. He made drawings of streaky markings on Mercury's surface, and comparing the drawings made at different times, he concluded that Mercury rotates on its axis once in 88 earth days, the same length of time that it requires for one orbital revolution. This meant that it always kept one face toward the sun, just as the moon always keeps one face to the earth. However, some of the other observers disagreed on this point, and it seems that they were right to question it, although it was eventually accepted and has appeared in textbooks until recently.

The modern technique for measuring Mercury's rotation is by *radar astronomy*. Unlike the radio astronomer, who measures whatever signals are coming in from the cosmos, the radar astronomer beams a pulse of radio waves of known properties at the target, then records the reflected signal. Comparison of the original and returned signals yields information about the target. For example, the time required by the pulse to travel to the target and return gives a measurement of the distance; the Doppler shift of its wavelength tells the relative speed at which the earth and the target are receding from or approaching each other; *Doppler spreading* of the pulse reveals the rotation rate of the target.

Doppler spreading appears in the following way. The initial signal produced by the radar transmitter is concentrated in a very narrow range of wavelengths, like an emission line in visible light. When it is reflected by the planetary target, the radar pulse is Doppler shifted to a new wavelength, depending on the relative velocity of approach or recession of the earth and the planet. But if the planet is rotating, one edge *(limb)* of it will be turning toward the earth, while the opposite limb will be turning away from us. The radar waves reflected from the approaching limb will have a slight Doppler shift toward the shorter wavelengths; those reflected from the receding limb will be shifted toward longer wavelengths. When the radar astronomer records the returned signal, he finds that it is spread over a range of wavelengths due to the planet's rotation, and the rotation velocity can be determined from the amount of this Doppler spread (Figure 8-4). The details of this procedure are more complicated

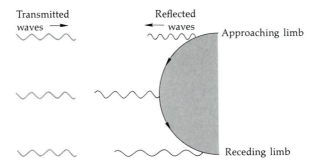

Figure 8-4. Reflection of radar waves by a rotating planet. The curved arrowheads indicate direction of rotation. The radar waves approaching the planet from the direction of the earth (transmitted waves) all have the same wavelength. The wave reflected from the upper limb, which is turning toward the earth, is Doppler shifted to a shorter wavelength; the wave reflected at the bottom limb, which is turning away from the earth, is Doppler shifted to a longer wavelength. The difference in the wavelengths of the upper and lower reflected waves is determined by the rotation velocity of the planet and by the direction in which its axis points. As shown in this sketch, the axis points out of the paper, toward the reader; an additional wavelength shift due to the relative motion of the planet with respect to the earth is omitted. (Adapted from *Radar Observations of the Planets* by Irwin I. Shapiro. Copyright © 1968 by Scientific American, Inc. All rights reserved.)

than we have indicated, and careful attention has to be paid to the geometry of the situation, but, by observing a planet over a length of time as it moves around its orbit, the radar astronomer can determine not only its rotation rate but also the orientation of its axis and the direction in which it is turning. (For example, if a radar astronomer on Mars observed the earth, he would find that the rotation rate is about 0.5 km/sec at the equator, or one turn per 24 hours, that the orientation of the axis is toward the star alpha Ursae Minoris (our "North Star"), and that the direction of the earth's spin is counterclockwise as seen from that star.)

In any case, radar astronomy observations do give precise measurements of planetary rotations, and in this way it was found that the old conclusions from visual drawings of Mercury were wrong. It does not rotate once per 88-day orbital period, but rather once in 58.7 days. In fact, Mercury makes *one and one-half rotations per orbital period.* As shown in Figure 8-5, this means that a certain diameter through the planet always points toward the sun *at the time of perihelion* (closest approach), although the opposite ends of this line (opposite faces of the planet) point toward the sun at alternate perihelia. A reasonable explanation of this phenomenon is that Mercury is not a perfect sphere, but is somewhat elongated along one diameter, which is the same one that lines up with the sun at perihelion.

According to the theory, the slightly greater pull of the sun on the near half of Mercury than on the far half controls this alignment, and hence the exact value of the rotation period. For if the planet turned slightly faster (or slower) than 1.5 times per orbit, the long axis would not point to the sun at perihelion time, and this gravitational effect would then act to slow or hasten the rotation, respectively. The relatively large ellipticity of Mercury's orbit is also involved, because this gravitational effect is much stronger at small distances (near perihelion) than at longer ones. If the orbit were a circle, the long diameter of the planet would point at the sun continuously, in which case the rotation and orbital periods would be equal.

Figure 8-5.   (a) Sketch of the rotation and orbital motions of Mercury. Both the orbit and the shape of Mercury are exaggerated for purposes of illustration. Suppose Mercury, rotating at the observed rate, with its orientation at perihelion as shown in (b), rotated slightly faster, so that it appeared as in (c) at the next perihelion passage; then the sun's greater pull on the near half than on the far half would retard the rotation. The opposite happens in (d), where Mercury has rotated slightly slower than the normal rate. If the orbit were circular, the long diameter of Mercury would always point to the sun (e).

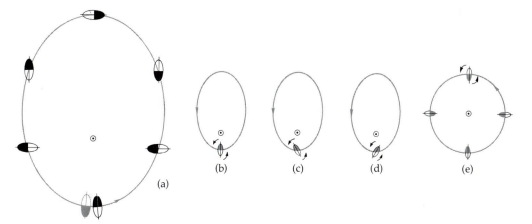

When the visible-light spectrum of Venus was observed from earth, the Doppler shift due to rotation was found to be too small for reliable measurement. It was only possible to conclude that Venus was a slow rotator, with a period of more than a few weeks. Earth-based telescopes revealed a thick, cloudy atmosphere, and some photographs made in blue light showed weak markings in the clouds that appeared to move around the planet with a period of only 4 days. This, of course, was inconsistent with the spectroscopic observations, and was mostly disregarded. Finally, radar observations showed that Venus' rotation period is 243 days, or 18 days more than its orbital period; furthermore, the rotation is *retrograde*. As seen from the direction of our North Star, the earth and Venus both move counterclockwise around their orbits, but whereas the earth also turns on its axis in the counterclockwise way, Venus rotates clockwise. On the earth, the difference between our solar and sidereal days is only about 4 minutes, an effect arising from the fact that the earth moves a short way along its orbit during the time that it rotates once on its axis. However, the orbital period of Venus is 225 days. Thus, during a *Venusian sidereal day* of 243 days, Venus moves completely around its orbit, and then some. As a result, the *Venusian solar day*[1] amounts to 117 days.

If we forget about the clouds for a moment, so that we can imagine seeing the surface of Venus, or imagine being on Venus and being able to see the stars, then the meaning of the two Venusian days is as follows: an observer on a distant star would see that Venus turns once on its axis every 243 days; an observer on Venus would see the sun rise once every 117 days.

The 4-day cloud pattern rotation (also retrograde, by the way) has now been verified through close-up photos (Figure 8-6) by the Mariner 10 spacecraft. It is not caused by the turning of the planet, however, but seems to be caused by a high-altitude wind of about 100 meters/sec (224 mph).

It is also possible to map the surfaces of planets by radar astronomy techniques. For example, as Venus rotates, it is found that some radar pulses take a little more or a little less time than expected to return from its surface. This indicates depressions and highlands, respectively. Further, the intensities of reflected signals corresponding to different areas of the planet are different, indicating differences in the smoothness (and hence the reflecting power) of the terrain. This method does not give us a

---

[1] Space engineers and planetologists have adopted the science fiction term *sol* to describe the interval between two successive sunrises on a given planet.

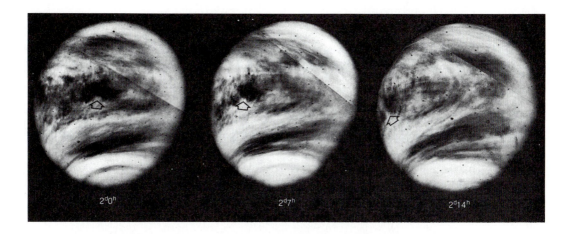

2^{d}0^h 2^{d}7^h 2^{d}14^h

Figure 8-6. Rotation of Venus' atmosphere. A series of mosaic photos taken by Mariner 10 shows that atmospheric clouds rotate with much higher speed than the (unseen) planetary surface below. The arrow shows the location of a feature in the clouds as it rotates from right to left. The period of 4 days for rotation of clouds above Venus' equator can be compared with the 243-day rotation of the planet itself, as measured by radar. Dark and light streaks are features of the atmospheric circulation system. (NASA.)

very detailed picture (see Figure 8-7) of the surface of Venus, but it is of value since the clouds of Venus prevent us from seeing or photographing its surface anywhere but in the immediate vicinity of spaceprobe landers. In particular, this method of mapping has revealed such features as a huge volcano, craters, and a long valley or tectonic rift, whereas lander photos show only the rocks in the immediate vicinity of the spacecraft.

PLANETARY TEMPERATURES

A simple theory has been used to calculate the expected temperatures of the planetary surfaces. The sun shines on a surface, providing an energy input. The input is equal to the total amount of solar energy received, minus the amount that reflects back into space. We can easily calculate the solar energy received by any planet, since we know its distance from the sun. The amount that is reflected back into space is determined by observ-

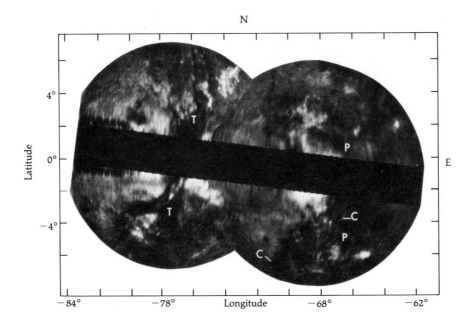

Figure 8-7. Radar observations of Venus obtained by Richard M. Goldstein and coworkers at the Goldstone Tracking Station. The area of the radar images is approximately 2,300 km by 1,400 km. The bright areas correspond to areas of high radar reflectivity. There is a trough (T) in the left image and a plateau (P) in the right image. Some large crater-like features (C) are also marked. These interpretations are supported by altitude data also obtained from radar. The landforms on Venus, particularly the trough, are interpreted as evidence for geological activity like the earth's. (Courtesy M. Malin, Jet Propulsion Laboratory.)

ing the planet with a photometer.[2] The solar energy that is not reflected is absorbed by the surface and heats it. The surface must radiate energy according to the Stefan-Boltzmann Law (Chapter 5) for its temperature. After a while, the amount of energy radiated by the surface will equal the amount of energy that it absorbs. Otherwise, if (for example) the surface radiated less energy than it absorbed, it would keep heating up indefi-

---

[2]The fraction of incident solar radiation that is reflected into space by a planet is called the *albedo*.

nitely. In fact, it heats up until it is just hot enough to radiate away an average amount of energy equal to the average input that it receives from the sun. It cannot heat up further, because then it would be radiating more energy than it receives, and this would violate the principle of conservation of energy. Thus, the energy output of the surface is equal to the energy input. We already know the input, and so we know the output, and we can use the Stefan-Boltzmann Law to find what the surface temperature must be in order to produce that output.

The average planetary surface temperatures calculated according to the simple theory outlined above are much lower than the temperature of the sun. (For example, the calculated temperature of Mars is 250 K, and the temperature of the visible solar surface is about 6,000 K.) We recall a property of the Planck Law discussed in Chapter 5: The wavelength at which the maximum radiation occurs is longer for cooler objects. Maximum solar radiation occurs in the wavelength range of visible light, but the planets, being cooler, radiate most of their energy in the infrared and radio wavelengths. For this reason, we use infrared and radio astronomy observations to measure the actual surface temperatures of the planets. The dependence of intensity on wavelength is determined by measuring the radiation from a planet at several wavelengths. This can be compared with a set of Planck curves (corresponding to various temperatures) to find the best agreement, and hence the observed planetary temperature. In studying the observations to deduce the average temperature, we have to take account of the fact that different parts of the planetary surface are at different temperatures; the part of the planet where the sun is overhead will be hotter than the other side, where it is night.

The observed average surface temperatures of Mercury, the Earth, the Moon, and Mars agree fairly well with the values calculated by the above theory. However, the radio and infrared observations of several planets show that they are significantly hotter than predicted by the theory. For Venus the temperature found by the radio technique is about 700 K, while the theoretical value lies between 325 and 375 K. The higher surface temperature apparently results from the atmospheric heat-trapping *greenhouse effect* discussed later in this chapter. A smaller greenhouse effect occurs on earth, where the theory predicts 246 K, but the average surface temperature is 290 K.

For Jupiter, the observed temperature is near 135 K, while the theory predicts 105 K; here the explanation is based on our ideas about Jupiter's interior. The best measurements and calculations indicate that Jupiter is radiating over twice as much energy per second than the planet is receiving from the sun. The excess energy must be generated at the planet; thus, Jupiter is *self-luminous* and in that sense could be considered a very cool, dim, small star. However, the temperature is not high enough for nuclear reactions to occur, so Jupiter's internal energy source is not the same as that of the sun or other normal stars. The thermal characteristics of Saturn resemble those of Jupiter; it too appears to radiate about twice as much energy as it receives from the sun.

EVOLUTION OF THE EARTH'S ATMOSPHERE

The earth grew by accretion of solid matter from the solar nebula. As the matter drew together, it was compressed and heated. Additional heat was generated by radioactivity in the rock, and perhaps also by the impact of falling meteorites and larger, asteroid-sized bodies called *planetesimals.* Some of this heat still remains in the earth's core today, accounting for its substantially molten state. The heat caused extensive vulcanism, much more than we find on the earth today. The early volcanoes released gases that had been locked up chemically in the solid matter of the earth. Thus the atmosphere was literally exhaled from the interior of the earth. Some gases of the solar nebula may also have been retained by the gravitation of the newly formed earth. The primitive atmosphere, however, contained almost no free oxygen, unlike the air that we breathe today, because oxygen is highly chemically reactive and forms compounds with various substances that it encounters in the environment. (The most familiar example is the rusting of iron to produce iron oxide.) Much water vapor was exhaled from the interior of the earth, and it soon cooled, condensing to form the seas. As the earth evolved, the atmosphere evolved along with it—a result of processes of evaporation, chemistry, geology, and biology.

Evaporation

The molecules of the air are in constant motion, their average speed increasing with the temperature. At the high atmospheric density (in terms of molecules per unit volume) near the earth's surface, the molecules travel only very short distances before colliding with each other. However, the thinner the air, the longer the average distance (*mean free path*) between collisions. Thus, the mean free path increases with height above sea level. Indeed, at heights of about 500 km (310 miles) and greater, the air density is so low that an atom or molecule moving upward will, on the average, not collide with another one. If it is going fast enough, it can escape into space. Recall our discussion of orbits in Chapter 2 and the fact that they can be thought of in terms of a balance of gravitational attraction and centrifugal force. The (always outward) centrifugal force increases with the speed. Three possible cases are of interest for atoms moving upward at heights above 500 km. (1) Particles traveling relatively slowly do not have enough speed to overcome gravity, and therefore they fall back (like a ball thrown upward). (2) Particles with an intermediate speed can achieve an orbit around the earth, but this orbit usually carries the atom back into the atmosphere. (3) Atoms traveling sufficiently fast will be subject to a centrifugal force too large to be balanced by the earth's gravity, and the atoms in question fly away, never to return. The minimum speed at which an atom moving upward can leave the earth in this way is called the *escape speed*, and it is about 11 km/sec (7 miles/sec). The escape speed does not depend on the mass of the object; it is the same for an atom of the air as for a space vehicle.

Atoms in the outer atmosphere have escaped continuously by evaporation. The hotter the temperature of the atmosphere, the faster are the speeds of the gas atoms and molecules—more of them can then exceed the escape speed. In a gas at a given temperature, the less massive molecules move faster than the heavier ones. Detailed calculations show that the earth is likely to have lost essentially all of its original hydrogen (hydrogen molecules are the lightest of all) and possibly all of its original helium (the second lightest gas) by evaporation. (Compare the concept of astronomical evaporation with the evaporation of water; the basic ideas are the same, but in the latter case the water molecules escape only from the body of water to the atmosphere.) According to one theory, the process was aided by heating due to an intense "wind" of particles from the young sun.

Biological, Chemical, and Geological Effects

The primitive atmosphere exhaled from the interior of the earth probably contained much carbon dioxide. Indeed, direct measurements on contemporary volcanoes have been made by brave geologists, who found that they exhale steam, carbon dioxide, nitrogen, and sulfur dioxide, among other substances. Further, carbon dioxide is the major constituent of the atmospheres of Mars and Venus. Unlike the present atmosphere of earth (see Table 8-2), the early atmosphere contained little or no free oxygen (O_2), and therefore had no ozone (O_3) layer. However, a variety of processes have operated to remove most of the carbon dioxide from the earth's atmosphere, and to liberate more oxygen than can be used up in

Table 8-2.   Chief Constituents of Dry Air at Sea Level

Gas	Percent by Mass
Nitrogen (N_2)	75.5
Oxygen (O_2)	23.2
Argon (A)	1.3
Carbon dioxide (CO_2)	0.05

chemical reactions.[3] One process that accomplishes both of these results is photosynthesis by plants, which produces oxygen and uses up carbon dioxide. Carbon dioxide also dissolves in sea water, where much of it is converted to other compounds that are laid down in sediments and eventually form rock. The best example is limestone, which is calcium carbonate.

Thus, the existence of the oceans helped to remove most of the carbon dioxide from the primitive atmosphere and convert it into rock. The presence of plant life produced the oxygen of the air, and from that oxygen came the ozone layer, which shields life from the harmful effects of solar ultraviolet radiation. According to some evolutionary theories, once the ozone layer formed, the simple life forms already present were able to evolve more successfully into advanced species. It is an incredible concept that primitive plant life was able to modify the environment many kilometers above the ground so that the earth became a more hospitable place for the descendants of those plants!

On the earth today, there is about 50 times more carbon dioxide dissolved in the seas than is present in the atmosphere, and there is more than a thousand times as much carbon dioxide transformed into sedimentary rock than is present in the seas. Thus, the presence of liquid water allowed the earth to evolve an atmosphere substantially free of carbon dioxide, in direct contrast to our sister planets, Mars and Venus.

Evolutionary Summary

Evaporation has been an ongoing process since the earth formed, and it (and/or an intense solar wind) probably removed all or nearly all of any

---

[3] A continuous source of oxygen is necessary, since estimates show that if the production of oxygen stopped today, the existing oxygen in the atmosphere would be exhausted within at most several tens of thousands of years, chiefly as a result of oxidation of rock.

light gases that may have been inherited from the solar nebula. Additional material for the primitive atmosphere was exhaled from the crust due to volcanic processes, probably about 4 billion years ago. The oceans developed to about their present volume around the time that the atmosphere was formed, or perhaps a little later. The combination of the primitive atmosphere and the water provided the raw materials for the primary broth from which life probably developed. Starting about 2 billion years ago, substantial amounts of oxygen began to be produced by plant life. By approximately 1 billion years ago, the oxygen content of the atmosphere had increased to its present value. The present atmosphere (except for pollutants) dates from that time.

CHARACTERISTICS OF THE EARTH'S ATMOSPHERE

The composition of the earth's present atmosphere was given in Table 8-2. A variable constituent not listed in the table is water vapor, which can amount to as much as 2 percent by mass under certain conditions at sea level. Solar radiation affects the bulk of the atmosphere indirectly; it heats the ground, and the atmosphere is then warmed by convection as air near the ground is heated and rises. Some parts of the atmosphere—for example, the ionospheric layers near 160 km (100 miles) altitude—are heated directly by the absorption of solar radiant energy. The pressure decreases with height in the atmosphere, as shown in Table 8-3; the variation of temperature is also shown. These are average values around which there are considerable fluctuations, depending on the time of day, the season, and (at the lower altitudes) the weather. The properties of the atmosphere at high altitudes were determined from many rocket and satellite studies.

A layer of ozone (O_3) molecules exists near 30 km (19 miles) altitude in the atmosphere. This layer is a highly effective absorber of solar ultraviolet radiation for wavelengths of 2,000 to 3,000 Angstroms. For wavelengths near 3,000 Angstroms, about 7 percent of the light passes through the ozone layer. At the absorption peak in this wavelength range (near 2,600 Å), only 10^{-32} of the incident sunlight passes through the ozone layer. The small amount of ultraviolet radiation that does reach the ground is responsible for suntans and sunburns. The intensity of this radiation at the earth's surface if the ozone layer were not present would be far more than uncomfortable—it would be lethal. On the other hand, in the early history of the earth, this radiation may have been an important source of energy for the chemical reactions in the primary broth before the ozone shield was formed from atmospheric oxygen.

Table 8-3. A Model Terrestrial Atmosphere

HEIGHT ABOVE SURFACE			Pressure (Percent of Surface Value)	Average Temperature (K)
Km	Miles	Remarks		
0	0	Surface	100	280
1.5	1	Altitude of Denver, Colorado	85	273
3	2		70	266
6	4	Tops of high mountains	50	255
10	6	Cruising altitude of jet-liners	30	238
16	10		10	210
30	20	Ozone layer	1	250
50	30		0.1	290
80	50	Aurora and meteors near 100 km	10^{-3}	170
160	100	Ionosphere	10^{-6}	600
480	300	Atmospheric evaporation takes place	10^{-13}	1,200

The absorption of solar radiation by ozone is an energy source for the atmosphere near 30 km and above, and this energy is responsible for the higher temperatures found at 30 and 50 km, as seen in Table 8-3. Similarly, the temperatures increase above 100 km because of the ionization[4] of atmospheric constituents by solar radiation. This region of the atmosphere, called the *ionosphere*, contains substantial numbers of free electrons and ions, although most of its atoms and molecules are neutral. It consists of several layers and reflects radio waves with wavelengths of a few meters or greater. Multiple reflections between the ionosphere and the earth can send radio signals all the way around the earth. The principal layers of the ionosphere are called the D region (altitude 90 km, or 56 miles), the E region (110 km, or 68 miles), and the F region (200 to 300 km, or 120 to

[4]The heating occurs because the energies of the photons absorbed in the ionization process are partially converted to kinetic energy of the electrons released from the atoms.

190 miles). The D layer has the smallest *electron density* (number of electrons per cubic centimeter) and the largest density of neutral gas molecules. This combination makes the D layer weakly reflecting and strongly absorbing, so far as radio waves are concerned. The electrons and ions in the D region recombine to form neutral atoms and molecules shortly after sunset, but particles in the higher ionospheric layers recombine more slowly. Therefore, the D region vanishes, and radio propagation by ionospheric reflection (Figure 8-8) is most effective at night. Since the ionosphere is produced by x-ray and ultraviolet radiation from the sun, variations in the intensity of this radiation due to solar flares and other events (Chapter 9) create temporary changes in the ionosphere and can thus affect radio communications.

Our own experience with the weather and the seasons tells us that the atmosphere is not in a static and uniform state. The combined effect of the heating of the earth by the sun and the rotation of the earth is to produce a global circulation pattern (Figure 8-9). Many facets of this pattern were known empirically to the navigators of sailing ships. The fairly steady *trade winds* are a result of large-scale atmospheric circulation patterns, as are the *horse latitudes* characterized by light and undependable winds. These examples are representative of the surface circulation pattern. The general atmospheric circulation is also largely responsible for the gross pattern of currents in the oceans.

Other features of the circulation are found at different heights. The most familiar example is the *jet stream*, a thin current of air that flows from west to east near an altitude of 10.5 km (35,000 feet). The wind speed in the jet stream can reach 320 km/hour (200 mph). As a result, for example, it takes longer to fly east–west across North America than it does to fly west–east; airline schedules reflect this fact, and the cruising altitude is normally chosen to avoid the stronger parts of the jet stream when going west and to take advantage of them when going east.

Figure 8-8. Sketch showing that reflection by an ionized layer in the atmosphere permits radio transmission over large distances on the earth.

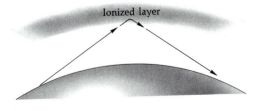

Ionized layer

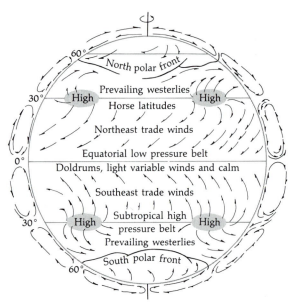

Figure 8-9. Major patterns in the circulation of the earth's atmosphere. (From *Principles of Geology*, 3d ed., by J. Gilluly, A. C. Waters, and A. O. Woodford. W. H. Freeman and Company. Copyright © 1968.)

Besides the gross changes of pressure, temperature, and circulation patterns, the atmosphere plays a vital role in the movement of water between the oceans and the land masses—the *hydrologic cycle*. Water moves through the cycle as illustrated in Figure 8-10, evaporating into the atmosphere principally from the oceans. It condenses into the clouds, which are transported by the wind, and is deposited as rain and snow on the continents. Eventually it reaches the ocean again, and the cycle is repeated. The same water flows continually through these stages, and we say the cycle is a closed one. Impurities in water blocked at some stage of the cycle will accumulate there; an example is the continual buildup of salt in the oceans as minerals are washed away from the land.

Clouds often contribute to a beautiful sunset, but certain features of the atmosphere, without clouds, can be seen in the setting sun and blue color of the sky. As the sun sets, its color turns first to orange and then red. This effect is caused by the molecules in the atmosphere. When the molecules are much smaller than the wavelength of light, the sunlight is affected by a process called *Rayleigh scattering*. We recall the reference to scattering in Chapter 5; the wavelength of the light remains unchanged in a scattering process, but the direction is changed. Blue light is scattered much more strongly than red light, and, as a consequence, the red light penetrates much farther through the atmosphere. The geometrical situation is shown in Figure 8-11. As the sun sets, the amount of air that its light passes through en route to the observer is continually increasing, and so we see a red sun. The blue light is scattered in all directions and gives rise to the blue color of the sky. Above the atmosphere, as seen by the astronauts, for example, the sky is black.

The effect of the wavelength dependence of the atmospheric penetration of light is illustrated by photographs of the same terrestrial scene taken with short (violet) and long (infrared) wavelength light, as shown in Figures 8-12a and 8-12b. The long wavelengths pass through, while the short wavelengths are scattered in the atmosphere. Figures 8-12c and 8-12d show a similar effect in the atmosphere of Mars.

The atmosphere is a barrier that shields us not only from the solar ultraviolet radiation but also from meteoroids, many of which burn up in the atmosphere near 100 km (62 miles) altitude from friction with the atmospheric gases. The atmosphere also stops some of the cosmic rays, and

Figure 8-10. The hydrologic cycle. (From *Principles of Geology*, 3d ed., by J. Gilluly, A. C. Waters, and A. O. Woodford. W. H. Freeman and Company. Copyright © 1968.)

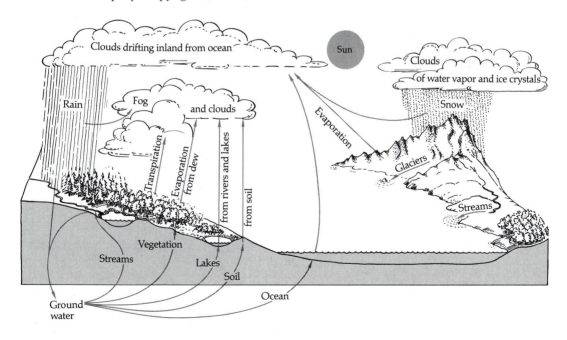

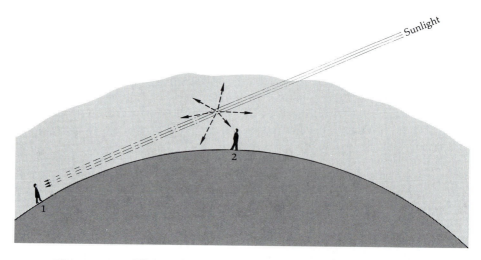

Figure 8-11. Effects of light scattering in the earth's atmosphere. Blue light is strongly scattered, unlike the red light, and so an observer at position 1 sees a red sun at sunset, while the scattered light gives the sky its characteristic blue color as viewed, for example, from position 2, where sunset has not occurred.

this effect, together with the tendency of the earth's magnetic field to deflect the slower-moving cosmic rays, keeps the genetic mutation rate of life down to a reasonably low value; biologists believe that a high mutation rate is undesirable.

Beyond the atmosphere, there is a region called the *magnetosphere*, where electrons and protons are trapped in the earth's magnetic field. This region was discovered with instruments carried on the first artificial satellites, and it is discussed in Chapter 15.

THE ATMOSPHERE OF VENUS

The surface of Venus is obscured by a thick, cloudy atmosphere (Figure 8-13). In 1932, spectroscopic observations showed characteristic absorption lines (as explained in Figure 8-14) of carbon dioxide. Nevertheless, for years the principal constituent of the atmosphere of Venus was assumed to be molecular nitrogen (N_2) by analogy with the earth's atmosphere (a convenient assumption, since molecular nitrogen has no strong absorption lines that could be searched for in the visible light spectrum). Radio observations of the planet indicated a temperature of about 700 K. This is surprisingly high, and there was some reluctance at first to interpret it as the temperature of the surface of Venus. One alternative interpretation was that the radio emission came from a thick ionospheric layer in the atmosphere of Venus.

The question raised by the radio observations of Venus was settled in 1962 by the deep spaceprobe Mariner 2. The intensity of radio emission would show a rather different variation across the disk of Venus, depend-

Figure 8-12. The penetration of light through the atmospheres of Mars and the earth. The photographs are: (a) San Jose, California (13.5 miles distant) in violet light; (b) San Jose in infrared light; (c) Mars in violet light; (d) Mars in infrared light. The greater penetrating power of the longer wavelength infrared light is clearly shown in both atmospheres. (Lick Observatory photographs.)

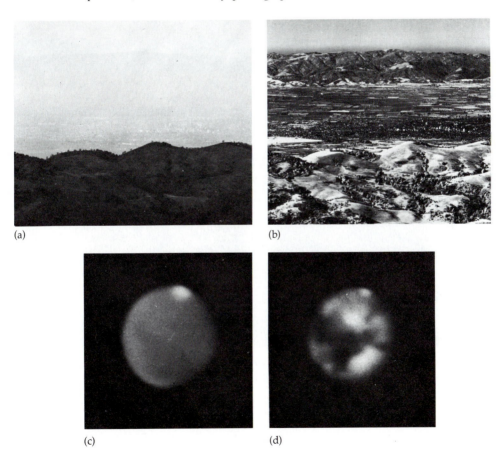

(a)

(b)

(c)

(d)

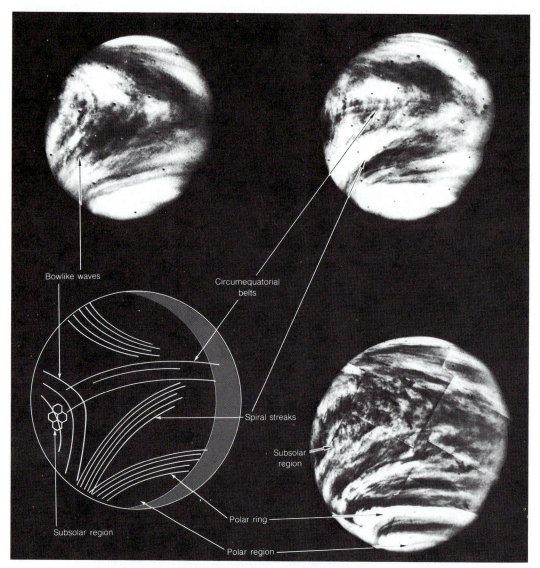

Figure 8-13.   The major features of the circulation pattern of Venus'
atmosphere are identified in the sketch and illustrated by examples in
the global pictures. (NASA.)

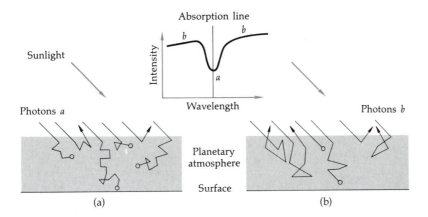

Figure 8-14. Schematic illustration of photon diffusion and absorption line formation in a planetary atmosphere. Photons (a) corresponding to a characteristic wavelength of an atom or molecule are scattered and/or absorbed more frequently than photons (b) in the continuum, where they are largely just scattered (not absorbed). Thus, fewer photons (a) than photons (b) diffuse back out of the planetary atmosphere, and an absorption line is found when the spectrum of the planet is recorded.

ing on which of the two likely models (Figure 8-15) was correct. The Mariner 2 observations at a wavelength of 2 cm clearly established that the high temperature observed by radio astronomers corresponds to the surface and not the ionosphere of Venus. This was directly confirmed on December 15, 1970, when the Soviet spaceprobe Venera 7 parachuted to a soft landing on the surface. It measured a temperature of about 750 K at an estimated atmospheric pressure 90 times greater than the sea level pressure on earth. The spectroscopic results give a temperature for the 25-km (40-mile) high cloud-tops of about 235 K. Thus the temperature of the atmosphere of Venus decreases rapidly with height above the surface. How can the very high surface temperature be maintained?

An answer to this question involves an understanding of how planetary atmospheres are heated. As mentioned earlier, it is a relatively simple matter to calculate the temperature of a planet made of rock (and lacking an atmosphere) as a function of distance from the sun. It is determined by a balance of energy input from the sun and the rate at which rocks radiate it away. However, the situation may be drastically altered when a thick atmosphere is present, as on Venus.

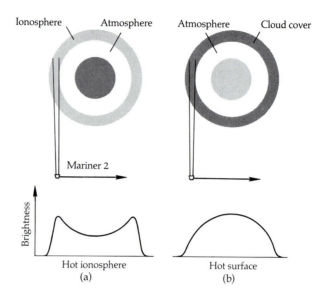

Figure 8-15. Predicted intensities of 2-cm wavelength radio emission that would have been observed by Mariner 2 as it scanned past Venus if (a) the hot ionosphere theory were correct or (b) the hot surface theory were correct. According to theory (a), Mariner 2 would measure maximum brightness near the edge of the planet because it would be viewing through the largest amount of hot ionosphere. According to theory (b), the maximum brightness would occur at the center, because radiation from the hot limb would be greatly reduced by the longer traverse through the relatively cool atmosphere. The Mariner 2 observations favored theory (b). (After Brandt and Hodge.)

The Greenhouse Effect

The basic explanation for the high surface temperature of Venus is the well-known *greenhouse effect*. On a hot, sunny day, anyone entering a greenhouse, or a car left with the windows rolled up, becomes conscious of this effect. The temperature at the surface of the earth is determined by a balance of net energy received due to incident solar energy during the day and net cooling at night by radiation into space. If the loss could be reduced, the result would be a higher temperature; the greenhouse at a botanic garden or nursery accomplishes just this. The visible sunlight passes readily through glass and heats the greenhouse floor, but the floor is much cooler than the sun, and instead of radiating yellow light, it radiates in the infrared. Most glass is not transparent to this infrared radiation, so it becomes trapped in the greenhouse. Thus, solar energy can get in, but the infrared radiation cannot get out. The result is a hot greenhouse.[5]

Measurements by spaceprobes indicate that the atmosphere of Venus is composed almost entirely of carbon dioxide; the old assumption that a substantial amount of nitrogen was present was wrong. The surface pressure as measured by Venera 7 and subsequent Soviet Venus landers is about 90 times the surface pressure on earth. To experience a comparable pressure on earth, a diver would have to descend more than 900 meters (3,000 feet) into the ocean depths! A thick atmosphere composed of carbon dioxide would allow the greenhouse effect to operate, because, although the solar radiation does not strike the surface directly, the atmosphere scatters (changes the direction of a photon but not its energy) in visual wavelengths, and thus the solar energy can diffuse down and heat the surface. The carbon dioxide is opaque (truly absorbing) to most of the infrared radiation from the surface; therefore the greenhouse effect and resultant high temperatures can be achieved. At some wavelengths, the infrared radiation may be trapped by other atmospheric constituents, such as sulfuric acid, but carbon dioxide is the principal absorber.

Atmospheric Circulation on Venus

The circulation of Venus' atmosphere is shown dramatically in the Mariner 10 photographs (Figures 8-6, 8-13). The features are quite different from the typical circulation pattern on earth (Figure 8-9). Clearly seen in the Venus photos is the fast-moving, east–west flow called *zonal circulation*, which proceeds at 100 meters/sec (224 mph) at the level of the cloud deck. But a pattern of cell-shaped (as opposed to layered) clouds surrounding the point where the sun is overhead on Venus shows that strong up-and-down convective motions occur there due to the direct heat of the sun. The subsolar point, of course, moves around the planet only once per solar day (117 earth days), much more slowly than the zonal flow. Thus, the convective activity at the subsolar point represents an obstacle to the adjacent zonal flow. As a result, a wave pattern (Figure

---

[5] Actually, there's an alternative theory. The atmospheric scientists R. G. Fleagle and J. A. Businger state that real glass-roofed greenhouses "are warmer than the surrounding air because the glass prevents the warm air inside from rising and removing heat from the greenhouse." They believe this effect is more important than infrared absorption by the glass. Thus, the greenhouse effect may explain the high temperature of Venus but not the warmth of a greenhouse!

8-13) is produced much like that seen around the bow of a ship in calm water *(bow waves)*. Recent spectroscopic observations indicate that Venus' clouds are composed largely of sulfuric acid droplets, although other substances are also present. Even at the cooler levels of Venus' atmosphere, the environment is distinctly hostile.

THE ATMOSPHERE OF MARS

Detailed studies of Mars, which were made both from spaceprobes and from observatories on earth, show that it has only a very thin atmosphere composed almost entirely of carbon dioxide. There are small amounts of nitrogen and argon, and just traces of water vapor and oxygen. There is no liquid water. There are thin clouds (Figure 8-16), and occasionally the surface is obscured by dust storms, which may occur over large areas of the planet. The surface pressure is about one percent of the terrestrial value. The temperature is about 250 K, and thus below the freezing point of water. It has been suggested that the rate of volcanic activity on Mars is

Figure 8-16. Clouds in the thin atmosphere of Mars are seen here as detached layers, about 25 to 40 km above the surface of the planet. The picture was taken by a Viking orbiter looking to the east at an altitude of about 19,000 km. The clouds are probably composed of tiny particles of frozen carbon dioxide (dry ice). (NASA.)

much less than on earth, and this may partially explain the much smaller total mass of the Martian atmosphere. Our ideas about the surface of Mars have been radically altered by spaceprobe results; they are discussed at length in connection with space exploration in Chapter 15.

EARTH, MARS, AND VENUS— THE KEY DIFFERENCE?

Three terrestrial planets, Earth, Venus, and Mars, are similar in many ways, but their atmospheres differ markedly. The atmospheres of Mars and Venus consist primarily of carbon dioxide, whereas the earth's has little carbon dioxide but a significant amount of oxygen. The oxygen was produced mostly by vegetation, and the carbon dioxide of earth is largely bound in the rocks of the crust and in sea water.

The physicist Ichtiaque Rasool has noted that the key circumstance leading to the separate evolution of these planetary atmospheres was *the different distances of Earth, Venus, and Mars from the sun.* According to Rasool's theory, most of the water on Mars would have remained in the frozen state due to the low surface temperature. On the other hand, essentially all of the water on Venus would have taken the form of water vapor (steam) due to the high temperature, and ultraviolet photons from the sun would have *dissociated* these water molecules into hydrogen and oxygen.[6] The hydrogen, being very light, would have escaped from Venus by evaporation. The oxygen would gradually have been absorbed into the crust of Venus through the oxidation of rocks and would not have been replenished by plants (as on earth), since the high temperature and lack of liquid water on Venus presumably prevented life from arising there. Liquid water on earth facilitates the absorption of carbon dioxide by chemical reactions in rocks of the crust, in addition to absorbing carbon dioxide directly. Thus, the absence of liquid water on Venus and Mars can account for the fact that their atmospheres are primarily composed of carbon dioxide, and the presence of liquid water on earth was crucial to the development of the present atmosphere and life.

This theory may represent an oversimplification of the processes that formed the atmosphere of the earth and its two neighbor planets. Nevertheless, Mars and Venus are, respectively, cold and hot desert

---

[6]Ionization is the process in which the absorption of a photon by an atom or molecule gives an electron enough energy to escape from the atom or molecule. Dissociation is the process in which absorption of a photon by a molecule gives one of its constituents atoms enough energy to escape from the molecule.

planets. If the earth were slightly closer to or farther from the sun than it happens to be, it would resemble one or the other of these planets. In that event, the earth would be uninhabited.

ATMOSPHERES OF JUPITER AND SATURN

Intensities of infrared and short-wavelength (a few centimeters) radio radiation from Jupiter and Saturn indicate atmospheric temperatures of about 135 K and 120 K, respectively. Although somewhat higher than expected, these temperatures are still so low that evaporative escape of atmospheric molecules is a very slow process, especially since the gravities of Jupiter and Saturn are much greater than that of the earth or the other terrestrial planets. The combined retarding effects of low temperature and high gravity make the escape of atmospheric gases from Jupiter and Saturn entirely negligible. Thus, these planets should have about the same composition as the sun and the original solar nebula.

Spectroscopic analysis shows that compounds of hydrogen, including molecular hydrogen (H_2), methane (CH_4), and ammonia (NH_3), are present in the Jovian atmosphere. Ammonia exists both as a gas and as ice crystals. On earth, droplets and ice crystals of water make up the clouds. On Mars, there are clouds of dry-ice (carbon-dioxide) particles (Figure 8-16). On Jupiter and Saturn, ammonia plays the same role. Helium is actually by far the most abundant element after hydrogen on Jupiter, which agrees with the assumption of solar composition. However, the helium went undetected until November–December 1973, when the Pioneer 10 spacecraft flew past Jupiter. Water is present in the Jovian atmosphere, but only as a trace constituent.

The composition of Saturn's atmosphere is much like that of Jupiter. However, as it is colder on Saturn, the ammonia is nearly all in the frozen state, and hence ammonia spectral lines (which are due to the gas) are not prominent in the spectrum, although the clouds are almost certainly composed of ammonia crystals.

Jupiter's atmosphere and internal structure are determined by its several remarkable properties, including low density, a chemical composition dominated by hydrogen, a major source of internal heat dating back to the time of planetary formation, and rapid rotation. Jupiter spins so fast (9 hours, 55 minutes, 30 seconds is the length of the sidereal day) that it is noticeably deformed from spherical shape, being flattened at the poles and bulging at the equator. The fast rotation has especially spectacular effects on the atmosphere, leading to the banded pattern of clouds (Figures 8-17 and 8-18). The internal structure of the planet, according to a

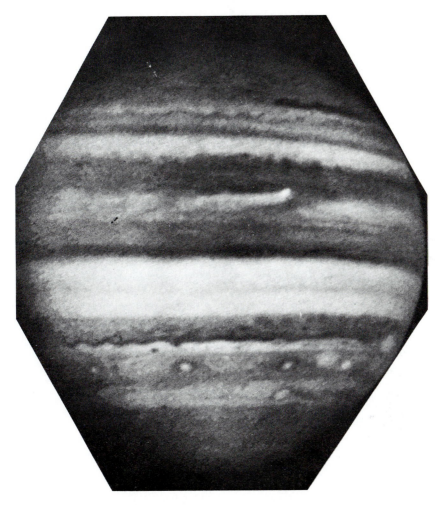

Figure 8-17. The prominent structure of Jovian cloud belts and zones, with many details of smaller cloud features, is seen in this image constructed from measurements by a scanning photometer device on the Pioneer 10 spacecraft. (NASA.)

leading theory, is shown in Figure 8-19. Note that the presence of a core of heavy substances is uncertain. Even if there is such a core, it must be molten or in an extremely compressed vapor state, as solid rock and iron cannot exist at the enormous pressure occurring from the weight of the overlying material. Thus, there is in no sense a solid rock surface any-

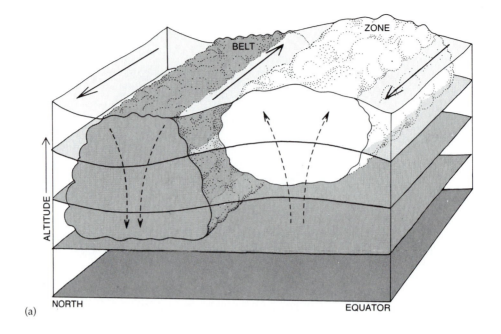

(a)

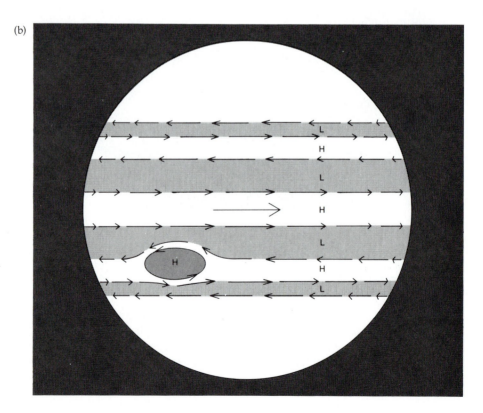

(b)

where on Jupiter. Surrounding the hypothetical core is a great region of *metallic hydrogen*, an electrically conductive gas under high pressure. The powerful magnetic field of Jupiter (measured by the Pioneer 10 and 11 space probes) probably is generated by electrical currents within this region.

The *great red spot* (Plate 12) is the most famous single feature of the Jovian atmosphere. It has been under observation from earth for more than three centuries. Generally orange-red, it changes hue and occasionally disappears, leaving a clearly recognizable oval region called the *red spot hollow.* Careful measurements show that it drifts slowly with respect to the surrounding atmosphere. Most planetologists believe that it is a meteorological disturbance, possibly resembling an immense hurricane, larger than the earth. Atmospheric features in general on Jupiter last much longer than they do on earth, so that the "weather" is much more predictable. On earth as on Jupiter, atmospheric disturbances arise basically from the differences in temperature of different parcels of air. On the earth, a warm parcel can cool down in a few weeks by radiating away its excess heat in the form of infrared radiation. However, the corresponding process takes much longer on Jupiter, and so features of the Jovian atmosphere often persist for years or decades. Since the great red spot is an extreme example, we may require a more involved explanation for it.

The general appearance of Saturn (Figure 7-17), setting aside the question of the rings, is similar to the appearance of Jupiter. The belts and

Figure 8-18.   Explanation of the appearance of Jupiter's belts (dark bands) and zones (bright bands). (a) The zones are at higher temperature than the belts, so at the same height the pressure in the zones is higher. (This situation is shown schematically by the up and down swings in the horizontal lines.) The interaction of this difference in pressure with Jupiter's rotation produces westward flow (toward the reader) on the edge of the zone toward the equator, and eastward flow (away from the reader) on the edge of the zone toward the pole. This flow is shown by the solid arrows. A secondary flow caused by the rising of warmer gases and the sinking of cooler gases is shown by the dashed arrows. (b) On a global scale, the banded appearance of Jupiter results. The zones are shown white, the belts are shown gray, and the symbols (H) and (L) denote regions of high and low pressure, respectively. The long, solid arrow indicates the planet's rotation. The short, solid arrows show the wind direction. Note the flow pattern around the great red spot. The entire banded pattern breaks down at latitudes greater than 45° from the equator. (After *The Meteorology of Jupiter* by A. P. Ingersoll. Copyright © 1976 by Scientific American, Inc. All rights reserved.)

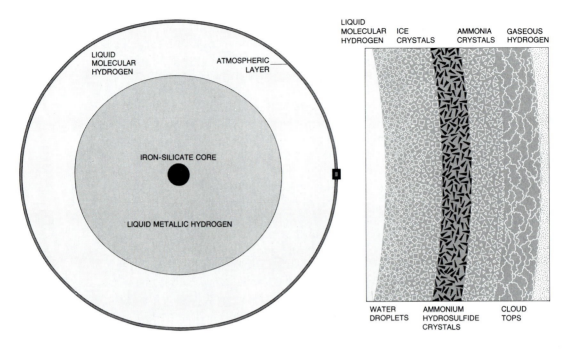

Figure 8-19. Model of Jupiter's internal structure. The upper layers (corresponding to the black box on the right side of the full model) are shown in expanded detail at right. This model is based on theory, and details are uncertain. (After *Jupiter* by J. H. Wolfe. Copyright © 1975 by Scientific American, Inc. All rights reserved.)

bands of Saturn are, however, more diffuse in appearance and somewhat harder to observe. Like Jupiter, Saturn shows considerable oblateness due to rapid rotation.

ATMOSPHERES OF URANUS AND NEPTUNE

The atmospheres of these planets are very cold as a result of their great distances from the sun; radio observations give temperatures of about 100 K. Methane is very prominent in the spectra of their atmospheres; it undoubtedly exists in both gaseous and frozen form. The visible clouds may well be composed of methane crystals. As yet undetected, ammonia clouds are thought to occur at lower depths. As on the two larger Jovian planets, the bulk of the atmospheres of Uranus and Neptune must be composed of hydrogen and helium.

Uranus seems to have very faint belts like those of Jupiter and Saturn, but features of these outer planets are very difficult to determine because of their great distance. Not much surface detail has been observed on Neptune. Both Uranus and Neptune show greenish disks, and Uranus has a slight polar brightening ascribed to some kind of aerosol haze layer over the poles.

Uranus' axis of rotation is nearly in the planet's orbital plane instead of being roughly perpendicular to it. Thus, we have the strange circumstance that, depending on Uranus' position in its orbit, sometimes the sun shines down from nearly overhead at the North Pole, sometimes the same occurs at the South Pole, and sometimes there is high noon at the lower latitudes! Imagine the incredible seasons we would have if this happened on earth. Can you diagram Uranus' motion as described here? Refer to Figure 2-7, which accounts for the earth's seasons. What kind of atmospheric circulation can we expect to discover when the first space probes reach this planet? If you know, you are the only one.

THE MAGNETOSPHERE OF JUPITER

Intense radio signals from the planet Jupiter at wavelengths of about 10 meters (33 feet) were detected in 1955 by Bernard Burke and Kenneth Franklin at the Carnegie Institute of Washington. Their discovery is a splendid case of serendipity. They meant to observe the Crab nebula (Chapter 16), but noticed a strong source of radio emission which passed through the stationary beam of their antenna about the same time on most days. The signal was so strong that it was considered to be some kind of local interference, possibly someone going home from work every day in a vehicle with a faulty ignition. After some time a curious fact became apparent, namely, if the noise were due to a commuter, he commuted according to the sidereal clock. Recall the discussion in Chapter 2 concerning the difference between the day reckoned with respect to the sun and the day determined with respect to the stars. The sidereal day is about 4 minutes shorter than the solar day; the stars rise and set by sidereal time. The fact that the hypothetical commuter was going home 4 solar minutes earlier each successive day clearly suggested an astronomical source. Jupiter was in the right area of the sky, but at first it was not considered seriously as a possible source of the radio bursts, because only continuous thermal radiation was expected from it. Further observations proved that the bursts do in fact come from Jupiter, and they are not thermal radiation.

Three fairly distinct types of Jovian radio emission have now been identified. One, occurring predominantly in the infrared and millimeter radio wavelengths, is in fact the thermal emission. At wavelengths of roughly 10 to 100 cm (4 to 40 inches), *nonthermal* (that is, not resembling the predictions of the Planck Law) emission of slowly varying intensity is received from a large area, as shown in Figure 8-20. These radio waves are polarized and are typical of the radiation emitted by energetic electrons trapped in a magnetic field. It was deduced that Jupiter has a magnetic field with a calculated strength at the visible "surface" of the planet about ten times that of the magnetic field on the surface of the earth. The electrons are analogous to the particles in the Van Allen Belts (Chapter 15) of the earth's magnetosphere. The third type of Jovian emission (the kind actually discovered by Burke and Franklin) occurs at longer wavelengths of, say, 10 meters. Here the radio emission only occurs in the form of intermittent bursts. After the announcement of the discovery, Australian radio astronomer C. Shain searched records of a radio sky survey carried out five years previously and found that Jupiter had been recorded (and unrecognized) many times.

Statistical analysis of the many observations of the radio bursts from Jupiter, dating back to 1950, has led to a theory of the burst-emitting region that would have stretched the imagination of any science fiction writer. The bursts are emitted *directionally*; they occur in three rather wide beams that rotate with the planet. However, we do *not* observe bursts

Figure 8-20. Contours of radio emission from Jupiter as observed at a wavelength of 10 cm. The strongest radiation comes from the two shaded areas located about two Jovian radii on either side of the planet. The circle represents the size of Jupiter as seen in visible light.

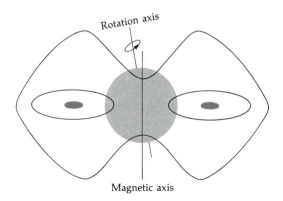

every time a beam crosses the earth. The likelihood of an occurrence has been found to correlate very strongly with the orbital position of the Jovian moon Io. Apparently the satellite triggers storms of radio bursts! Their origin apparently involves charged particles trapped in the magnetic field, perhaps in localized concentrations, which are disturbed by the presence of this moon.

All of these conclusions were verified when Pioneers 10 and 11 flew through the Jovian magnetosphere. It was found to be endowed with intense concentrations of magnetically trapped particles, which caused radiation damage to some electronic components of the spacecraft. Indeed, an astronaut could not survive such a trip, as the dose of radiation would cause death. Radiation biology specialists at the University of Rochester's School of Medicine and Dentistry concluded from the Pioneer 10 measurements that the radiation dose experienced within the spacecraft as it flew past Jupiter "would be lethal to man and to most multicellular biological organisms."

9

Our Sun

The sun, our closest star, supplies the energy that maintains life on earth. Because of its proximity, it is a unique target for astronomers; it is the only star for which we can observe the detailed features of the surface. Since serious solar study began in 1610, long before the advent of modern physics, early investigations were confined largely to observations of the positions and sizes of the solar surface features that are most conspicuous as viewed in visible light—the sunspots.

SUNSPOTS

Dark areas on the sun were noticed occasionally by observers before the advent of the telescope. The idea that these sunspots were actually planets seen against the bright background of the sun was ruled out by Galileo, who found that the spots moved across the sun in the same direction and at the same rate; in this way he discovered the rotation of the sun. Taking this solar rotation into account, we can, by analogy to the earth, define the solar equator, axis, and poles (Figure 9-1).

Watching spots move across the sun was the first of several methods used to study the sun and solar rotation (Figure 9-2). In view of the early date of discovery of sunspots and the relative ease of observation, it is surprising how slowly the basic facts were accumulated. The number of spots and spot groups changes from day to day, sometimes going slightly up, sometimes slightly down. Over longer periods of time, however (such as a few years or more), distinct trends are evident, as shown in Figure 9-3. The sunspot number increases gradually, reaches a peak, declines to a minimum value (sometimes there are no sunspots at all), and then increases again. The interval between peaks (they are called *sunspot maxima*) averages eleven years, the length of the *sunspot cycle*. By definition, a sunspot cycle begins at *sunspot minimum*, when a few new spots appear.

These basic facts were established in 1851, when the work of the German amateur astronomer Heinrich Schwabe received wide attention. He had searched for a new planet between Mercury and the sun, and instead discovered a fundamental fact of solar physics—the periodic variation of the number of sunspots. (In fact, Schwabe had published these results about seven years earlier, but they had been largely ignored until 1851, when Baron von Humboldt mentioned them in his book *Kosmos*.)

Watching sunspots form (sometimes as individual spots and sometimes in groups), move across the solar disk, and gradually fade away, it was natural to mark down their positions each day. These measurements have shown that the rotation is actually *not* uniform. In 1863, the English amateur astronomer Richard C. Carrington reported that the spots at low

Figure 9-1. The solar coordinate system.

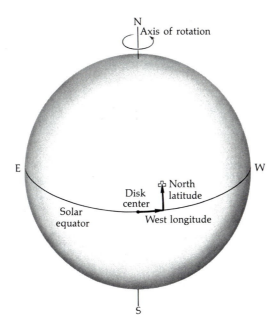

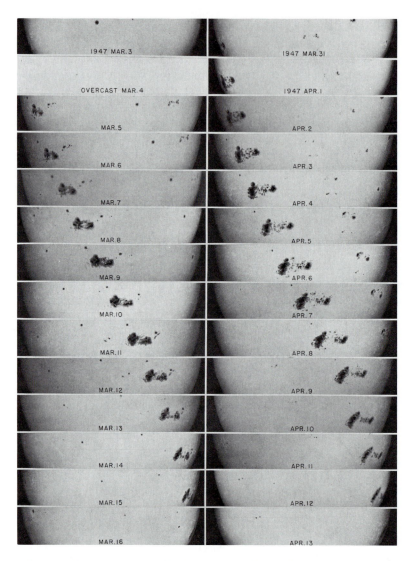

Figure 9-2. Sunspots and solar rotation. A very large sunspot group is shown during two solar rotations in 1947. A solar rotation takes 27 days, so we see the spot group for about two weeks at a time. (Courtesy of the Hale Observatories.)

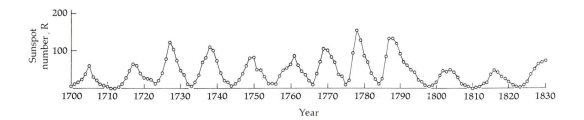

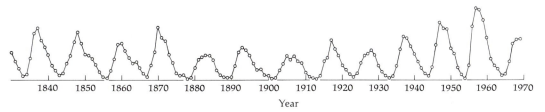

Figure 9-3. The solar cycle during the years 1700 to 1969 as shown by the annual average sunspot number. This quantity is defined as the number of individual spots plus ten times the number of sunspot groups; it is adjusted to allow for the fact that individual observers with different ability, equipment, and viewing conditions tend to count more or fewer spots than the average observer. The data for the years 1700–1715 actually are *false*. It now seems certain there was neither a detectable solar cycle nor many sunspots at all from 1645 to 1715.

Table 9-1.   Differential Rotation of the Sun

Latitude	Rotation Period
0° (equator)	25 days
20° N and S	26 days
40° N and S	28 days
60° N and S	31 days

latitudes take less time for a complete rotation about the axis than do spots located closer toward the poles. This was a remarkable result. If the equivalent situation existed on the earth, it would mean that northern Canada would take longer to rotate about the earth's axis than does, say, Mexico City. Fortunately for the people who draw maps, this does not happen here, because the earth is a solid body. Clearly, the sun is not solid, and thus the *differential rotation* of sunspots at different latitudes confirms that the sun is a gaseous object. Table 9-1, summarizing the differential rotation, was determined from newer methods in addition to sunspot observations, since the spots do not occur at all locations on the sun.

Carrington also noted that the spots do not appear in the same locations on the solar disk throughout the sunspot cycle. Near the beginning of the cycle, the new spots form mostly between the latitudes of 20° and 30° in each hemisphere. As time passes and the spots form in larger numbers, they appear at lower latitudes, and near the end of the cycle, approaching sunspot minimum, the few spots that form usually occur below latitude 15°. When the latitude positions of spots are marked on a graph that shows how they change during the cycle, a striking wing-shaped pattern appears (*butterfly diagram*, Figure 9-4). Individual spots from a cycle that has just ended may persist on the sun briefly while the first spots of the next cycle are forming. Thus, one occasionally sees spots at low latitudes from the previous cycle, and other spots near latitude 30° from the new cycle.

In addition to Carrington's two major discoveries about sunspots, he also observed the first recorded solar flare (see Figure 9-19) on September 1, 1859. These achievements boded well for Carrington's future scientific work, but unfortunately, the responsibility of running the family brewery when his father died effectively ended his scientific career.

Sunspots are found in the *photosphere* (Greek, sphere of light), the region of the solar atmosphere that produces the visible light of the sun. Although it is gaseous, the photosphere is frequently referred to as the surface or visible surface of the sun. Even ignoring the sunspots, the photosphere does not appear to be smooth and uniformly bright when examined with a telescope under conditions of good seeing. In particular, the *granulation*, a grainy pattern of small bright markings, spaced by narrower dark lanes, covers the photosphere (Figure 9-5), and motion pictures show that the individual granulation cells exist for only a few minutes. In fact, movies of the photosphere reveal that the whole granulation pattern is in seething motion, reminiscent of a pot of boiling water. A typical granule is only about 1,100 km (700 miles) in diameter, which is

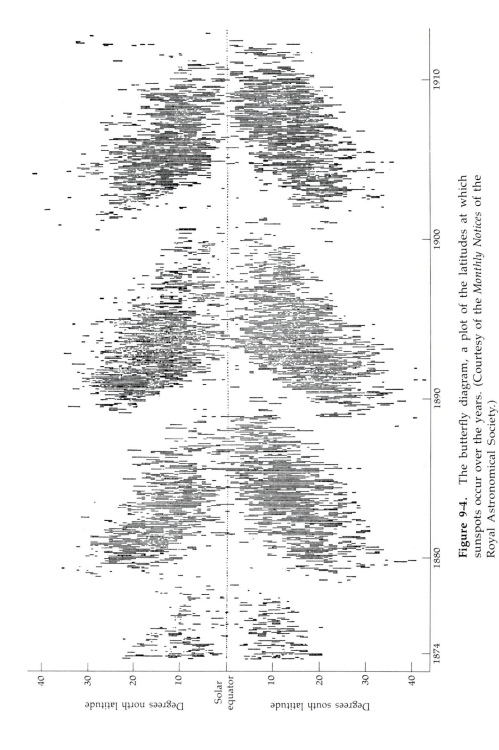

40   30   20   10   Solar equator   10   20   30   40

Degrees north latitude   Degrees south latitude

1874   1880   1890   1900   1910

Figure 9-4. The butterfly diagram, a plot of the latitudes at which sunspots occur over the years. (Courtesy of the *Monthly Notices* of the Royal Astronomical Society.)

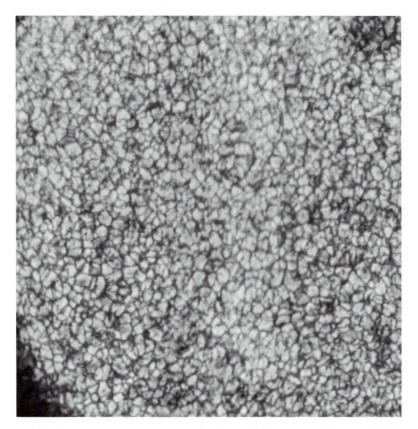

Figure 9-5. The granules on a small part of the sun, as photographed through a telescope carried aloft by a balloon. (Courtesy of Project Stratoscope of Princeton University, sponsored by the Office of Naval Research, the National Science Foundation, and the National Aeronautics and Space Administration.)

very small compared to the sun's diameter (measured at the photosphere) of about 1,400,000 km (864,000 miles).

The life of a sunspot begins when two or more narrow, dark features (called *arch filaments*) form in a small region of the photosphere, and then grow larger and merge to become a recognizable spot. It is believed that the arch filaments correspond somehow to the emergence of curved lines of magnetic force from below the visible surface. This process has been observed to occur in less than one hour. Such a small spot (with a diameter of only about 2,500 km, or 1,600 miles) is called a *pore*. The pore stage lasts from a few hours to a few days, after which most sunspots appear to fade or dissolve into the background pattern of photospheric granulation. A small fraction of the spots, however, grow larger and develop a surrounding, less dark region called a *penumbra;* the central dark part is now called an *umbra.* High-resolution photographs of penumbrae show that they consist of filamentary structures that point outward from the central umbra. A photograph of a sunspot is shown in Figure 9-6; the umbra and penumbra are clearly visible, along with the adjacent granulation. Finally, sunspots that survive the pore stage may occur together in *sunspot groups* (Figures 9-2 and 9-7). These can achieve lengths of over 100,000 km (62,000 miles).

Before discussing the physical properties and origin of sunspots and other aspects of solar activity, we will summarize the basic properties of the underlying or *quiet sun* and the source of its energy.

Figure 9-6.   A sunspot, photographed by a balloon telescope. (Courtesy of Project Stratoscope of Princeton University, sponsored by the Office of Naval Research, the National Science Foundation, and the National Aeronautics and Space Administration.)

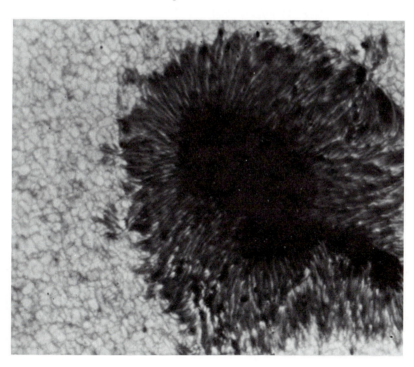

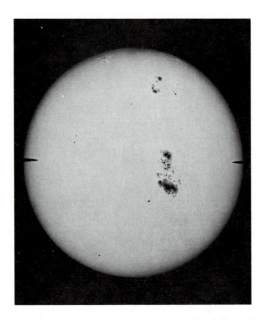

Figure 9-7. A photograph of the sun in visible light, showing sunspot groups and limb darkening. (Courtesy of the Hale Observatories.)

THE SOLAR SPECTRUM
AND THE PHOTOSPHERE

The visible light of the sun is emitted from the photosphere. By studying the spectrum of this light we can determine the physical conditions, such as temperature and pressure, under which it is formed. The most prominent features of the solar spectrum are the absorption lines that are named for the Bavarian physicist Joseph Fraunhofer (1787–1826).[1] Fraunhofer first noted the lines in 1817. He counted more than 500 of them (note the dark lines in Plate 7) in the solar spectrum and used letters to name the strongest ones. Now we know the origins of these lines; for example, the dark lines in the yellow region of the spectrum that he referred to as the "D lines" are produced by sodium atoms.

The solar spectrum can be given a simple interpretation on the basis of Kirchhoff's Laws (Chapter 5). Photons are supplied to the photosphere

---

[1] Four of the strongest Fraunhofer lines were actually discovered by the British chemist W. H. Wollaston, in 1802. Wollaston thought they represented the boundaries of different colors of sunlight.

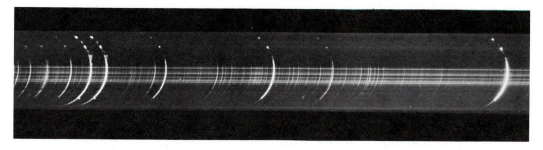

Figure 9-8. The flash spectrum observed at the solar eclipse of January 24, 1925. The crescent shape of each individual emission line corresponds to the shape of the emitting region above the moon's limb. (Courtesy of the Hale Observatories.)

from the interior, which can be regarded as an incandescent source. The photospheric gases are at a lower density, hence the disk spectrum shows the characteristic absorption lines corresponding to the various elements in the atmosphere. This interpretation can be corroborated by observing the spectrum of the next outermost layer of the solar atmosphere (*chromosphere*) on the edge of the sun, where the incandescent interior is not behind it. The chromosphere is hard to observe under ordinary conditions, but this can easily be done during a total eclipse of the sun.[2] This spectrum (Figure 9-8) of the chromosphere consists of emission lines, as expected from Kirchhoff's Laws. It is called the *flash spectrum* because it is only observable for a brief time during the eclipse. In fact, the strong emission lines in the flash spectrum correspond to lines seen in absorption in the solar disk (photospheric) spectrum. Thus, we find a simple demonstration of Kirchhoff's Laws in the sun.

Kirchhoff's Laws are rather general statements. Looking at the process of radiation transfer in more detail, we find that photons generated in the solar interior diffuse upward through the atmosphere, being acted upon by the two different processes of pure absorption and pure scattering (discussed in Chapter 5). Pure absorption in the solar atmosphere is mostly due to H^-, the negative hydrogen ion. In this atom, the proton completed its outermost shell of two (Figure 9-9) and became helium-like by simply adding another electron. However, the singly charged positive proton cannot really balance the negative charge of two electrons, so the second electron is bound only loosely to the proton. A photon of visible light has enough energy to knock off the second electron. This effect pro-

---

[2]Or, with specialized instrumentation, from space.

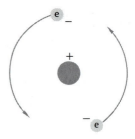

Figure 9-9.   Schematic diagram of the H⁻ ion.

vides a source of *continuous absorption* of photons of various wavelengths.[3] A photon of entirely different energy may be emitted later when another second electron combines with the hydrogen atom to form a new H⁻ ion. Thus, radiant energy migrates upward through the photosphere via photons that are continually being absorbed and re-emitted by H⁻ ions. If an atom (for example, sodium) that scatters photons at its characteristic wavelengths is present, then the upward migration of photons with these wavelengths is impeded, and the chances of being absorbed by H⁻ and re-emitted at other wavelengths are increased. This process, illustrated schematically in Figure 9-10, produces the Fraunhofer lines. The model that we have just described for the formation of the photospheric absorption lines differs from the simple picture of Kirchhoff's Laws, because the incandescent (continuum producing) region and the selective absorption region overlap, a situation not covered by Kirchhoff's investigations.

What do the Fraunhofer lines tell us about the photosphere? First, by comparing wavelengths measured for these lines in the solar spectrum with the wavelengths in the spectra of known elements as recorded in the laboratory, we can identify elements present in the sun. Lines of one element were actually observed in the chromospheric emission spectrum in 1868, before the element was discovered on earth. It was therefore called helium (after the Greek word for the sun, *helios*). It was subsequently recognized on the earth, in the atmosphere, in minerals, and in gas wells. Not all the elements can be detected in the solar spectrum, but it is a sensitive indicator for many of them.

The strength of a particular absorption line changes, depending on the amount of the element that occurs in the sun or other stars. The line will be darker when more of the element is present.[4] Thus, a careful

---

[3]Continuous, because photons of different wavelengths, not just those at or near one wavelength (as in an absorption line), are affected.

[4]The strength of an absorption line depends on other factors as well, especially on the temperature, and this is taken into consideration when a spectrum is analyzed to determine the abundances of the elements.

analysis of the spectrum can show not only what elements are present but also their amounts. This is the basic method used to determine the compositions of the stars. When we speak of stellar abundances, we refer to the compositions of the photospheres of stars, as it is those atmospheric regions that give rise to most of the light we can observe. The interiors of stars are so hot and dense that nuclear reactions occur, such as the transformation of hydrogen to helium, and so the abundances of the elements inside a star can differ from the atmospheric abundances. The continuous spectrum *(continuum)*, namely the general dependence of light intensity with wavelength, disregarding the Fraunhofer lines, enables us to determine the temperature of the photosphere by comparing it with Planck's Law. The solar continuum is a reasonably good match for an

Figure 9-10. Schematic diagram of photon diffusion in the solar atmosphere. Photons *(a)* with wavelengths close to the characteristic wavelength of an atom are scattered relatively frequently, find it harder to diffuse through the photosphere, and are preferentially absorbed. Scatterings are denoted by bends, and absorptions by circles. Photons *(b)* with wavelengths different from the characteristic atomic wavelength diffuse through the surface. The term *scattering* is used loosely in describing the movement of photons *(b)*. In fact, the process is the ionization and recombination of H⁻ (described in the text), but continuum photons *(b)* are involved in both the ionization and recombination. Thus, the observer outside the sun sees fewer of photons *(a)* and more of photons *(b)*—a situation which produces an absorption line.

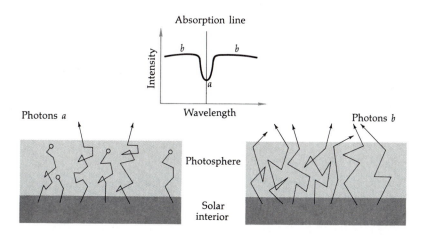

ideal radiator curve at 6,000 K, as given by Planck's Law and as shown in Figure 5-20.

The photospheric temperature can also be inferred from the *total* radiant energy emitted by the sun (called the *solar luminosity*), the solar surface area, and the Stefan-Boltzmann Law. If we measure the *solar constant* (the rate at which solar radiation of all wavelengths is received by a surface of 1 cm^2 at a distance of 1 a.u. from the sun), then we can calculate the amount of radiation passing through the surface of a sphere of radius 1 a.u. Since we know the surface area of the sun, we know how much energy each square centimeter must emit to produce the amount observed here at 1 a.u. Then, the temperature needed follows from the Stefan-Boltzmann Law, and a value of about 6,000 K is also found.

Thus, the temperature of the solar photosphere of about 6,000 K (11,000° F) is found by two different methods. (There are other, more complex methods as well.) The agreement is not exact, nor should it be; the sun is not quite an ideal radiator as assumed in the formulae used to determine the temperature.

Does the temperature in the photosphere increase or decrease as we go into the sun? Most people would expect an increase, but can we prove this? In visible-light photographs, the photospheric regions (such as granules) with the higher temperatures appear brighter, and regions with the lower temperatures appear darker. The solar *limb darkening* as shown in Figures 9-7 and 9-11 provides the evidence needed. When we look across a big city from the top of a very tall building, we can see to a certain distance, depending on the amount of dust and smog in the air. Likewise, when one looks into the photosphere one sees light originating from a depth determined by the continuous absorption of the H$^-$ ion. Because of the geometrical effect shown in Figure 9-11, one sees to a greater depth at the center of the disk than at the limb. If the limb intensity is higher or lower than the disk center, the temperature would decrease or increase inward, respectively. The limb actually appears darker than the disk center, and hence the temperature in the solar photosphere increases inward as expected. Surprisingly, this trend reverses in the chromosphere and corona.

THE SUN'S ENERGY

The interiors of the sun and stars cannot be observed directly at present. However, we can use the total energy output of the sun, the *luminosity*, to probe the solar interior indirectly. Since the basic energy source of the sun is in the interior, the conditions there must be capable of producing the solar luminosity.

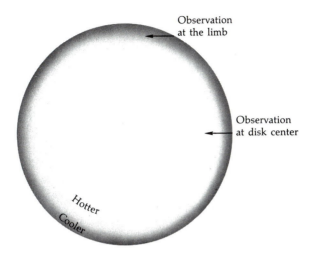

Figure 9-11. The geometry of solar limb darkening.

The amount and source of energy radiated by the sun was an astronomical mystery not solved until the late 1930s. It was an important problem because it is the sun's energy that sustains all life on earth. The basic datum, the *solar constant*, is difficult to measure because it includes energy radiated over all wavelengths of the electromagnetic spectrum. Thus, one problem is that the best detector for one region of the spectrum is not necessarily the best for another wavelength region (Chapter 6). Furthermore, not all wavelengths pass through the earth's atmosphere. However, most of the radiation from the sun can reach the surface of the earth, and by making theoretical estimates for the small amount of energy that does not penetrate the atmosphere (or more recently, by observing these wavelengths from space vehicles), the solar constant has been determined to be about 2 calories/cm²-minute (note the unusual units; the calorie was defined in Chapter 5). The total energy radiated by the sun (*solar luminosity*) can be found by multiplying the solar constant by the number of square centimeters on the surface of a sphere of radius 1 a.u. This yields the enormous value of 9×10^{25} calories/sec, or 4×10^{26} watts.

Studies of rocks and fossils suggest that the sun has been emitting roughly the same amount of energy during at least the major part of the earth's lifetime. The argument for this is rather simple: if the earth received very much more solar energy than it does now, water would exist mostly as vapor, rather than as a liquid, and if the earth received much less solar energy than it does now, water would exist primarily as ice. But the fossils of aquatic life forms, and the existence of very old sedimentary rocks formed by the deposition of eroded material in ancient seas, show that the earth has had extensive bodies of liquid water for at least several billion years. This does not rule out fairly small changes in the solar constant, which some scientists believe have occurred and caused the ice ages on earth and/or the climatic changes on Mars.[5]

It is possible that a greenhouse effect occurred during the primitive-atmosphere stage of the earth. In that case, a loophole exists in our argument, since the sun might have emitted somewhat less energy than it does now, and yet the greenhouse effect might have kept the earth warm. However, our argument for the rough constancy of the solar luminosity would still hold true for the time since the present atmosphere formed. Thus, an energy source must exist to have produced the solar luminosity of 4×10^{26} watts over at least the last few billion years. The road to the discovery of this energy source was a long one, and several wrong turns were taken on the way.

Attempts to understand the source of the sun's energy were made by Julius R. Mayer, a young German physician, in 1848. He was among the first to note that the sun was the ultimate source of all heat and power on earth. (Even when we burn fuel we remove from "coal storage" the sunshine energy of past geologic times.) Some of the first ideas considered by Mayer can easily be discarded. If the sun were simply a big lump of hot gas radiating at the sun's present luminosity (but without any internal source for continuing to heat the gas), its temperature would drop severely in 5,000 years. If the entire mass of the sun were coal, it could sustain its present luminosity for only 4,600 years. (If the earth were a solid lump of burning coal, it could emit at the rate of the solar luminosity for only five days.) Of course, the calculations ignore the problem of obtaining oxygen for the combustion and the problem of getting rid of the ashes, both of which would shorten the lifetime, but they do illustrate the magnitude of the problem.

Clearly a better source of energy was needed, and Mayer suggested that the sun's atmosphere was bombarded by a continuous rain of meteoroids. By friction with the solar gas, the meteoroids would heat the gas, just as they heat their immediate surroundings when they enter the earth's atmosphere. Unfortunately, the mass of meteoroids required was huge; in particular, it was inconsistent with the known rates of meteor occurrence at the earth. Moreover, the required mass of meteoroids impacting the sun would increase the solar mass at a rate that would be

---

[5]There is no proven theory of the ice ages. This theory is mentioned only for purposes of discussion.

detectable in the motions of the planets, since the gravitational force of the sun on each planet would be increased; this is not observed and Mayer's hypothesis must be discarded.

The next attempt was made by the German physicist Hermann von Helmholtz in 1854; the idea was also discussed by Lord Kelvin and is called the Helmholtz-Kelvin hypothesis. According to this idea, the sun is shrinking under the force of its own gravity, and the solar gases are thereby compressed. We know from experience that compression heats a gas and expansion cools it (refrigerators cool their contents by expanding a gas). So the sun, if compressing itself, would produce heat. The Helmholtz-Kelvin hypothesis greatly expanded the length of time over which one could imagine the solar luminosity being maintained. The observed solar luminosity according to this theory would be produced by a shrinkage of about 60 meters (200 feet) per year in the diameter of the sun. At this rate, the sun would last some 15 million years, since its diameter is about 1.4×10^6 km. (These numbers do not check exactly, because the yearly rate would change as the sun shrank.)

Even a time scale of millions of years, however, is not sufficient to explain the geological evidence, which requires times of billions of years. Hence, the Helmholtz-Kelvin process does not furnish the energy to support the solar luminosity over the necessary time scales. Nevertheless, we will see that such a process probably was important in the early stages of the sun's history.

The actual source of the sun's radiant energy[6] is now known to be nuclear energy, which is released according to Einstein's equation, $E = mc^2$. Deep within the sun a series of nuclear reactions called the *proton-proton chain* converts hydrogen into helium by merging sets of four hydrogen nuclei (protons) into one helium nucleus (alpha particle). The mass of the helium is slightly less than the sum of the four hydrogen masses, and the missing mass is released as energy. Hotter stars than the sun mostly derive their energy from another chain of reactions (the *carbon cycle*) in which hydrogen is also converted to helium, but carbon, nitrogen, and oxygen atoms are involved in the reactions.

The central temperature of the sun is about 15 million K. At temperatures as high as this, atomic particles are highly energetic, and upon collision can penetrate the nuclei of other atoms. If the sun's central temperature were too low, the collisions would not be energetic enough, and nuclear reactions could not occur. In the early stage of solar formation, the Helmholtz-Kelvin type contraction played an important role in raising

---

[6]Based on the work of the nuclear physicist Hans Bethe in 1938.

the central temperature of the sun above the threshhold value where the nuclear reactions could begin. At the currently estimated temperature of about 15 million K, nuclear reactions can supply the sun's entire luminosity. By heating the solar interior, the reactions also raise the pressure there to the point where the difference in pressure between the interior and surface regions balances the gravitational force on the gas so that the contraction no longer occurs.

There is enough hydrogen in the sun to generate the present solar luminosity by the proton-proton chain for another 100 billion years. However, the theory of stellar evolution predicts that, long before this time, changes in the internal conditions of the sun will lead to a different set of energy-generating reactions.

THE SOLAR MODEL

The sun's energy is liberated by nuclear reactions in its central regions. Energy is transported outward by photons that are scattered, absorbed, and reemitted (as discussed above) throughout the bulk of the solar interior. The gas must be reasonably transparent to the radiation in order for this energy transport to be efficient. This condition is no longer met at a point about 85 percent of the way from the center to the surface of the sun. In this region, the conditions of pressure and temperature are such that He$^+$ (singly ionized helium) ions form, and they can be photo-ionized; thus the solar material becomes relatively opaque. Since radiation becomes an ineffective method of energy transport at this point, the energy finds other ways to travel outward. From this point to just below the photosphere, the mechanism of convection becomes important for the energy transfer. In the higher parts of the convective region, the solar material is made opaque by the existence (and ionization) of, first, H (neutral hydrogen) atoms, and then H$^-$ ions in the region just below the photosphere. Currents of solar material circulate between the hotter regions below and the cooler regions above.

Convection currents take the form of cells with regular flow patterns, as shown in Figure 9-12. The hotter gas is brighter than the cooler gas, and the corresponding pattern influences the photosphere above, as can be seen in Figure 9-5. Thus, the photosphere shows this regular cell structure, called the *granulation*, because energy transport takes place by convection just below it.

The photosphere is a skin 300 km (about 200 miles) thick, forming the visible surface or *disk* of the sun. The temperature there is about 6,000 K. Energy travels through the photosphere by radiation. This occurs because the density of H$^-$ is greatly reduced in the photosphere (compared to the upper convection zone) so that, although H⁻ is the major absorbing constituent in the photosphere, it is not plentiful enough to almost cut off the flow of photons, as it does in the convection zone.

Photons traveling outward from the photosphere have little additional interaction with the solar atmosphere above the photosphere. Because the outer atmosphere is a rarified gas, a photon collides with an absorbing atom only relatively rarely. Hence, the sunlight we receive on earth is literally from the photosphere. The solar model and energy flow are shown schematically in Figure 9-13.

Just as the atmosphere of the earth thins out with altitude above the surface, so the solar gas extends outward an enormous distance above the photosphere, growing more and more tenuous with altitude. The region just above the photosphere, reaching to heights of several thousand kilometers, is called the *chromosphere* ("color sphere"), named for its intense red color seen at solar eclipse. Above this is the solar *corona*, which occupies such an enormous volume of space that its size is usually expressed in units of the solar (photospheric) radius (about 700,000 km). During total eclipses, astronomers on the ground have observed the coronal light out to as far as 10 solar radii from the photosphere (Figure 9-14); beyond that point the coronal light is too weak to be distinguished from stray light in the earth's atmosphere, even during the eclipse. However, observations with radio telescopes and space instruments show that the corona extends even farther. Eventually, as the corona thins out with height, it must merge into the background gas of the interplanetary medium. Matter, in the form of electrons and protons, streams away from the corona and can be detected as the *solar wind* in the vicinity of the earth.

Figure 9-12. Convection schematic (shown in vertical cross-section). Hot gas masses rise and cool; cool masses sink and are warmed again. Arrows indicate direction of gas motion. As seen from above, rising masses are brighter than falling masses, since they are hotter.

Hotter gases   Cooler gases   Hotter gases

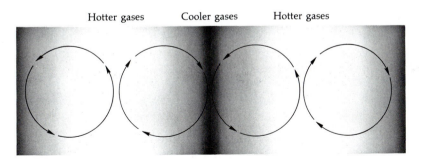

Figure 9-13. Schematic diagram of solar structure and energy flow. Energy is generated in the core (1) of the sun by nuclear reactions and transported through most of the interior (2) by photons. Although photons carry energy throughout the sun, convective motions (3) transport most of the energy in the region located between about 0.85 solar radii from the center and a level just below the photosphere (4) where radiative transport by photons is dominant again. Above the photosphere the sun's radiation does not interact strongly with the remaining material, and light emitted from the photosphere travels outward with little hindrance. Acoustic energy generated in the convection zone (3) is deposited in the low corona (5), heating it. Throughout most of the corona (6), energy transport is by conduction. Far from the sun, the energy is still carried outward in the bulk motions of the solar wind.

Although the mass of the solar atmosphere above the photosphere is trivial compared to the photosphere mass, and its interaction with the bulk of the solar radiation is totally negligible, the chromosphere and corona are nevertheless of unusual scientific interest. This arises from the surprising variation of temperature in the solar gas. The central region of the sun is about 15 million K, and the temperature decreases continually out to the photosphere, where it is about 6,000 K. But the temperature does *not* continue to decrease above the photosphere—it actually increases to 10,000 K in the chromosphere and to 2 million K in the corona. Why does this happen?

The convection cells beneath the photosphere are noisy and are a source of energy in the form of sound waves. A common analogy to the generation of sound waves in the sun's convection zone is the sound generated in a boiling pot of water. You do not have to see the water boil, you can hear it. The convective motions in the boiling water are noisy. Thus, energy in the form of sound waves is generated.

Figure 9-14. The solar corona on November 12, 1966, photographed with a filter which compensates for the sharp decrease in coronal brightness with distance from the sun. The planet Venus is at upper left. (Courtesy of High Altitude Observatory.)

Although the sun's outer atmosphere is transparent to light, sound waves are absorbed by the corona. Since the corona has a very small mass, a little acoustic energy goes a long way, and the corona is heated to a temperature of 2 million K. This hot gas is illuminated by photospheric radiation, and this scattered light gives the corona its pearly appearance when seen at the time of total eclipse (Figure 9-14). The same effect is visible in photographs made from satellites above the earth's atmosphere.

The high temperature of the corona and the relatively low gravity at these distances from the photosphere lead to a continual process in which the corona boils off, producing the solar wind. Thus, physical conditions in the corona cause particles to escape from the sun to interplanetary space. Equilibrium exists, however, for as the coronal material boils off, it is replaced from below.

The corona is a hot, fully ionized gas; that is, virtually every atom has lost at least one electron. (A gas in this state is called a *plasma.*[7]) Since the solar atmosphere is mostly composed of hydrogen, the coronal plasma consists primarily of electrons and protons. As the corona expands into the space between the planets, its heat energy (corresponding to a temperature of 2 million K) is converted into the energy of bulk motion of the solar wind. The solar wind temperature drops to about 100,000 K at the earth,[8] but the speed at which the wind moves rises from a very small value near the sun to about 500 km/sec (300 miles/sec) at the earth. Typically, there are 5 to 10 solar wind particles per cubic centimeter in space near the earth.

The solar wind is very important in the physics of the solar system. It (1) shapes the magnetospheres of earth and Jupiter, (2) pushes the ionized comet tails away from the sun, and (3) strikes the unmagnetized objects in the solar system. For example, the impact of solar wind particles on the lunar surface probably darkens the rocks and hence may be responsible for the relatively low fraction of sunlight reflected from the moon (small lunar albedo).

---

[7]Plasma is regarded as a fourth basic state of matter (after solid, liquid, and gas), and it appears that most of the visible matter in the universe may be in this form, although it is virtually absent on the earth.

[8]If the solar wind has this high temperature as it encounters the earth, why doesn't it burn us up? The answer is that it is hot, but it doesn't have much heat. The temperature is high but the number of calories is low because the total mass of hot particles colliding with the earth is very small. Also, the magnetic field of the earth shields us by deflecting the charged particles of the solar wind.

THE NEUTRINO PROBLEM

There is a basic test of our overall understanding of solar energy generation and the internal structure of the sun. Unfortunately, thus far our theories flunk the test! Although we cannot see into the core of the sun, we can calculate that the nuclear energy generation process which occurs there should release a large number of outward streaming particles, the *solar neutrinos*. The neutrino is remarkable among particles known to the physicist in that it has *no mass* and no electric charge, but it spins and it travels at the speed of light. Further, unlike a photon of light, a neutrino released at the center of the sun can pass unhindered through all the layers above it, reaching the earth about eight minutes later. Deep underground (the ground above screens out unwanted cosmic ray particles) in a South Dakota gold mine, physicist Raymond Davis, Jr. used a tank of 100,000 gallons of tetrachlorethylene (dry cleaning fluid) to trap neutrinos. Since 1970, careful measurements with this very complex and sensitive experiment showed that the number of solar neutrinos of the predicted type is about three times smaller than expected. It is conceivable that the experiment is defective in some way, but careful scrutiny by many physicists has failed to turn up the defect. The result stands as a serious challenge to our understanding of energy generation and internal structure in the sun, and by implication challenges our understanding of these matters in other stars as well. Surely the basic ideas of nuclear reactions, core, convection zone, and so forth must be correct, but the detailed theories for them may contain significant errors. Since 1976, solar astronomers have studied what appear to be oscillations of the sun as a whole. The study of these oscillations may reveal facts about the sun's internal regions, much as the study of earthquake vibrations at the earth's surface unveiled the internal layers of the earth.

SOLAR ACTIVITY

Although sunspots are the most obvious solar features, there are many other kinds of temporary structures or disturbances on the sun, most of which seem to be associated with the sunspot cycle (Figure 9-15). These occurrences are collectively called *solar activity,* and the complete description of their variations or events over the roughly eleven-year period is called the *solar cycle.* A large sunspot with its penumbra is just one aspect of a *center of activity.* Above the sunspot in the chromosphere is a much larger disturbed region called a *plage.* Since the photosphere is much

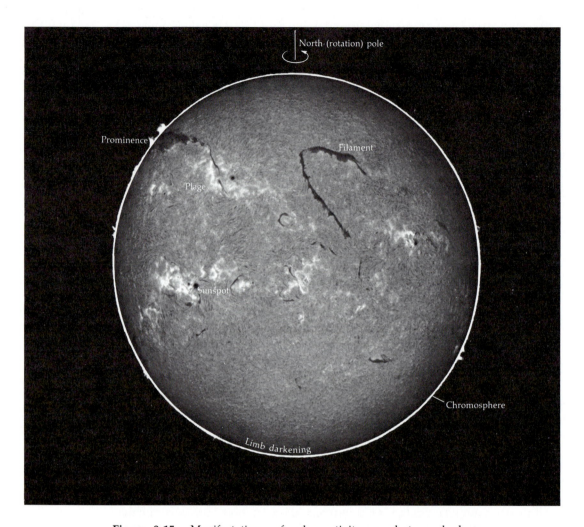

Figure 9-15. Manifestations of solar activity as photographed on March 7, 1970, through a filter that isolates light of the hydrogen alpha line wavelength (6,563 Å). The illustration is a composite of a short exposure to show disk features and a longer exposure to show the fainter prominences and the chromosphere. (Courtesy of H. Caulk and R. W. Hobbs, Goddard Space Flight Center.)

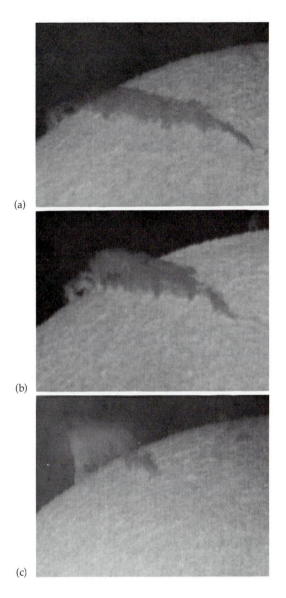

(a)

(b)

(c)

Figure 9-16. A series of photographs made on April 9, 10, and 11, 1959 (a, b, c, respectively), showing that filaments (dark features on the solar disk) and prominences (bright features on the solar limb) are physically identical. (Courtesy of Observatoire de Paris—Meudon.)

brighter than the chromosphere in ordinary light, the latter region is studied with special filters and spectrographic instruments to isolate light of specific wavelengths coming primarily from the chromosphere. This includes lines of Ca^+ ions (calcium atoms that have lost one electron). In pictures of the sun taken at these wavelengths, the plages appear as bright patches. (These photos are called *calcium filtergrams* and *calcium spectroheliograms*, depending on whether they are obtained using a filter or a spectrograph to isolate the wavelengths of interest.)

A major plage may last for several months. Occasionally, a brief, very intense burst of light, primarily from hydrogen atoms, is observed in a plage; often this *flare* (Figure 9-19) occurs above a sunspot. Above the chromosphere the center of activity extends into the corona in the form of a *condensation*, which appears to be denser and brighter than its surroundings. Straight, curved, and loop-shaped gas structures extend from the plage up into the corona. These *prominences* (Plate 3) appear as bright objects when they are located at the limb but are dark by contrast to the photosphere when they are seen on the disk (Figure 9-16), in which case they are called *filaments*. The fact that the filaments look dark when we know that they are bright prominences reminds us that "bright" and "dark" are relative terms. Actually, sunspots are bright objects, as are the "dark parts" of the granulation, and only look dark by contrast to their even brighter surroundings. Sunspots have a temperature of about 4,500 K and are thus some 1,500 K cooler than the surrounding photosphere.

The presence of a magnetic field can affect the emission lines produced by an atom. This was discovered in 1896 by the Dutch physicist Pieter Zeeman, who observed the change in the spectrum of a sodium flame when it was placed between the poles of a magnet. Depending on the direction of the magnetic field, an atomic line is observed to split into two or three individual lines; the separation of the lines depends on the strength of the field and the properties of the atom. The *Zeeman effect*[9] (illustrated for a sunspot in Figure 9-17) allows detection, measurement, and mapping of the magnetic fields on the sun with an instrument called

---

[9]The principles of the Zeeman effect were applied to the spectra of sunspots by George Ellery Hale, and the magnetic fields in sunspots were confirmed in 1908. Hale's contributions to astronomy are legion. He founded the Yerkes, Mount Wilson, and Palomar Observatories, and also the *Astrophysical Journal*. The 200-inch telescope on Mount Palomar was named for him, and later the Mount Wilson and Palomar Observatories were renamed the Hale Observatories.

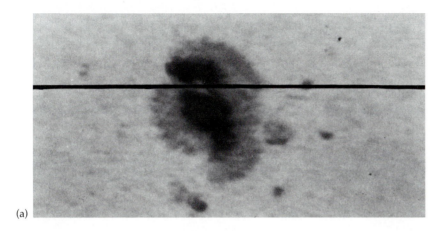

(a)

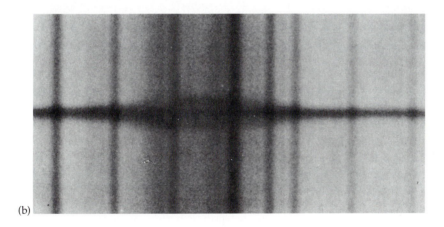

(b)

Figure 9-17.   Effects of a magnetic field on spectral lines. (a) The dark line across the photograph marks the position on the sun from which light is admitted to the spectrograph. (b) An absorption line in the spectrum of the admitted light. In the sunspot, strong magnetic fields split the line into three components through the Zeeman effect. (Courtesy of the Hale Observatories.)

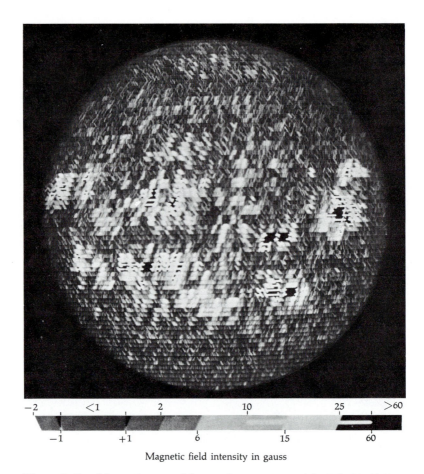

−2   <1   2   10   25   >60

Magnetic field intensity in gauss

−1   +1   6   15   60

Figure 9-18.   Magnetic map of the sun (magnetogram) for July 21, 1961. The intensity in gauss is shown on the scale; polarity is shown by the slant to the right or left. One gauss is about three times the magnetic field strength at the equator of the earth. (Courtesy of the Hale Observatories.)

a magnetograph. A solar *magnetogram* is shown in Figure 9-18. Comparing magnetograms with filtergrams and spectroheliograms shows that sunspots, plages, and other features of the center of activity correspond to regions of strong magnetic fields that may be hundreds or thousands of times more powerful than the background magnetic field of the quiet sun. This background field is about as strong as the earth's magnetic field.

In fact, strong magnetic fields seem to be fundamental to the origin of the centers of activity, as such fields are seen to emerge on magnetograms before the sunspots appear. As the sunspots and the plages grow, so does the magnetic field, and the flares may occur in the chromosphere above the spots. The flares are accompanied by bursts of radio emission observed with radio telescopes and by *x-ray bursts* detected by satellites. The spectrum of an x-ray burst sometimes includes emission lines from iron atoms which may have lost as many as 24 of their 26 electrons due to the temperatures in the flares. These temperatures may reach 50 million K. They confirm that the flares involve the sudden release of enormous amounts of energy. The explanation of the great energy releases in flares is one of the outstanding unsolved problems in astronomy. Most theorists believe that the strong magnetic fields are in some way responsible for flares.

Even as the sunspots in a center of activity are shrinking from their maximum stage, the magnetic field region and the associated plage continue to grow, and above the plage the prominences (filaments) increase in size, reaching lengths greater than 160,000 km (100,000 miles). Although the plage eventually fades from view, prominence activity continues, and long streamers may form in the corona above the weakening magnetic field region. This whole process of birth, growth, and death of a center of activity can take place over periods of 10 solar rotations (270 days) or even longer.

An equally remarkable development occurs in the background magnetic field of the sun. The two solar hemispheres have fields of opposite polarity (north and south), as do the earth's hemispheres. These are somehow related to the much stronger fields of the sunspots, because a sunspot group tends to evolve as two subgroups of opposite polarity, separated slightly in solar longitude. These are called the p and f spots (p for preceding, f for following; the p spots are defined as those that lead the others as the sun rotates). It is found that when the p spots in the northern hemisphere of the sun have north magnetic polarity, the p spots in the southern hemisphere have south polarity. But, in the next 11-year solar cycle, the polarities are reversed, with the northern hemisphere p spots having a south magnetic polarity. This seems to be related to the fact that the background magnetic polarities of the two solar hemispheres reverse every 11 years! If this happened on the earth, one would find a north-seeking magnetized compass needle pointing toward the Southern Hemisphere.

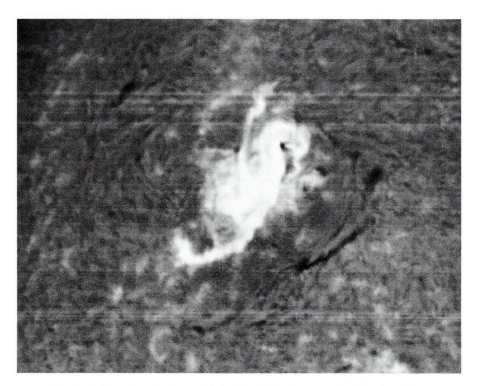

Figure 9-19. A solar flare of July 16, 1959, photographed in the light of the hydrogen alpha line. (Courtesy of the Hale Observatories.)

The effect of the magnetic field on solar material is apparent in several ways. The material in the outer corona (Figure 9-14) is confined into narrow streamers. Flares (Figure 9-19) are often very directional in terms of their effect on solar features, as if the flare particles were channeled by the solar magnetic field. This confinement or inhibition of ionized material by a magnetic field stems from the effect shown in Figure 5-4; both positively and negatively charged particles are deflected by a magnetic field.

Prominences are made possible by the solar magnetic fields, which support condensed coronal gas against the solar gravity. Material that condenses[10] from the corona to form a prominence may leave a void visible as a dark place in the corona above the prominence, as illustrated in

---

[10]By condensation, we mean here a localized increase in density of the gas, not a change from the gaseous to the liquid state.

Figure 9-20. It appears that matter is continuously condensing from the corona into the prominence and "slides" down the magnetic field into the photosphere.

The fact that magnetic fields inhibit the motions of plasma particles provides an explanation for the observation that sunspots are about 1,500 K cooler than the surrounding photosphere. As we noted in our discussion of the solar model, the region just below the photosphere is heated by convection currents. This process involves the circulation of material, and this circulation is inhibited by the magnetic field because the plasma is constrained to move along (and not across) the lines of magnetic force. Thus, part of the energy in the convective motions must be used to overcome the resistance of the magnetic field. There is then less energy available to be carried upward to heat the photosphere, and a relatively cool region, the sunspot, is formed.

Solar activity is of interest because post-flare effects often occur on earth. A particularly large release of flare radiation by the sun can cause anomalous ionization conditions in the earth's ionosphere and disrupt radio communications; this is called a radio fade-out. In addition, coronal gas particles can be ejected as a result of a flare. One or two days later this material strikes the earth's magnetosphere, leading to the aurora as a side effect. The ability to predict flares would enable us to anticipate these events; the forecasts would be of value to astronauts in orbit or in space, where they generally would be outside the protection of the earth's atmosphere and/or magnetosphere. Unfortunately, our knowledge in this area is still unsatisfactory, and flare forecasts, although improving, still are not very reliable.

Figure 9-20.   Schematic diagram showing the relationship of coronal structures and prominences.

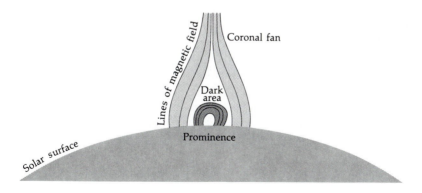

Solar activity is also of interest because similar events must occur on other stars. Although we cannot see the surfaces of other stars, indirect evidence suggests that they must have "starspots" and other fascinating phenomena of their own.

THE MISSING SUNSPOTS

One remarkable aspect of the sunspot cycle was forgotten or overlooked for many years and has been rediscovered only recently. During a particular interval of about 70 years *there was no sunspot cycle*. In fact, there were very few sunspots at all, and the astronomers who occasionally managed to see one treated it as a significant event! The missing sunspots did not cause great puzzlement, because the years were 1645–1715, and although sunspots already had been discovered, the sunspot cycle itself was not yet known. No one had any idea of how many sunspots to expect. During this time, many of the terrestrial effects of solar activity also were absent. For example, the astronomer Halley mentioned that he personally did not see an aurora borealis until 1715, when he was nearly 60 years old. This period of few sunspots and no cycle is now called the *Maunder minimum*, after the British astronomer E. W. Maunder, one of the persons who called attention to it.

In nearly all astronomy textbooks published prior to 1977, you will find a diagram of the sunspot cycle that shows the cycle extending all the way back to the year 1700. Evidently, the first person to draw this particular diagram wanted it to start neatly at the beginning of a century and therefore faked the first 15 years worth of data (see Figure 9-3). He or she never dreamed that the falsification would be detected, let alone that there was no sunspot cycle during those years!

What significance does the Maunder minimum, as rediscovered around 1975, have? It means that theories of the sun's magnetic field and the nature of the sunspot cycle must now be revised to explain why sometimes the cycle and the sunspots may disappear. Further, if there was one Maunder minimum, perhaps there will be another one some day, so that a whole future generation of people may be born, live and die without ever (or hardly ever) seeing a sunspot or an aurora. According to John Eddy of the University of Colorado, there may be a causal relation between the Maunder minimum and a cool period known as the "little ice age." This controversial suggestion, if verified, would indicate a direct connection between solar activity and climatic change.

10

The Stars—A Census

The distance of the sun is 1 a.u., and the distance of the star Proxima Centauri is about 260,000 a.u. We know that the intensity of light decreases as the square of the distance increases. Thus, if the intrinsic brightness of Proxima Centauri happened to be exactly equal to that of the sun,[1] it would appear to us as 260,000 squared (almost 7×10^{10}) times fainter than the sun. Yet Proxima Centauri is the *next closest star*.

We cannot see the surface of Proxima Centauri or any other star except the sun. Despite the obstacles imposed by the great distances of the stars, we can determine some of their characteristics, including temperatures, motions, masses, sizes, chemical compositions, and luminosities. Stars are found sometimes as isolated individuals, and sometimes in pairs, small groups, and clusters. We will consider both the individual characteristics and the "social behavior" of the stars, but the first item of our cosmic census is their distances.

DISTANCES—PARALLAXES

Distances are a subject of the greatest interest in astronomy. No attempt to understand the solar system, the galaxy, or the universe in terms of the forces that regulate their structure, the sources of their energy, or the processes that may have given rise to them seems possible without an understanding of the scale of each system.

---

[1] It is actually much fainter.

In Chapter 2, we discussed the evidence that the earth moves in orbit around the sun and mentioned that a direct consequence of this motion is that the nearby stars should show a *parallax* (shift back and forth during the year) with respect to the distant stars. Measurement of this effect permits a direct determination of the distance to a star. The same principle is used for measurements on land by surveyors.

For stellar distances, we measure the parallax angles as shown in Figure 10-1a; the baseline is the known diameter of the earth's orbit. The measurement is entirely straightforward in principle, and the heliocentric view of the solar system requires that such a parallax angle exist—otherwise the earth must be at rest. Unfortunately, the early attempts to measure parallaxes were unsuccessful. All the principles used were correct; however, the distances to the stars were vastly underestimated, and so the expected parallax angles were overestimated.

The first certain measurements of stellar parallaxes were obtained independently at three different observatories during the 1830s. The observers were Friedrich Bessel at Konigsberg, Germany; Wilhelm Struve at Dorpat, Estonia; and Thomas Henderson at the Cape of Good Hope. Their measurements are given in Table 10-1, along with more accurate values from modern observations. The favorite unit of interstellar distance in popular science writing is the *light-year*, defined as the distance that light traveling through the vacuum at 300,000 km/sec (186,000 miles/sec) will traverse during one year. This distance is about 9.5×10^{12} km (5.8×10^{12} miles). On the other hand, astronomers usually express distances in units of the *parsec*, which is the distance of a star having a parallax of one second of arc, as seen from the earth. (One parsec equals 3.26 light-years, or 3.1×10^{13} km.) In Table 10-1, the distance of each star, as determined from the modern parallax measurements, is given in parsecs. Note the simple relation between the last two columns: once the parallax angle (in seconds of arc) is measured, the distance is found by dividing 1 by the parallax.

Table 10-1.   The First Parallaxes

Observer	Star	Early Parallax Measurement (Arc Seconds)	Modern Parallax Value (Arc Seconds)	Distance to Star (Parsecs)
Henderson	alpha Centauri	1.16	0.75	1.33
Struve	Vega (alpha Lyrae)	0.26	0.12	8.3
Bessel	61 Cygni	0.31	0.29	3.4

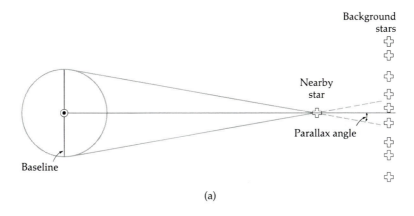

(a)

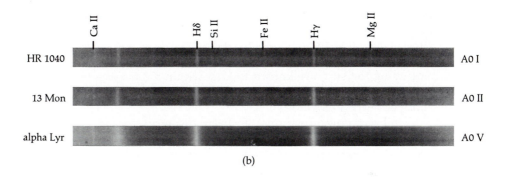

(b)

Figure 10-1. The determination of stellar distances. (a) Direct method; the geometry of stellar parallaxes. (b) Indirect method; spectral lines are analyzed to determine intrinsic brightnesses of stars. The three spectra shown here were produced by stars with the same photospheric temperature (about 10,000 K). However, the supergiant (indicated by I) is much brighter than the giant (II), which is brighter than the main-sequence star (V). Note that the hydrogen lines (H) are weaker in the brighter stars, but that certain ionized metal lines (Mg II, Fe II, Si II) are stronger in these stars. These effects in the giants and supergiants are related to the fact that the lines are formed in less dense, larger atmospheric regions than in the main-sequence stars. The spectra are reproduced here as negatives, just as a working astronomer would examine them on the original photographic plates or film. Thus the spectral lines which appear white here are due to *absorption* of light. (Yerkes Observatory photograph.)

The star alpha Centauri, which is listed in Table 10-1, is actually a triple star system; that is, it consists of three stars in a common orbiting system. One of the three, called Proxima Centauri, travels in an enormous orbit with a period of several hundred thousand years. At its present location it is closer to the sun than the other two stars, and it is in fact the closest known star in space to the sun. Its parallax is 0.763 seconds of arc, corresponding to a distance of 1.31 parsecs.

With the exception of Bessel's value quoted in Table 10-1, the errors in the original parallaxes are considerable. The small size of stellar parallaxes makes them difficult to measure (note that even the parallax of Proxima Centauri is less than one second of arc). As more distant stars are considered, the parallaxes get smaller. Beyond 50 parsecs, that is, for parallaxes smaller than one-fiftieth of a second of arc, the uncertainty in a parallax measurement is comparable to the measurement itself, and so these parallaxes are unreliable. Such small parallaxes, although too crude to give reliable distance values, do at least show that the stars involved are far away. On the other hand, accurate (to about 10 percent) parallaxes have been measured for about 700 stars closer than 20 parsecs. These directly measured quantities are called *trigonometric parallaxes*. Sixty stars, including the sun, are known to exist within 5.2 parsecs of the earth. It is possible that other, extremely small and dim stars are near us but have not yet been discovered.

Indirect determinations of stellar distances (necessary for stars beyond about 50 parsecs) are mostly based on techniques for estimating the *intrinsic* brightnesses of stars. How bright a star appears in the sky *(apparent brightness)* depends on (1) its intrinsic brightness, (2) the distance to the star, and (3) complicating effects that we ignore for now (such as absorption of starlight by dust in space and by the earth's atmosphere). If we somehow know the intrinsic brightness, we can figure out how distant the star must be to have the observed apparent brightness.

A simple example of the principle of indirect distance determination can be given for the case of a light bulb. Suppose we see a light bulb shining through the night some distance away, and we wish to measure this distance. We could make direct measurements, such as by pacing off the distance, or by sighting on the bulb from two different spots with a surveyor's transit and measuring its parallax. However, we are too lazy for all that—we prefer to just point our telescope at the bulb. By attaching a photometer to the telescope we can measure the apparent brightness of the bulb. If we also knew its true or intrinsic brightness, we could compare this with the apparent brightness and compute the distance with the inverse square law (which says that the apparent brightness of an object decreases as the square of its distance, so that if you move the bulb twice as far away, it will seem one-fourth as bright). But how can we find the intrinsic brightness? A clever trick would be to look through the telescope and read the wattage label on the bulb.[2] Then, knowing the true brightness, and having measured the apparent brightness with the photometer, we can calculate the distance of the bulb.

There also are "labels" that astronomers have found in their observations of stars, and we can read these to find the intrinsic brightnesses of stars too distant to be measured by the parallax method. The astronomical labels of course are not written in English, but are coded in terms of the properties of starlight. The most important example is the class of pulsating stars called *Cepheid variables*. These stars are seen to get brighter and fainter and brighter again in a periodic fashion. The number of days that elapses between one time of maximum apparent brightness and the next maximum is called the star's *light period*. As described later in this chapter, the period depends on the average intrinsic brightness; in particular, the longer the interval between successive times of maximum brightness, the greater the average intrinsic brightness. Thus, looking through the telescope and counting the number of days between successive times of maximum brightness, we can determine the average intrinsic brightness of a Cepheid variable. Measuring the average apparent brightness and comparing it with the intrinsic value, we can determine the distance at which the star must be for its known intrinsic brightness to be reduced to its measured apparent brightness. The "label" on the star is its light period. No matter how far away the Cepheid, if we can see it, we can find its period. The intrinsic brightness of another class of pulsators, the *RR Lyrae stars*, also has been determined, so their distances also can be estimated.

By studying the absorption lines in the spectrum of a star, we can determine the physical conditions of temperature and density in the photosphere of that star. Among the many hundreds of stars for which accurate distances are known, we may find a similar spectrum and thus similar physical conditions. It is therefore a good guess to say that the two stars have similar intrinsic brightnesses as well, and by comparing them we can estimate how far away the first star must be. This enables us to calibrate a method of indirect distance determination, called the *spectroscopic parallax* technique, which uses the spectrum of the star as its label. When two stars of the same photospheric temperature have different diameters, their photospheric densities, and thus their absorption line

---

[2]For a given type of light bulb, the wattage is an approximate indicator of the brightness, if the lamp is operated at the standard voltage for which it is designed.

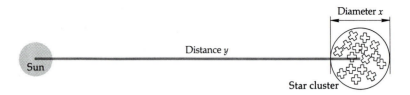

Figure 10-2. Every star in the cluster is at distance y within an accuracy of $\frac{1}{2} \cdot x$. Distances to different stars in a cluster are all nearly the same.

patterns, are different. Therefore, the details of the absorption line pattern constitute a label we can use to estimate the intrinsic brightness, and hence the distance, of a star (Figure 10-1b).

Astronomers are always on the alert for other ways to determine distance. Several methods apply to groups of stars. Stars in a cluster[3] are all at essentially the same distance (Figure 10-2) from the earth, and some of the group properties (Chapter 11) allow the distance to be determined. A few clusters (such as the Hyades, Plate 4) are close enough so that we can accurately measure their speed and direction of motion, or *space velocity*. By comparing the observed motion in angular units (seconds of arc per year) with the space velocity in linear units (parsecs per year) we can determine how far away the cluster is. This technique has yielded fairly accurate distances for hundreds of stars located beyond the reach of the trigonometric parallax method.

Finally, *statistical parallaxes* are based on a given star's proper motion (see below) across the sky. Part of the star's apparent change in position with time is due to the motion of the sun with respect to the nearby stars. As the earth travels through the galaxy with the sun, we view the stars from a constantly changing position. The nearer stars seem continually to move with respect to the background of distant objects. Thus, the nearby stars have, on the average, larger proper motions, caused by the sun's motions, than those stars far away.

MOTIONS IN SPACE

The motion of a star in space with respect to the sun can be determined from knowledge of its distance, its *proper motion*, and its *radial velocity*. The

---

[3]The various kinds of star clusters are described on page 302.

radial velocity, or velocity along the line of sight, is found (in units of kilometers per second) by measuring the wavelength shifts in the spectrum caused by the Doppler effect (Chapter 5).

The velocity at right angles to the line of sight is much more difficult to determine. It can be measured as the proper motion in angular units (seconds of arc per year) by observing the rate of change in the position of a star with respect to background stars that, because of their much greater distance, hardly seem to move. This can be done by comparing photographs taken many years apart. The star with the largest known proper motion was discovered by the American astronomer E. E. Barnard (1857–1923). Known as *Barnard's star*, it has a proper motion of 10.3 arc seconds per year. When the distance to a star is known from one of the methods discussed previously, then the observed proper motion in angular units can be converted to the *tangential velocity* in units of kilometers per second. When the radial and tangential velocities are combined by trigonometry, as shown in Figure 10-3, the *space velocity* of the star relative to the sun is known. We say "relative to the sun" because the sun has its own motion through the galaxy. At different angles to the sun's direction of motion, we see an effect in the space velocities of the stars. This is analogous to the apparent velocities of other cars when we are driving on the highway; everyone may be going at roughly 55 mph, but oncoming

Figure 10-3. A star's total space motion is determined by its radial and tangential motions.

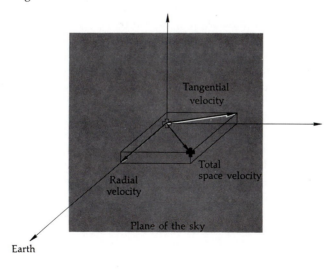

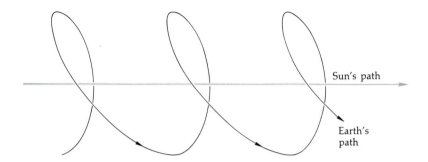

Figure 10-4. The earth's corkscrew motion through the galaxy. As the earth orbits around the sun, the sun is also moving through space.

traffic seems to pass our car at 110 mph. A house on the side of the road seems to be left behind at 55 mph, but distant objects such as mountains on the horizon several miles away seem to move very slowly as we cruise past them.

When the space velocities are determined for the stars near the sun, some regular patterns are observed. These can be explained in terms of the above analogy by postulating a space motion of the sun of about 20 km/sec (12 mps) with respect to the nearby stars, directed toward a point in the constellation Hercules. This phenomenon shows why part of the proper motion of stars is caused by the sun's motion through the nearby stars, and why proper motion depends on the star's distance from the sun. Stars close to the sun have large proper motions; those farther away have small or unobservable proper motions.

The direction of the sun's motion relative to the nearby stars was first determined by William Herschel in 1783. The combination of the sun's motion and the earth's motion around the sun produces a corkscrew path for the earth with respect to the nearby stars (Figure 10-4).

INTRINSIC PROPERTIES OF STARS

In this section we consider some stellar properties that are independent of distance or motion. Much of this information is learned from photometry (measurements of light intensity) and spectroscopy (study of spectra).

In photometry, one usually measures the *relative brightnesses* of stars with respect to some *standard stars* whose apparent brightnesses are already known; this allows us to determine the apparent brightnesses of stars on a common scale of *magnitudes*. Magnitudes are defined by the statement that a star that is one magnitude brighter than another star is 2.512 times brighter. The filters and photomultiplier tubes used in a photometer are chosen and standardized to produce measurements in a specific range of wavelengths. One such system, the so-called UBV, gives magnitudes in the U (ultraviolet), B (blue), and V (visual or yellow) wavelengths. The sensitivity of a UBV photometer to light of different wavelengths is illustrated in Figure 10-5. The zero point of the UBV scale is defined in terms of standard stars whose magnitudes have been carefully studied and agreed upon by several astronomers. Because the magnitude scale is defined so that brighter objects have smaller magnitudes, a star with B = 6.3 is 2.512 times brighter in the blue wavelength range than a star with B = 7.3. Other photometric systems, including some with a large number of narrow wavelength bands, also are used.

Differences in magnitudes can be used to study the wavelength distribution of light from stars. A blue star with a surface temperature of 25,000 K produces more blue light than yellow. Thus, the B (blue) magnitude will be small (bright) and the V (yellow) magnitude will be large (faint). Similarly, an orange star with a temperature of 5,000 K emits more light in the yellow than in the blue, with a corresponding result in the magnitudes. We can subtract the V magnitude from the B value for a star, forming the quantity B – V, which indicates the photospheric temperature of the star by telling how its blue and yellow light compare in intensity. A quantity such as B – V (U – B is another example) is called a *color index*, or for short, a *color*. For example, a fairly blue star such as Alcyone (the brightest star in the Pleiades, Figure 6-19) has B – V = −0.1; a yellow star such as the sun has B – V = +0.6, and a red star such as Antares in the

Figure 10-5.   Photometer sensitivity in the wavelength bands of the UBV system.

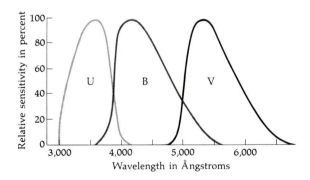

constellation Scorpius has B – V = +1.8. The existence of different colors in stars is shown in Plate 4.

All observed magnitudes must take into account the effects of absorption of light by the earth's atmosphere, as well as absorption by dust in interstellar space. Corrections for these effects are especially important in determining the *bolometric magnitudes,* which measure the light emitted at *all* wavelengths. Most of the energy radiated by the sun is carried by light of wavelengths that can pass through the earth's atmosphere. But, for the hot, blue stars, a large fraction of the radiation is absorbed in the air and does not reach our telescopes; therefore, substantial corrections must be applied to the observed magnitudes and colors of these stars.

When it is possible to determine the intrinsic brightness of a star, it is expressed in terms of an *absolute magnitude,* which is the magnitude that the star would have in a certain wavelength range[4] if it were at a standard distance from us and no absorption occurred in the intervening space. This distance has been arbitrarily adopted by astronomers as 10 parsecs. Absolute *bolometric* magnitudes tell us the total energy emitted per second, or the luminosity. Since the luminosity corresponds to the energy output of a star, it also tells physicists how much energy must be produced inside the star by nuclear reactions.

The photospheric temperature of a star can be estimated by comparing the wavelength distribution of its light with the Planck curves for different temperatures; the same procedure was used to determine the temperature of the solar photosphere in Chapter 9. Photometric measurements of stellar color indices are used to infer temperatures. For greater precision, allowance is made for the fact that the stars are not perfect radiators.

The temperature and luminosity can also be estimated from a detailed analysis of a star's spectrum. Spectra of stars with different temperatures are shown in Figure 10-6. (Also recall the spectra of stars with the same temperature, but different luminosities, in Figure 10-1.) Stellar spectra were originally classified without the benefit of a physical theory. Similar-appearing spectra were assigned the same code letter or *spectral class* at a time when the real significance of differences among spectra was not clear. Later, it was determined that stars of different spectral classes have different temperatures, and the classes into which most stars fit were arranged in a new order, according to decreasing temperature. This order is: O B A F G K M. The O stars are hot and blue (temperatures around 30,000 K or more), the A stars look white to the eye (temperatures around 10,000 K), and the M stars are cool and red (temperatures around

---

[4]For example, the U, B, or V wavelength ranges.

O6
B3
A0
F2
G2
K5
M5

Figure 10-6. Some examples of the basic types of stellar spectra, arranged in order of temperature, with the hottest stars at the top. The sun has spectral class G2, and its spectrum resembles the fifth one from the top in this illustration. (Courtesy of the Hale Observatories.)

3,000 K or less). The sun is a G-type star. There are numerical subdivisions of these classes; for example, K3 designates an orange star cooler than K1 or K2 and hotter than K4. The sun is G2 on this system. People who wish to memorize the sequence of spectral classes from O to M can use the mnemonic device, "Often, Beetles And Fleas Give Kangaroos Malaria."

A rather reasonable idea which led the original spectrum classifiers astray was the possibility that the main differences in appearances of spectra (as in Figure 10-6) were due to the presence of different relative amounts of the elements in the various stars, so that one star, having a lot of iron, would have a spectrum dominated by iron absorption lines, whereas another, being mostly hydrogen, would just show hydrogen lines. Although there are composition differences among stars, it turns out that they rarely have a strong effect on the overall appearance of the spectrum. The problem was solved after 1920 through the work of Indian astrophysicist M. N. Saha, who analyzed the influence of temperature and pressure on the ionization of atoms. His results, when applied to the stars, show that the presence of absorption lines of (for example) neutral iron in one star's spectrum, and the absence of the same lines in a second star's spectrum, does *not* mean that the latter star lacks iron. This latter star may be hot enough so the iron is singly ionized, that is, has had one electron knocked out of the atom. An iron ion such as this absorbs light at a different set of wavelengths than does a neutral iron atom. In fact, we can calculate what fraction of the iron atoms should be neutral or ionized at various temperatures[5] and densities by using Saha's theory. We indeed find that when the theory predicts (say) neutral iron, we find absorption lines of neutral iron. Extensive work of this nature has shown that the general appearance of the great majority of all stellar spectra is determined by temperature. Spectral differences due to chemical abundance variations among the stars (see Figure 10-8) or other variations (such as pressure or magnetic field) do occur, but the effects of temperature usually dominate.

When we have found both the photospheric temperature and the luminosity of a star by the above methods, we are in a position to determine its size (at the photosphere). The Stefan-Boltzmann Law tells us how much energy per square centimeter is produced at a given temperature. Then we simply calculate how many such square centimeters must exist in order to produce the total luminosity of the star. In that way we

---

[5]Remember that we can estimate the temperature of a star from its color index measurements.

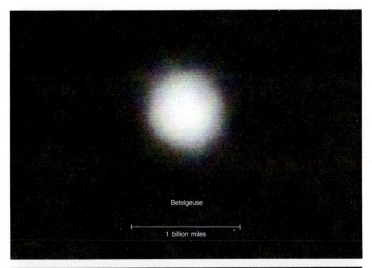

Betelgeuse

1 billion miles

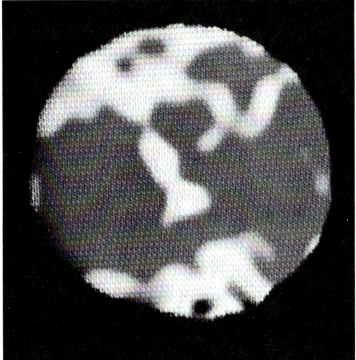

Figure 10-7. Photograph made with the 4-meter telescope at Kitt Peak seems to show some features on the red-supergiant star Betelgeuse (angular diameter about 0.05 arc seconds). If real, the "starspots" may correspond to gigantic convection cells, larger than our sun. Lower picture has been enhanced by computer to bring out detail. (Kitt Peak National Observatory.)

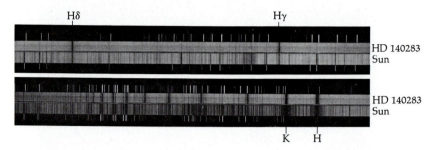

Figure 10-8. Comparison of the same region of the spectrum of the sun and a metal-poor star (HD 140283). Notice the difference in the strengths of the lines. (Lick Observatory photograph.)

learn the surface area of the stellar photosphere, and by a simple calculation (the surface area of a sphere is 4π times the square of the radius) we can find the stellar radius. As a result of such studies, we know that the more luminous stars are usually also the bigger ones.

The determination of stellar brightnesses and sizes revealed the existence of *giant stars* far larger than the sun. The bright red supergiant, Betelguese (Figure 10-7), in the constellation Orion, has a radius about 800 times that of the sun. If placed in the sun's position, Betelguese's atmosphere would extend beyond the orbit of Mars. It is also found that giant stars have lower pressures at their photospheres than do smaller stars. The differences in pressures cause slight differences in the spectrum. By looking for these pressure effects in a stellar spectrum, we can judge whether the star is a giant, normal size, or even smaller, and by combining this spectroscopic size estimate with the temperature judged from the spectrum, we can make a rough estimate of a star's luminosity and its intrinsic brightness in a given wavelength range. Comparing the intrinsic and apparent brightnesses, we can estimate how far away it is. This is how the *spectroscopic parallax* method of distance determination works.

Finally, the most difficult basic property of a star to determine is its mass. The methods used to measure mass require that the star have another object in orbit with it; fortunately such pairs or *binary stars* are common. This method is analogous to the way (Chapter 2) in which the mass of the sun was found from the motions of the planets and Kepler's Third Law. Let us briefly return to a discussion of the earth's motion around the sun and consider some of the differences in the two methods of solar and stellar mass determinations. Kepler's Laws are only an approximation of the true situation. The sun, for example, is not at rest at a focal point of the planetary ellipses. Unknown to Kepler, it too performs a small elliptical motion with respect to the *center of mass* of the solar system, although this solar orbit is smaller than the sun itself. The center of mass is, in essence, the average position of the mass in a given system. For objects of equal mass this position is exactly half-way between them. If one object is more massive, the center of mass would then be closer to the more massive object (Figure 10-9), and if one object is very much more massive (as is true in the case of the sun and the earth), then the center of mass may even be within the bigger body. Thus, if the radii of the respective orbits of binary stars can be determined with respect to the center of mass of the binary system, the *ratio* of the masses of the two stars can be obtained. For example, if one orbit radius is half the size of the other, then the star with the smaller orbit is twice as massive as its partner.

In addition, Kepler's Third Law (as it was originally stated by Kepler) is not a very precise description of the way in which the orbital periods of the planets depend on the sizes of their orbits, because the period of a planet also depends slightly on its mass. The variation is such that the period of revolution of the earth around the sun (for example) depends on the *sum of the masses* of the sun and the earth. In this case, since the sun is much more massive than the earth, the error involved in neglecting the earth's mass is small (the mass of the sun is 2×10^{33} grams; the earth's is 6×10^{27} grams). For visual binary stars the period of revolution

Figure 10-9. Illustration of the center of mass in binary star systems by considering the analogous problem with children on a see-saw.

Center of mass   Center of mass

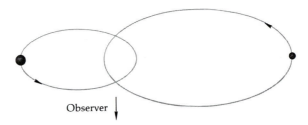

Observer

Figure 10-10. Schematic diagram of motions in a spectroscopic binary system. (From *Life Outside the Solar System* by Su-Shu Huang. Copyright © 1960 by Scientific American, Inc. All rights reserved.)

can be observed easily. If the distance is known, the sizes of the orbits can be calculated, and Kepler's Third Law allows a determination of the sum of the masses after a correction is made for the effect of the inclination of the orbital plane to our line of sight.

Once the *ratio* of the two masses of a binary star has been determined from observations of the relative sizes of the orbits around the center of mass, and the *total mass* of the two stars is found from Kepler's Third Law, we can solve for the individual masses. The study of visual binary stars has yielded about 15 fairly accurate stellar masses; these happen to be mostly of stars less massive than the sun.

Other masses can be determined from *spectroscopic binary stars*. Here, the stars are not visible separately in the telescope, as they are in the case of visual binaries. The resolution is achieved not by the telescope, but by the spectrograph. If the stars are in orbit around each other, then at a given moment one star will be approaching the sun while the other recedes (Figure 10-10). Because of the Doppler effect, the absorption lines from the receding star are shifted to the red, and those of the approaching star are shifted to the blue. This situation can produce two sets of lines that move back and forth periodically, as shown in Figure 10-11, and then we call it a *double-lined spectroscopic binary*.[6] The mass ratios can be inferred from the observed velocities, because the more massive star will move more slowly than the less massive star. The larger star moves at a slower rate than its companion because the gravitational force on each star is the same (action-reaction), and a given force will produce a smaller acceleration of a more massive object than of a lesser one (Newton's Second Law, Chapter 2).

---

[6]Sometimes, one member of a binary system is much fainter than the other, and we can only record the spectral lines of the brighter star. However, they shift back and forth as the star orbits, and so we can identify the system as a *single-lined spectroscopic binary*.

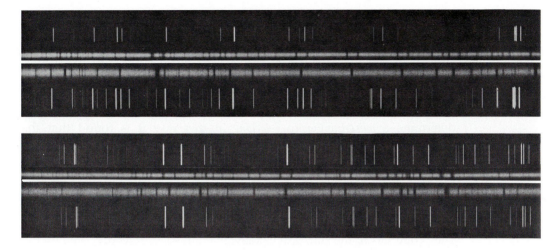

Figure 10-11.   Spectra taken at two different phases of the spectroscopic binary star kappa Arietis, showing the doubling of lines (upper spectrum) due to the Doppler shifts arising from the motions of the two stars. The same lines are single in the lower spectrum. (Lick Observatory photograph.)

The period of a spectroscopic binary can easily be determined by observing the changes in the positions of the spectral lines, since the cycle of changes will repeat once per orbital period. The separation between the stars cannot be determined, however, when the inclination of the orbits to the line of sight is unknown. In some cases, one star eclipses the other. This tells us that the plane of the orbits must contain our line of sight, and so we know the inclination is close to zero degrees.[7] If this is the case, we can use the velocity of each star, as determined from the Doppler shift of the lines in the spectrum, and multiplying the velocity by the period gives the distance traveled in one period.[8] If we neglect the details of ellipses and consider this distance to be the circumference of a circle (equal to 2π times the radius), the orbit radius can be determined. When the period,

---

[7]In this text, we have defined the inclination of an orbital plane with respect to the line of sight. However, astronomers usually quote the inclination with respect to an imaginary plane at right angles to the line of sight (plane of the sky), and so you may find a different definition in other books.

[8]If the binary does not undergo eclipses as seen by us, the inclination is greater than zero degrees, and the observed Doppler shifts only correspond to a fraction of the full orbital velocities.

orbit radii, and mass ratios are known, the masses of the stars in such systems can be determined. As it happens, these *eclipsing spectroscopic binaries* have provided a few dozen accurate masses, mostly for stars more massive than the sun. Thus, they complement the data on smaller stars in visual binaries. Altogether, masses have been determined for less than 100 binary systems.

The known masses can be plotted against the stars' luminosities to determine a *mass-luminosity relation*. In this way it is found that the luminosity of normal stars (called *main-sequence* stars) varies roughly with the mass of the star cubed, so that a star twice as massive as another will be about eight times brighter.

THE SUN AND STARS COMPARED

The properties of the stars are usually given in solar units (denoted by the subscript $\odot$) for convenience, and the comparison is interesting. Most stellar masses are in the range from 0.1 to 50 times the mass of our sun (denoted by 0.1 $M_\odot$ to 50 $M_\odot$). The radii of ordinary stars range from 0.1 $R_\odot$ (one-tenth the solar radius) up to hundreds of solar radii for the giants, although two kinds of unusual, very small stars (white dwarfs and neutron stars) are outside this range. Photospheric temperatures of normal stars range from 2,000 K to above 30,000 K; the sun's value at 6,000 K is not near either extreme. In terms of the solar luminosity, stars that are 10^4 times fainter than the sun are known, and stars 10^5 times brighter than the sun are also known. For these reasons, the sun is often described as an average star. This is really a little misleading; its properties fall midway in their ranges among normal stars, but most of these stars are smaller, cooler, and less massive than the sun. A further distinction is that the sun is a single star, whereas it appears that most of the stars in the galaxy are found in binary, triple, or multiple systems.

HERTZSPRUNG-RUSSELL DIAGRAM

In this chapter, we have been discussing a variety of techniques used to determine masses, radii, luminosities, and atmospheric temperatures of stars. The chief differences in the appearance of stellar spectra were explained as the result of temperature differences. A convenient way of illustrating the observational data will be helpful for our further discussions of the stars and how they change with age. Such a method is a diagram of luminosity (or absolute magnitude) plotted against temperature (or color index), as shown in Figure 10-12. This type of chart was first used by the Danish astronomer Ejnar Hertzsprung in 1911, and indepen-

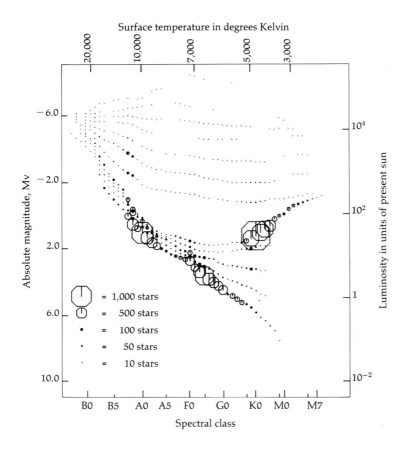

Figure 10-12. A modern observational H-R Diagram showing over 24,000 stars down to a rough limiting magnitude for a given area of the sky. The number of stars of a particular type is proportional to the symbol size; examples are given in the key. The main sequence and the red-giant region are conspicuous. Approximate surface temperature and luminosity scales have been added; they are not part of the observational diagram. See Plate 1, which shows the location of the major types of stars in the H-R Diagram. (Courtesy of N. Houk and R. Fesen, University of Michigan.)

dently by the American astronomer Henry Norris Russell in 1913; it is called the Hertzsprung-Russell Diagram or *H-R Diagram*. Note that the stars are not uniformly spread over Figure 10-12, but in fact seem to fall mostly in certain restricted locations or bands (note that temperature increases to the left on the horizontal scale).

Many of the principal classes of stars are readily identified on the H-R Diagram. Most stars are found on the wide band (called the *main sequence*) that runs from the upper left portion of the diagram to the lower right. This main sequence was not predicted, it was simply found when the measurements of stellar luminosities and temperatures were plotted on the diagram. Note that since the stars at the left end of the main sequence are brighter (higher up in the diagram) than the ones at the right end, they must be more massive, according to the mass-luminosity relation. Since most stars are observed to fall on the main sequence, and since we find compelling evidence for evolution of stars over the H-R Diagram (that is, the position of a star in the diagram changes with time; see Chapter 11), we conclude that stars spend most of their lifetime on the main sequence. The sun, for example, is a main-sequence star. Most stars near the sun are main-sequence stars that are fainter and redder than the sun.

The *giant stars* are above and to the right of the main sequence in the H-R Diagram. Since giants are much brighter than main-sequence stars with comparable temperatures, the Stefan-Boltzmann Law tells us that they are also much larger, by factors ranging from 10 to several hundred. The biggest giant stars, such as Betelgeuse (Figure 10-7), are often called *supergiants*. On the other hand, a given *white dwarf* star is much fainter than main-sequence stars of the same temperature. The white dwarfs have radii about 100 times smaller than that of the sun, and are thus comparable in size to the earth.

The Hertzsprung-Russell Diagram and the various measured quantities discussed in this chapter provide the fundamental data for our discussion of the birth and death of stars in Chapter 11. In fact, the principal task of the science of *stellar evolution* is to explain the facts of the stellar census as illustrated by the H-R Diagram.

BINARY STARS

Studying the stars through telescopes and on photographs, it has been found that most of them occur in pairs, triples, or multiples. We have already indicated that binary stars are of great interest because studies of them can reveal the masses of stars.

In the latter part of the eighteenth century, many double stars were found with telescopes. The two stars of a pair, however, were believed to be widely separated in space and were thought to be seen near each other only in projection on the celestial sphere. William Herschel carefully observed many double stars in the hope of detecting their parallaxes, but he

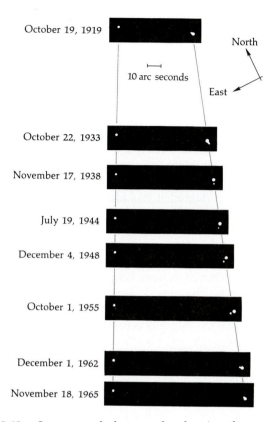

Figure 10-13. Sequence of photographs showing the proper motion and orbital motion of the components (right) A and B of the visual binary star Krüger 60. A more distant star appears at the left. (From *Principles of Astrometry* by Peter van de Kamp. W. H. Freeman and Company. Copyright © 1967.)

never found them because his instruments were not precise enough. However, he observed that for some double stars the positions changed in such a way that they appeared to be revolving in orbits (they are now called *visual binaries;* an example appears in Figure 10-13). After about a quarter century of careful study, Herschel concluded in 1803 that many double stars "are not merely double in appearance, but must be allowed to be real binary combinations of two stars, intimately held together by the bonds of mutual attraction." Thus, Newton's laws of motion and universal gravitation had been extended to the stars.

The fact that most double stars must be physically connected had already been deduced by the Reverend John Michell. He noted that there were far too many of them to be merely chance occurrences or perspective effects. Although Michell's statistical arguments preceded Herschel's results, it was the latter's observations of orbital motions that put the matter beyond dispute.

Another demonstration of the reality of physically associated stars followed from measurements of the star Algol in 1783 by the Englishman John Goodricke. It had already been known for more than a century in Europe that the brightness of this star changes. (In fact, a story persists in the astronomical literature that the early Arabs were aware of its light variations and therefore named it Ras al Ghul, the Demon's Head, from which we get Algol. There is no evidence for this origin of the name. In any case, there was considered to be a constellation of the demon or monster in this area of the sky in ancient times, according to the records of several civilizations.) Goodricke observed that the brightness of Algol changed by about a factor of 3 from brightest to dimmest, but his great discovery was that these changes occurred *regularly*, with a period of 69 hours. The *light curve* of Algol is shown in Figure 10-14, along with the explanation for its behavior. Algol is actually a binary system composed of

Figure 10-14.   The light curve and schematic diagram for Algol, an eclipsing binary star. The relative separation of the stars, their size, and the size of the sun are also shown. (After *Matter, Earth, and Sky,* 2nd ed. By George Gamow. Prentice-Hall. Copyright © 1965. Used by permission.)

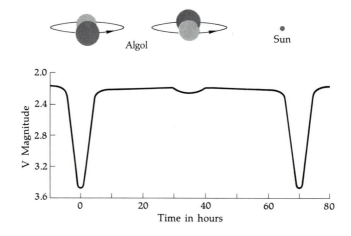

a bright star and a faint star in orbit around their center of mass. The plane of their orbits contains our line of sight from the earth, so they are an *eclipsing binary.* Primary eclipse (the major brightness decrease) occurs when the brighter star is obscured by the fainter star, as seen from the earth. A secondary eclipse occurs when the bright star hides the faint star. The graph of brightness plotted against time is called the *light curve.* Eclipsing binaries are most valuable because the inclinations of their orbital planes to the line of sight are definitely known to be close to zero degrees (from the simple fact that we can see the eclipses), and their masses can be calculated in the way described earlier in this chapter. Eclipsing binaries are of additional value because the observations of the changes in brightness produced by one star gradually cutting off the light of the other reveal information on the sizes of the stars and even their limb darkenings. To estimate the sizes we (1) get the orbital velocities from Doppler shifts in the spectrum, and (2) get the length of time it takes for one star to eclipse the second star from the light curve. Knowing the speed and how long it takes to go across the diameter of the second star, we can calculate the size of the second star.

STARS AND PLANETS

A curious fact concerning binary stars is that the average separation between the members of a binary is about 20 a.u., which is roughly comparable to the distance of the Jovian planets from the sun. Could we detect a planet the size of Jupiter revolving around a star near the sun?

The answer is yes, even though Jupiter emits no visible light of its own; stars or planets with masses of less than about 0.07 $M_\odot$ have negligible luminosities, because the amount of Helmholtz contraction that they undergo is not enough to raise their internal temperatures to the point where nuclear reactions can occur. Nevertheless, the existence of such an object can be inferred, because its gravitational attraction causes the associated visible star to execute an orbit around the common center of gravity. Thus, as seen from the earth, the visible star moves through the galaxy along a wiggly path, almost a sure clue to the existence of a dark companion. If the spectrum of the bright star can be obtained, its mass can be estimated by the techniques mentioned previously; then the mass and orbital parameters of the dark body required to produce the observed wiggle can be determined. In some cases these dark bodies may be comparable to the Jovian planets, as suggested, for example, by results obtained by Peter van de Kamp of Swarthmore College. Van de Kamp concluded that Barnard's star has a companion that is a large planet. If van de

Kamp is correct, this is the first known case of a planet outside the solar system. He believes there is also some evidence for a second planet of Barnard's star.

The basic difference between the two interpretations of the Barnard's star system (one planet versus two planets) is in the orbital shapes. If there is only one companion, then its orbit must be a highly elongated ellipse to account for the observed motion of the star. If there are two planets, the orbits are nearly circular. If we can use our solar system as a basis for judgment, then the model involving nearly circular orbits is to be preferred. The parameters of the models are compared in Table 10-2, along with some data on Jupiter.

The existence of dark companions to a few other nearby stars has also been inferred, although the evidence is not conclusive. If correct, this suggests that the occurrence of planets is not very unusual. However, it must be pointed out that in no case has the existence of planets of stars other than the sun been definitely proven. Even the planet(s) of Barnard's star has been challenged by some experts.

Barnard's star is of interest for another reason; its motion in space is carrying it closer to the sun. By the year A.D. 11,000 it will be closer than Proxima Centauri, with a parallax of about 0.87 arc seconds and a distance of 1.15 parsecs.

STAR CLUSTERS[9]

The social organizations of people and animals include the lone wolves, the family groups, and the tribes or herds. Affiliation with such a group is generally related to the circumstance of birth. Photographs of the sky show that in addition to the many isolated stars that seem to follow their own paths through space *(field stars)*, there are pairs of stars like the binary star Algol, triple stars like the alpha Centauri system (of which Proxima is a member), *multiple stars*, consisting of several stars orbiting around a common center of mass, and larger groups, the *star clusters* and *associations*. These groups include the *open clusters* like the Pleiades (Figure 6-19) and Hyades (Plate 4), and the *globular clusters* like the Hercules cluster (Figure 10-15). It seems reasonable to assume that the members of a

---

[9]The field stars and clusters that are discussed in this section, which we can see or photograph when we observe the sky at night, are members of our own Milky Way galaxy, as discussed in Chapter 12. Other galaxies, like the one in Andromeda (Plate 8), are also seen to contain both field stars and clusters when they are observed or photographed through telescopes.

Table 10-2. Comparison of the Planet(s) of Barnard's Star with Jupiter

	COMPANION(S) TO BARNARD'S STAR			
	One Companion Case	*Two Companion Case*		*Jupiter*
Orbital period	25 years	12 years	26 years	12 years
Radius of orbit	4.4 a.u.	2.8 a.u.	4.7 a.u.	5.2 a.u.
Mass	1.5 M_J	0.8 M_J	1.1 M_J	1.0 M_J

M_J represents the mass of Jupiter, 1.9×10^{30} grams.

Figure 10-15. The globular cluster M13 in the constellation Hercules. (Courtesy of the Hale Observatories.)

binary or multiple star system share a common origin, and the members of a star cluster are likewise related by birth. A large globular cluster may have more than one million member stars and a diameter of 50 parsecs or more. The term "globular" is derived from the round shape, evident in Figure 10-15. The membership of open clusters ranges from less than 100 to more than 1,000 stars; their diameters range from about 3 to 20 parsecs. The members of an open cluster are not organized into the compact spherical shape of a globular cluster, hence the cluster is called "open."

Associations resemble open clusters in their lack of an obvious geometrical shape, but they are distinguished from open and globular clusters by several properties: (1) stars visible in associations consist primarily of the young, hot, main-sequence O and B types, whereas clusters include stars of several spectral types; (2) the member stars of an association are not closer together on the average than other stars in the sky,[10] whereas the property of clusters that allows astronomers to discover them easily is that the clusters are obvious as concentrations of stars; (3) an individual association is expanding; that is, the member stars are moving away from the center of the group. We see only young stars in the associations, and the associations themselves are expanding. Thus these groups are of relatively recent origin.

PULSATING STARS

For years, light variations of all periodic variable stars were considered to be caused by eclipses in binary systems, since no other mechanism, such as pulsations, was known. John Goodricke, the observer of Algol, also discovered another kind of variable star, named *Cepheids* after the first one found, delta Cephei. Hundreds of Cepheids are now known, and a typical light curve is shown in Figure 10-16. Compare this light curve with that of Algol shown in Figure 10-14. They definitely are not alike. In fact, because of the difference in shapes of the rising and falling parts of the Cepheid light curve, it is hard for us today to imagine any way that it could be produced by eclipses. Observations of Cepheid spectra showed variable Doppler shifts, and the pattern of spectral lines seemed to change as well as shift as the brightness of a star changed.

Harvard astronomer Harlow Shapley investigated the Cepheids and in 1914 struck the death blow to the eclipsing hypothesis. Cepheids were clearly giant stars (because of their brightness), and he calculated that the

---

[10]The existence of an association has to be checked by measuring the distances of O and B stars in a given region of the sky and seeing whether a group of them are all roughly in the same part of the galaxy.

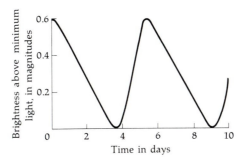

Figure 10-16. The light curve of the pulsating variable star delta Cephei, as measured in yellow light. The period is 5.37 days.

likely dimension of a Cepheid would exceed the orbit size of any companion, as deduced from the observed Doppler shift and period. If the companion were inside the Cepheid, it would find it hard to keep orbiting, and, in any case, it could hardly eclipse the Cepheid! Shapley pointed out that all the observations of Cepheids (light curves, spectra, and radial velocity changes) are consistent with the idea that these stars are pulsating. The Doppler shifts observed are literally due to the photosphere moving toward us as the star expands and away from us as the star contracts; of course, we only observe the light from the side of the star that faces the earth.

The remaining obstacles to the pulsating star hypothesis were theoretical; that is, *why* should stars pulsate? Apparently, pulsating stars are unstable, and the meaning of this word is illustrated in Figure 10-17. An object in a stable configuration returns to its ordinary position or arrangement after a small disturbance occurs. An object in an unstable configuration does not necessarily return to the rest position, but may collapse or oscillate back and forth around it. The latter situation applies to pulsating stars such as the Cepheids. As a star ages, its changing internal structure may become unstable and start to oscillate. This problem was tackled by the English astronomer Sir Arthur Stanley Eddington (1882–1944), who established the theory of pulsating stars as a branch of astrophysics. As an example of the reasonable accuracy that was attained, the theory predicts that the increase or decrease in radius compared to the average should be less than 10 percent; the value for delta Cephei is observed to be 6 percent. As the star increases and decreases in size, the gases of its outer layers are alternately expanded and compressed; its surface temperature also changes, and this causes the changes in the pattern of lines in its spectrum.

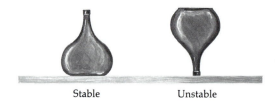

Stable   Unstable

Figure 10-17.   Simple examples of (a) stable and (b) unstable configurations.

The previous discussion is just a small part of the Cepheid story. These stars are very bright and easily recognizable by means of their light variations; as such, they are good candidates for studies of distances in astronomy. The brightest Cepheids have luminosities that are 10^4 times that of the sun. They can be seen in other galaxies, such as the Magellanic Clouds. In 1912, American astronomer Henrietta Leavitt studied the Cepheids in the Small Magellanic Cloud and noticed that the average brightnesses are related to their periods. The Cloud is so far away that all its Cepheids are effectively at the same distance from us (remember Figure 10-2). Hence, the relation between periods and apparent brightnesses was in fact a relation between periods and absolute magnitudes (although the *values* of the absolute magnitudes were uncertain, since the distance of the Cloud had to be determined). What a discovery! If the *period-luminosity relation* could be calibrated, the identification of a star as a Cepheid and the simple determination of its period of light variation would immediately give its absolute magnitude and ultimately its distance. (Such a calibration was obtained from the statistical parallaxes of the Cepheids in our galaxy; none of the Cepheids is close enough for a trigonometric parallax.) The use of Cepheids as distance indicators by this method became a prime tool in surveying our galaxy and measuring the distances to other galaxies.

Care must be exercised to establish the correct type of variable star, as a few different types of Cepheids are known, and each has a different period-luminosity relation. Stars such as delta Cephei are called *Classical Cepheids*; another group is named after the star W Virginis. The period-luminosity relations for Classical Cepheids and W Virginis stars are shown in Figure 10-18. Thus, the classification and period determination for Cepheids enables us "to read the label on the light bulb" and calculate the distances to these stars.

The pulsating stars used as distance and population indicators are the two types of Cepheids just discussed, along with the so-called cluster variables, or *RR Lyrae stars*. These stars are named for their frequent occurrence in globular clusters. However, some of them are field stars, including RR Lyrae itself. The RR Lyrae stars are rapid pulsators, with periods around one-half day. Observation of RR Lyrae stars in the same cluster shows that (unlike Cepheids) they all have about the same absolute magnitude. There is no close globular cluster, and so the calibration comes from statistical parallaxes of the few relatively nearby RR Lyrae stars found outside of clusters. These determinations can be checked by the cluster parallaxes, as described in Chapter 11. The RR Lyrae stars have an absolute V (yellow) magnitude close to +0.6. Hence, if a star can be shown to be an RR Lyrae variable, its absolute magnitude (about 50 times that of the sun) is known, and, comparing this quantity with its average apparent magnitude, we can find the star's distance.

There are still other kinds of variable stars, notably the *long period variables*. These are red giant stars with periods of roughly a year. A typical example is the star omicron Ceti, also called Mira (Latin, "The Wonderful"). It has a period of eleven months, during which its brightness typically changes by a factor of 160 (about 5.5 magnitudes), so that sometimes it is conspicuous to the unaided eye, and at other times a telescope is needed to see it. In some of Mira's cycles, the brightness changes are greater than in other cycles. Thus, its behavior is not as accurately reproducible as that of a Cepheid. It appears that additional, complicating phenomena (perhaps shock waves) as well as pulsations are present.

Figure 10-18. The relationship between absolute magnitude and period for Classical Cepheids (dark line) and W Virginis stars (gray line). (From *Pulsating Stars and Cosmic Distances* by Robert P. Kraft. Copyright © 1959 by Scientific American, Inc. All rights reserved.)

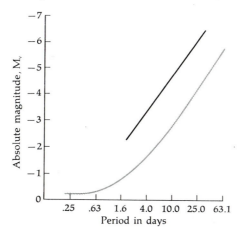

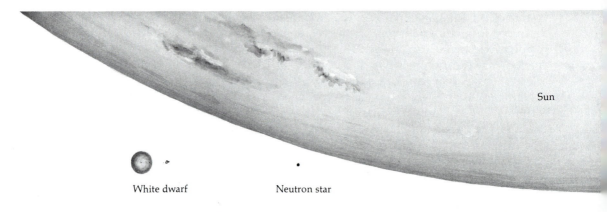

White dwarf   Neutron star

Figure 10-19. Relative sizes of the sun, a white dwarf, the earth, and a neutron star. Neutron stars are discussed in Chapter 11. (Adapted from *X-Ray Astronomy* by Herbert Friedman. Copyright © 1964 by Scientific American, Inc. All rights reserved.)

On the basis of average temperatures and luminosities of the pulsating variable stars, we find they occur in specific zones of the H-R Diagram. It is an important task of astrophysics to explain why stars with these particular temperatures and luminosities tend to be unstable.

WHITE DWARFS

Inspection of the H-R Diagram shown in Plate 1 reveals the existence of a group of very faint stars with typical radii only about one percent of the sun's radius (Figure 10-19); thus, these stars are comparable in size to the earth. *White dwarfs* are common among the nearest stars, but none are visible to the naked eye because they are so dim. Figure 10-20 shows a famous white dwarf called Sirius B (also known as "The Pup" since it is the binary companion of Sirius A, known in ancient times as the Dog Star). Note that Sirius A is a main-sequence star, not even a giant, but (thanks to its distance of only 2.7 parsecs from the solar system) it is the brightest star other than the sun as viewed from earth. Yet the white dwarf Sirius B, at virtually the same distance, is a few magnitudes fainter than the *dimmest* star visible to the naked eye.

The remarkable fact about white dwarfs appears when we consider their masses. A few of these stars have been observed in binary systems, and therefore fairly accurate masses have been found for them. These

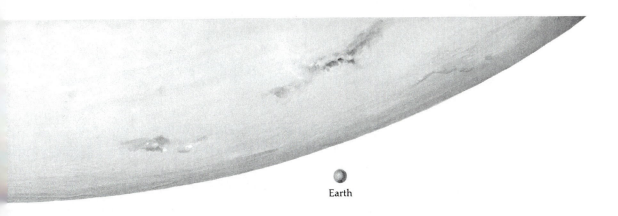

Earth

masses range from roughly one-half to roughly one solar mass. But, since the white dwarfs are so much smaller than the sun, their average densities must be several hundred thousand to a million times greater than the average density of the sun (which is about equal to the density of water). The state of this material can be illustrated by recalling the Rutherford model of the atom (Chapter 5), in which the electrons orbit the nucleus and the volume of the atom is, to a large extent, empty space. The densities in the interiors of white dwarfs are so high that the electron shells around the nuclei of atoms are broken down, and the nuclei and electrons are pushed closer together. This state is called *electron degenerate* matter.[11] It is believed that the white dwarfs were once normal stars, generating energy by nuclear reactions in their interiors. The pressure of this radiation helped support the gaseous matter of the star against its tendency to contract under its own gravitation. However, the nuclear fuel was eventually exhausted, and the star contracted. Now it is radiating away its internal heat, slowly fading and cooling.

White dwarfs are probably a terminal phase of stellar evolution. If so, the theoretical deduction that they cannot have a mass in excess of about 1.4 M⊙ becomes of interest, because many young stars have masses above this limit. These massive stars must die in some other way; this happens through a supernova explosion, as discussed in the next chapter. Some stars, however, may shed enough mass by means of *stellar winds* (similar to but more intense than the solar wind) or by ejecting shells of

---

[11]It is a fifth state of matter, in addition to the gas, liquid, solid, and plasma states that we previously described.

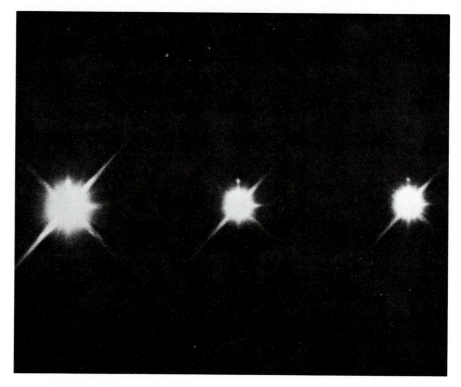

Figure 10-20. Three exposures with the 120-inch reflector of the Lick Observatory showing the white-dwarf companion, Sirius B, just above the primary star, Sirius A. (Lick Observatory photographs.)

gas, so that they can reduce themselves to the point where they are light enough to end up as white dwarfs.

EXOTIC STARS

The studies of binary stars have turned up some unusual cases. Some binaries are so close together that the surfaces of the two stars are actually in contact. These *contact binaries* are distorted by gravitational effects, and there is the possibility of matter passing from one star to the other. Mass transfer in a binary system is another method by which a massive star can reduce its mass sufficiently so as to alter its final state.

The variable star beta Lyrae is another kind of exotic binary system. It cannot be resolved with a telescope, but the spectrograph shows that the brighter star is a B star, and its lines shift back and forth in the way that characterizes orbital motion. The other star is too faint to show up in the spectrum, but there are unusual spectral lines that have been ascribed to different streams of gas around the stars. Three different streams are

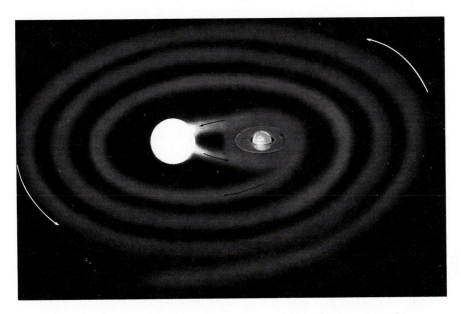

Figure 10-21. Schematic diagram showing the arrangement of gas clouds in the binary system beta Lyrae. The white arrows indicate the direction of the stars' orbital motion, and the black arrows indicate the direction of gas flow.

needed to explain the observations. (1) A stream of gas flows from the more massive to the less massive star. (2) A lower density stream circles around the less massive star and returns to the source. (3) Part of the gas flow forms a spiral pattern outside of the binary system. This complex system is shown schematically in Figure 10-21.

Complicated star systems like beta Lyrae remind us that there are some really intriguing problems in astronomy. The noted astrophysicist Otto Struve (1897–1963) devoted a substantial part of his professional career to studying the changing spectrum of beta Lyrae and trying to explain it in terms of models such as that shown in Figure 10-21. The problem is still not completely solved. Not many years ago a paper entitled "A Final Model for Beta Lyrae" was read before a meeting of the American Astronomical Society; a few minutes later another speaker announced a new "Preliminary Model for Beta Lyrae!"

11

Birth and Death of Stars

The title of this chapter may at first sight appear presumptuous. In Chapter 3 we established that the age of the earth was about 5×10^9 years; humans have life spans of at most a hundred years or so. Even records kept from one generation to the next do not tell us much, because recorded history covers only a few thousand years. How then can we infer that stars are born, live out their lives, and die?

The answer involves a philosophical adjustment to a difficult problem, and the correct attitude toward gathering the relevant observations. The situation can be illustrated by considering the growth of trees in a forest based on one afternoon's observations. Even if you watched a few trees closely during the afternoon you would not detect their growth. But if you noticed that trees of the same species were present in different sizes, that many of the larger trees had approximately the same size, and that some of these larger trees had fallen over and lay dead, you might reach some useful conclusions. One reasonable hypothesis might be that trees begin small, grow to a maximum size at which they spend most of their life, and then die. A similar approach to the stellar census (as organized in the H-R Diagram) allows us to discuss the birth and death of stars.

LIFETIMES

The discussion of the various proposed energy sources for the sun (Chapter 9) centered around the sun's lifetime. From the geological evidence, we believe that the sun has had about the same luminosity for billions of years, hence any acceptable source of energy must be capable of supporting the solar luminosity for the sun's lifetime. We found that nuclear energy was an adequate source, but it is not an unlimited one. If all the hydrogen in the sun were available for conversion into helium (that is, if it were all in the correct state of high temperature and pressure), the sun would radiate at its present rate for about 10^{11} years.

In practice we are interested in the star's main-sequence lifetime, or the time that the star spends on the main sequence of the H-R Diagram. This lifetime is shorter than the hypothetical total lifetime by about a factor of 10. The reason for this stems from the fact that the burning of hydrogen to helium occurs in the hotter, denser, central region of a star. The tendency is therefore to create gradually a central core of helium with a surrounding shell where burning of hydrogen to helium still takes place, and an outer region, where nuclear reactions do not occur, composed primarily of hydrogen. The star thus becomes very inhomogeneous, as the atomic nuclei in the core have about 4 times the mass of the nuclei in the outer parts. The larger the core grows, the more important the inhomogeneity becomes with regard to the stability of the star. When approximately the inner 10 percent of the star's mass of hydrogen has been burned, the star is no longer stable. Calculations show that when the star reaches this state, the outer layers rapidly expand, so that the photosphere reaches 100 times its original radius and a red-giant star is formed. This will happen to the sun when it reaches the age of 11×10^9 years. So far, the sun has lived out about one-half of its main-sequence lifetime, and still has about 6×10^9 years to go before it expands to fill a substantial volume of the present solar system.

This happy situation (which gives NASA 6 billion more years to perfect interstellar travel) does not hold true for the massive, hot, and very bright stars near the upper end of the main sequence. This difference arises from the mass-luminosity relation (Chapter 10), which states that the luminosity increases very rapidly with mass. Consider a very bright (blue) O star with a mass of 35 $M_\odot$. Its luminosity is about 10^5 times that of the sun. The main-sequence age for this O star (as compared to the sun's main sequence age of 11×10^9 years) is *increased* 35 times because of the higher mass (greater fuel supply), but it is *reduced* by 10^5 because of the higher luminosity (or greater rate of burning). A simple arithmetical calculation then gives a main-sequence time for an O star of about $11 \times 10^9 \times 35 \div 10^5$ years, or only 4×10^6 years. A bright blue star therefore exists on the main sequence for a time less than 0.1 percent of the sun's present age. If these bright blue stars had been created at the same time as the sun, they would have long since used up their nuclear fuel. This conclusion holds true even if all their hydrogen were available, instead of only the inner 10 percent. Any bright blue stars that were born contemporaneously with the sun have already passed off the main sequence. Therefore, the O stars that we see at present must have been created no more than 10^6 to 10^7 years ago—a very short time by cosmic standards. The very existence of the bright blue stars with huge luminosities and resultant very short lifetimes shows that stars are not static creatures but are born and evolve. We will also see that there is evidence that stars die.

H-R DIAGRAMS AND THE AGES OF STAR CLUSTERS

The basic ideas of the preceding section predict certain consequences that can be tested. The H-R Diagrams of clusters are a nearly ideal aid, because a cluster consists of a group of stars presumably formed from the same cloud of material at about the same time. Thus, an individual cluster represents a group of stars of similar age and chemical composition, although these properties can and do vary from one group to another.

A composite H-R Diagram is shown in Figure 11-1. Except for the adjustment of the vertical scale, which involves a determination of distance, the preparation of the H-R Diagrams for individual clusters is straightforward; a sample with the data points is shown in Figure 11-2. The distances for some clusters can be determined independently, and for others we utilize the fact that evolution is very slow and main-sequence lifetimes very long for stars fainter than the sun. Thus, the lower main-sequence regions of the various clusters' H-R Diagrams should be approximately the same. This idea has been used in the preparation of Figure 11-1, where the main sequences of stars fainter than the sun have been made coincident. This was achieved by sliding the individual H-R Diagrams vertically; the amount of slide determines the distance to the cluster (because it corresponds to the difference between the apparent and absolute magnitudes of the stars on the lower main sequence). A *cluster parallax* is thus determined.

The study of clusters and stellar evolution was actively pursued during the 1950s. The interpretation of the composite H-R Diagram (Figure 11-1) was made by the Hale Observatories astronomer Allan Sandage. The fact that the very bright stars are still on the main sequence in the cluster NGC 2362 immediately implies that this is a very young group. Could this composite diagram, with a common main sequence at the left and segments leading off to the right, be an age sequence with the youngest cluster at the top and the oldest at the bottom?

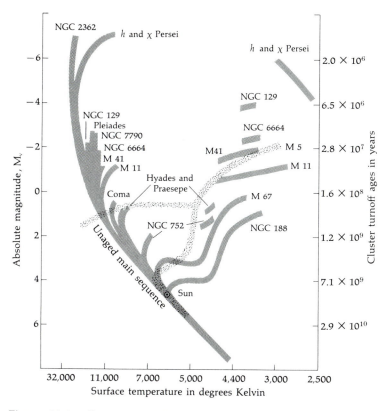

Figure 11-1. Composite H-R Diagram, showing the absolute magnitudes and surface temperatures of stars in a variety of clusters. The cluster names or numbers are given on the diagram. The age of a cluster, defined by the position along the main sequence at which its stars turn off to the right, is marked on the vertical scale at the right. For example, the cluster NGC 188, which turns off near the position of the sun on the main sequence, has an age of about 1×10^{10} years. (Adapted from *Pulsating Stars and Cosmic Distances* by Robert P. Kraft. Copyright © 1959 by Scientific American, Inc. All rights reserved.)

This last idea is entirely consistent with our discussion of lifetimes in the preceding section. The most massive stars are found at the upper left part of the H-R Diagram, if they are still on the main sequence. The more massive the star, the shorter is its time on the main sequence; hence, as time passes, an initial main sequence would evolve schematically as shown in Figure 11-3a. As the age increases, the point where the actual H-R Diagram turns off from the initial main sequence moves down the

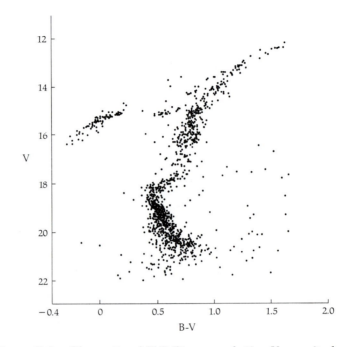

Figure 11-2.   Observational H-R Diagram, plotting V magnitude versus B–V color for the globular cluster M5. The data points correspond to individual stars. (Courtesy H. C. Arp, Hale Observatories.)

remaining main sequence to the fainter, less massive stars. The clusters can actually be dated on Figure 11-1 by the main-sequence lifetime of the stars that have just left the main sequence; these *turnoff points* are converted into ages in the right-hand scale of Figure 11-1. The stars in the very old clusters, such as M5, M67, and NGC 188, have evolved off the main sequence all the way down to the sun's position in the H-R Diagram. Their ages are about 10^{10} years or perhaps a little older, and thus they must be representative of the oldest stars in the galaxy.

The H-R Diagrams of clusters contain few stars between the main-sequence segments and the giant stars to the right. Stars do not spend much time in this area, that is, with temperatures and luminosities intermediate between those on the main sequence and those of red giants, or more would be seen there. The evolution of a cluster up to an age of 4 billion years is shown in Figure 11-3b.

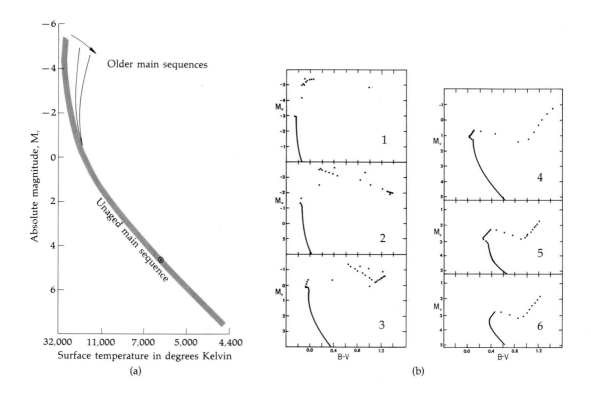

Figure 11-3. (a) Schematic evolution of the upper main sequence of a galactic cluster. As time passes, stars of lesser mass (lower absolute magnitude) move toward the red-giant region, and the main sequence moves away from the initial or *zero-age* main sequence. (b) Theoretically predicted color diagrams, showing stars on and beyond the main sequence for a cluster as it ages. Solid lines represent regions of many stars; dots indicate regions of fewer stars; the farther apart the dots, the fewer stars are present at a given position. Ages corresponding to these diagrams are (1) 22.5 million years; (2) 70 million years; (3) 250 million years; (4) 600 million years; (5) 2.25 billion years; (6) 4.0 billion years. (Courtesy of B. M. Schlesinger and the *Astrophysical Journal*, published by The University of Chicago Press. Copyright © 1969, The University of Chicago. All rights reserved.)

AN OUTLINE OF EVOLUTION

The cluster H-R Diagrams provide us with a fund of knowledge against which we can check our calculations and theories of stellar evolution. The theoretical and mathematical aspects of stellar evolution are extremely complex. The development of the large, fast electronic computer enabled astronomers and physicists to carry out the calculations necessary to investigate them. These calculations could then be compared with the observed H-R Diagrams of clusters for confirmation or improvement. In order to be valid, the calculations must place stars in the parts of the H-R Diagram where they are observed.

Star Formation

Stars are born as *protostars* in dark clouds of gas and dust in interstellar space. The theory of this first stage of evolution has been developed by Richard B. Larson of Yale University, among others. A typical scenario goes like this: We begin with a clump of cold interstellar matter at a temperature of only 10 or 20 K with relatively uniform density. The clump is contracting under the influence of its own gravitation, although an external influence, such as a shock wave from some kind of cosmic disturbance, may have caused the contraction to begin. As the clump contracts, the density gets higher at the center, and soon a much denser and warmer zone, the *core* or "embryo star" forms there. The protostar now consists of this core and the remaining infalling material or *envelope*. The luminosity of the core is supplied by gravitational contraction, as discussed earlier in connection with possible energy sources for the sun; the energy available increases as more mass is added to the core. The envelope is in turbulent motion, swirling randomly as it falls into the core. The protostar also is rotating, turning faster as it contracts. The rotation causes the envelope, which may have begun with a spherical shape, to gradually flatten at the poles. The flattening is greater for material distributed deep inside the envelope; right around the core, where the envelope is thickest, the material assumes a flat distribution that may resemble a thick pizza centered on the core. This structure is the *accretion disk*. Material from the envelope falls onto the accretion disk, where it spirals around the core in an ever-shrinking orbit until it drops into the core itself. The orbit gets smaller because the revolving material suffers friction in the thick, viscous disk, and hence loses energy.

As the contracting core gets hotter, it emits an increasing amount of radiation. At first, this is mostly in the form of infrared light, but as the temperature rises, it is in the form of visible light. The radiation is absorbed by the thick dust of the envelope, and the protostar as a whole gets hotter. As seen by an outside observer, no visible light escapes, and so the protostar is regarded as an infrared source. Over time, the wavelength of maximum radiation detected by the observer gets shorter, because the protostar is getting hotter. A very young protostar may emit its maximum energy at wavelengths of about 50 to 100 microns (one micron equals 10,000 Ångstroms, or one millionth of a meter); an older and hotter one may be most intense at 5 to 10 microns.

The core of a small protostar—one that becomes a star either less massive than the sun or at most a few times more massive—will eventually accrete all or nearly all of the envelope material. It will then be observable in visible light, providing that it's bright enough. However, shortly before this occurs, gaps may open in the envelope, and we may get a glimpse of the new star within, either directly or by reflection from a neighboring dust cloud. Certain curious types of young stars, known as *Herbig-Haro objects* and *T-Tauri stars*, may be examples of protostars in or near this stage of evolution.

Protostars much more massive than the sun—say with 5 or more solar masses—evolve differently. The core gets very hot and bright before much of the envelope has fallen into it. Nuclear reactions may even begin while the core is still accreting. As a result, the protostar core produces intense light. The pressure of this radiation drives the dust grains of the envelope outward, just as the pressure of sunlight drives the dust tail of a comet away from the sun. The dust grains collide with the gas atoms and molecules of the envelope, and thus tend to drag the envelope gas outward as well. Thus, the envelope stops falling onto the core, and the star that finally forms is less massive than the initial protostar. The more massive the initial protostar, the more pronounced the effect and the larger the fraction of mass that is lost in the escaping envelope. Theory predicts a certain mass above which stars will not form. This value is 60 solar masses, according to one calculation, and indeed stars more massive than this are extremely rare.

Another interesting effect occurs in the envelope of a protostar as the core gets hot. The dust of the inner envelope absorbs radiation from the core and increases in temperature. At about 1,500 K (depending on the makeup of the dust particles), the dust begins to evaporate. Thus, the inner envelope is cleansed of dust, while the outer, cooler portions retain it. Now the envelope dust (as distinct from the gas) has the form of a thick shell, rather than having a continuous increase in density toward the center. As a result, it resembles an insect's cocoon, and indeed such objects are called *cocoon stars*.

The preceding discussion is based on a great deal of theory and a relatively small number of confirming observations, since the necessary techniques of infrared and radio astronomy are still in the experimental stage. Further, it is often difficult to relate a particular observed object to a particular evolutionary stage predicted by the theory. The theory itself is controversial, with different astrophysicists deriving somewhat different results. However, some of the individual objects studied by infrared astronomers seem to fit with the theory, even though they don't necessarily prove it. For example, there are numerous objects in dark clouds of the Milky Way with infrared emission similar to that expected from protostars. An infrared source, called "BN" for its discoverers (the astronomers Eric Becklin and Gerry Neugebauer), seems to be a massive pre-main-sequence star that has begun driving its envelope outwards. It may possibly become visible to the naked eye 50,000 to 100,000 years from now, although it can't even be photographed with the 200-inch telescope at present. Another infrared source, known as MWC 349, has characteristics that suggest it may be a protostar with a well-formed accretion disk, although other possibilities exist.

Despite the observational difficulties, much work has been devoted to predicting the track of the protostar core on the H-R Diagram. The astrophysicist A. G. W. Cameron realized some years ago that at a certain stage the gradual contraction of the core would suddenly accelerate, and its rapid collapse would produce so much heat that the object would begin to shine as a star. A detailed theory that supports this idea was calculated in Japan by Chushiro Hayashi, who found that this was indeed the way in which evolution proceeds toward the main sequence (Figure 11-4a). If the contraction to the main sequence occurs in this way, some stars in very young clusters should be found above and to the right of the main sequence (above, because they are very large and hence fairly bright; to the right, because they are cool). This is apparently confirmed by the observations.

The Main Sequence and Beyond

The contraction to the main sequence continues until the central temperature reaches about 10×10^6 K, and nuclear reactions begin generating energy at a fairly stable rate. The basic result of nuclear burning on the main sequence is the conversion of hydrogen into helium. The star then spends most of its life on the main sequence, a conclusion in agreement with the large number of stars found there. The time spent on the main sequence is determined by the mass and luminosity of the particular star; bigger (more massive) stars burn brighter and faster and spend less time

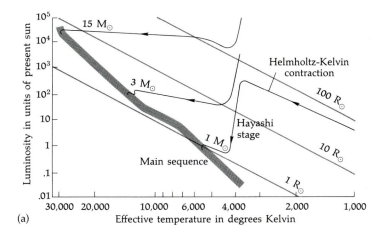

(a)

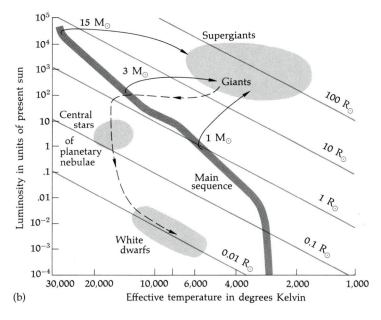

(b)

Figure 11-4. Schematic evolution of stars on the H-R Diagram, showing stars of 1, 3, and 15 solar masses. (a) The fairly well understood contraction to the main sequence beginning with the Helmholtz-Kelvin contraction and continuing through the Hayashi stage (vertical part of track). Lines of constant radius run diagonally from upper left to lower right and are shown for sizes of 1, 10, and 100 solar radii. (b) The less understood post-main-sequence evolution beginning with movement into the giant region and continuing back toward the main sequence. The evolution probably then proceeds along a sequence of faint, hot (blue) stars to the white dwarf region. (Adapted from *The Youngest Stars* by George Herbig. Copyright © 1967 by Scientific American, Inc. All rights reserved.)

on the main sequence than smaller stars. When the inner 10 percent of the star's mass has been converted from hydrogen to helium, the star moves rapidly up and to the right of the main sequence to the region of red giants. The evolution between the main sequence and giant regions is calculated to be rapid, and this agrees with the fact that few stars are found there on the cluster H-R Diagrams.

The departure of a star from the main sequence occurs because of the growing helium core and the resultant inhomogeneity. The star changes structurally in two ways. Its outer part expands, and we recognize the star as a giant. At the same time, the central region contracts and is heated to even higher temperatures. At temperatures of about 10^8 K, it is hot enough so that a nuclear reaction producing carbon from three helium nuclei takes place. This event in the life of a star is called the *helium flash;* it appears to stop the star (already a red giant) from moving farther upward and to the right in the H-R Diagram. Additional reactions involving helium and carbon can produce oxygen, neon, magnesium, other elements up to the atomic weight of iron, and some additional energy. The net effects of the helium burning and its consequences are to reverse the direction of evolution in the H-R Diagram and to send the star back toward the main sequence.

Some observational clues to the path of a star that is leaving the giant stage are given by the H-R Diagrams of globular clusters (Figure 11-2). Notice the horizontal band of stars reaching back toward the main sequence. Stars move along this band after the giant stage. The evolution of the star, as it moves back toward the main sequence, probably takes it through one or more unstable states, as we observe that some pulsating variable stars occur in this part of the H-R Diagram.

The internal structure of globular-cluster stars in three evolutionary stages—main-sequence star, red giant, and horizontal-branch star—is illustrated in Figures 11-5, 11-6, and 11-7, respectively. Note that these models include zones where radiative and convective energy transport dominate. Recall the explanation of such zones given in the discussion of the solar interior (Chapter 9).

Final Stages of Evolution

Regardless of the exact path taken (still a matter under investigation), many stars reach the white dwarf stage, where they are composed of electron degenerate matter. In this degenerate state the star cannot contract further, and therefore has no energy input from gravitational contraction. It also has no nuclear energy source because all the nuclear fuel

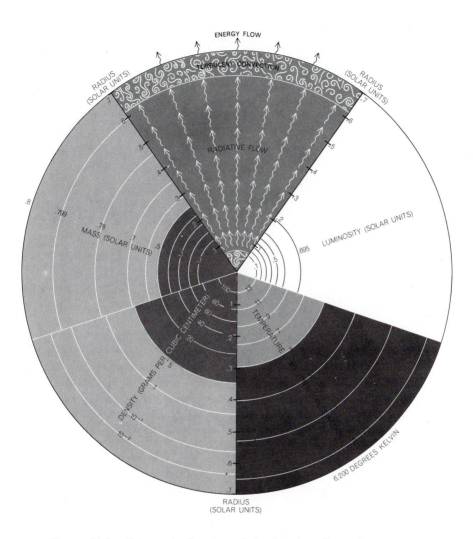

Figure 11-5. Schematic drawing of the interior of a main-sequence star. This model represents a typical globular cluster star, with radius smaller than the sun, mass of 0.8 M$_\odot$, luminosity of 0.7 L$_\odot$, and a surface temperature of 6,200 K. Energy production takes place in the star's core, and, in the very center of the core, matter is mixed by turbulent convection. Energy transport throughout most of the star is by radiative flow. Turbulent convection also occurs in the outer layer of the star. (From *Globular-Cluster Stars* by I. Iben, Jr. Copyright © 1970 by Scientific American, Inc. All rights reserved.)

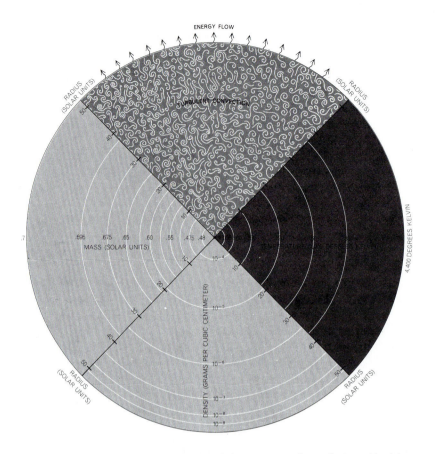

Figure 11-6. Schematic diagram of the interior of a red giant. In this case, the star began as a main-sequence star quite similar to the star in Figure 11-5, but with a slightly lower mass of 0.7 M$_\odot$. After exhausting the hydrogen in its center, the star has left the main sequence and now has a luminosity of nearly 1,000 L$_\odot$. The surface temperature has dropped to 4,400 K. The central helium core of the red giant cannot be properly shown to the same scale; it is represented by the dot at the center, although it is much hotter and denser and contains most of the mass. In particular, although the star has a radius of more than 50 R$_\odot$, 64 percent of its mass is located within a core whose radius is only 5 times that of the *earth*, but within which the temperature rises to 72 × 10⁶ K. The star's energy is produced by nuclear burning of hydrogen on the outskirts of the helium core, in a region only 3,000 km thick. In the region surrounding the core, which makes up the bulk of the red giant, the energy is transported by turbulent convection. In effect, almost the entire volume of the star is a giant convection zone. Contrast this with the interior of the star in Figure 11-5 and the sun (Figure 9-13), in which the convection zone is much smaller. (From *Globular-Cluster Stars* by I. Iben, Jr. Copyright © 1970 by Scientific American, Inc. All rights reserved.)

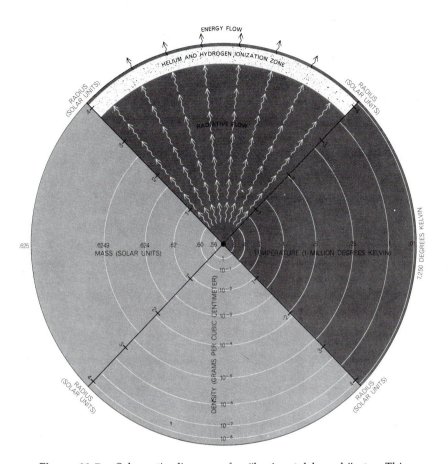

Figure 11-7. Schematic diagram of a "horizontal branch" star. This star was once a red giant similar to the star in Figure 11-6, but this particular model has a slightly smaller mass of 0.625 M$_\odot$. The luminosity is 43 L$_\odot$, the radius is about 4 R$_\odot$, and the surface temperature is 7,250 K. The central core of this star cannot be properly shown to the same scale; it is represented by the dot at the center. Over 80 percent of the star's mass is contained in a volume with radius 0.1 R$_\odot$. Within the core, energy production takes place in a helium burning core and in a hydrogen burning shell. Energy transport throughout the bulk of this star is by radiative flow. This flow tends to be blocked by the gases in the "helium and hydrogen ionization zone," which are only partially ionized. When this zone becomes deep enough, it induces periodic variations of brightness and radius. Such stars are observed as RR Lyrae variables. (From *Globular-Cluster Stars* by I. Iben, Jr. Copyright © 1970 by Scientific American, Inc. All rights reserved.)

in the hot central region has been used up. Hence, the white dwarfs can only radiate away their thermal energy. The star then gets cooler and cooler, fainter and fainter, and approaches death as a *black dwarf*. This is probably a common form of terminal evolution; however, such stars have not been observed, presumably because they are too faint. The path to the white dwarf region of the H-R Diagram may be along the sequence of faint blue stars (Figure 11-4b), which includes the central stars of *planetary nebulae* (Figure 12-3).

The stellar life story discussed above is appropriate for a star of the sun's mass and is illustrated schematically in Figure 11-4. Bear in mind that the exact path taken after the red giant stage is uncertain, but that the final stage is probably among the white dwarfs.

Other terminal phases are also possible, however. We have noted that white dwarfs with masses greater than about 1.4 solar masses probably cannot be formed. What happens to the stars with masses greater than 1.4 $M_\odot$? If they are to become white dwarfs, they must lose mass, and this can be accomplished in a variety of ways. A steady flow of gas (like the solar wind) might remove the mass, although the rate of loss would have to be millions of times greater than the present solar rate. There is a considerable amount of evidence for heavy mass loss in red giants. For example, some of the absorption lines in the spectra of these stars seem to be produced by matter surrounding the star in a shell far outside its photospheric surface, moving outward from the star.

Figure 11-8. Nova Herculis 1934. Photographs taken with the same exposure on March 10, 1934 (a), and May 6, 1935 (b), showing the large change in brightness as the nova faded. (Lick Observatory photographs.)

(a)   (b)

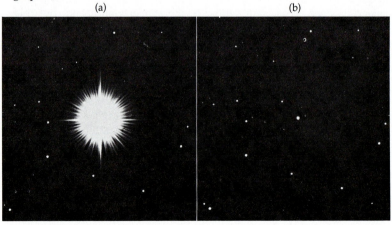

Figure 11-9.   Expanding nebulosity around Nova Persei 1901, photographed with the 200-inch telescope. (Courtesy of the Hale Observatories.)

It is also thought that the planetary nebulae found around certain
relatively small, hot stars represent shells of gas that were somehow
ejected from these stars at an earlier stage of evolution, when they were
red giants. Mass loss also occurs in explosive events, as observed in the
novae and *supernovae*.

The term *nova* is based on the Latin for *new*. When novae were first
observed, they were thought to be new stars that somehow appeared and
gradually faded away. Now we know that when a nova is discovered, it is
possible to find the *prenova* (the much fainter star that exploded and
produced the nova) on previously recorded photographs of the sky region where the nova is observed. After a few years, the star fades (Figure
11-8) to roughly its prenova brightness, indicating that the explosion did
not disrupt the bulk of the star. However, the spectrum of a typical nova
shortly after the explosion occurs does show that some matter is ejected.
The presence of emission lines shows that the star is surrounded by a hot
radiating gas,[1] and interpretation of the emission lines in terms of the
Doppler effect indicates that the ejected gas is expanding away from the
star at speeds of typically a few hundred to over 2,000 km (1,240 miles)/

---

[1] Absorption lines with large Doppler shifts are also found in novae spectra. The shifts are
toward the shorter wavelengths, indicating that the absorption lines are produced in the
part of the ejected gas that is moving toward us and is observed against the incandescent
background of the star.

sec. In some cases (see Figure 11-9) the nova was close enough to us that the expanding gas cloud eventually could be resolved on telescopic photographs. Recently, such clouds have been detected with a radio telescope shortly after the novae exploded. It is believed, but not known, that a given star that becomes a nova is likely to undergo this explosive process again and again, perhaps at intervals of several thousand years, each time losing a small percentage of its mass. In recent years, studies of the *old novae*, the stars that have already been novae, have revealed that they are actually binary stars. Theorists have concluded that the presence of a companion star induces the nova explosion. It appears that matter lost by one star falls onto the other star, producing a hot surface layer that explodes.

Supernovae are produced by much more catastrophic explosions than novae. A typical nova increases by a factor of 10,000 in brightness as it explodes and ejects a gas cloud, and at peak brightness it must be one of the brightest stars in its galaxy.[2] But a supernova is sometimes brighter than *all or most of the stars in its galaxy put together* (Figure 11-10), and there may be 10^{11} stars in a large spiral galaxy. Furthermore, it is possible that much of a supernova's energy is emitted in the form of ultraviolet light, so that its bolometric luminosity is even more impressive. A substantial fraction of the mass of the star is disrupted and flung into space by the explosion, resulting in an expanding gas cloud, the *supernova remnant*. The expansion velocities of remnants range from less than 1,000 km (620 miles)/sec to more than 10,000 km (6,200 miles)/sec. The Crab nebula (Chapter 16) is a supernova remnant, as is the Veil nebula (Plate 9). Supernova remnants are observed to be strong sources of radio emission (over 100 of them have been found in this way), and several of them have also been found to emit x-rays.

Theoretical computations suggest that a typical supernova explosion occurs when a star of several times the sun's mass has evolved so far that the atomic nuclei available as fuel for the star's energy-generating nuclear reactions are used up. There is no longer a powerful source of radiant energy inside the star, and thus the radiation pressure is not adequate to support the outer layers. The star collapses—what we really have is an *implosion*—and this catastrophic event releases an enormous amount of energy. This takes several forms, including the radiation emitted by the supernova and the kinetic energy of the matter in the expanding remnant that is ejected as a "splash" from the implosion. The implosion compresses the star enormously, perhaps to a diameter of 10 or 20 km, and

---

[2] Both novae and supernovae have been observed in our own and other galaxies, but unfortunately, no supernova has been seen in our own galaxy since telescopes were invented.

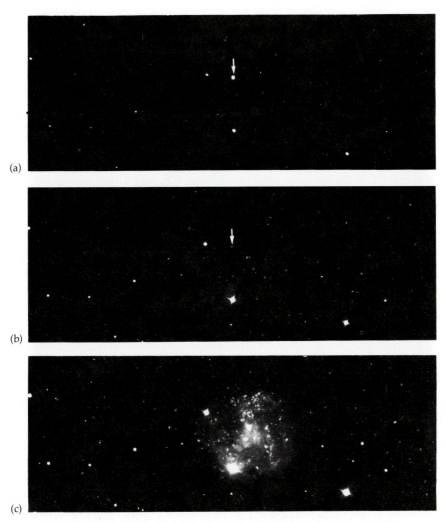

Figure 11-10. Sequence showing brightness changes of a supernova in the galaxy IC 4182. The supernova reached maximum brightness (a) in August, 1937, was much fainter in November, 1938 (b), and was invisible by January, 1942 (c). (Courtesy of the Hale Observatories.)

the resulting highly condensed object is called a *neutron star* (Figure 10-19). Until 1968, this discussion, and especially the suggested existence of neutron stars, was entirely hypothetical. Now it appears that the discovery of pulsars, as told in Chapter 16, has revealed the existence of neutron stars produced by supernova events. Supernovae probably also produce many of the cosmic ray particles that we observe at the earth, some of which, as mentioned previously, may be responsible for biological mutations.

Prior to the supernova event, nuclear reactions in the stellar interior probably produce most of the elements up to iron (atomic weight 56). In the course of the supernova implosion, the interior temperature may reach 10^9 K, and nuclear reactions take place that produce the heavier elements. *The ejection of a part of the supernova into space is believed to represent the major source of the heavy elements that we find on the earth, in the sun, and elsewhere in the universe.* (We recall from Chapter 1 that it is generally believed that the explosion of the Primeval Fireball at the birth of the universe only produced hydrogen and perhaps some helium.)

We have discussed the white dwarfs, black dwarfs, and neutron stars as final states of stellar evolution. It is also believed by some physicists that some massive stars can implode beyond the neutron star point (the central density of a neutron star is about 10^{14} to 10^{15} grams per cubic centimeter). The gravity of these hypothetical highly condensed stars would be powerful enough to prevent light waves from escaping from them. This property has led to the term *black holes.*

STELLAR EVOLUTION AND THE INTERSTELLAR MEDIUM

We know that stars are being formed at present and that many stars have already lived out their lives. Some of the original matter of these stars is still locked up in them in their final configuration, whatever it might be. Another part of the original stellar material, as well as heavier elements produced by nuclear reactions inside the stars, has been returned to the interstellar gas and dust by mass loss processes such as explosions, stellar winds, and the ejection of gas shells to produce planetary nebulae.

Nuclear reactions that power stars proceed in the direction whereby elements of low atomic weight are "burned" to form an "ash" consisting of elements of heavier atomic weight. The elements found in a star are conveniently divided into three categories: (1) hydrogen (by far the most common substance in the universe), (2) helium, and (3) all of the other elements, which astronomers group under the heading of "heavy elements." Recall the basic nuclear reactions that power the stars. On the main sequence, hydrogen is converted into helium; in giants, helium is converted into carbon, and the combination of carbon and helium can produce elements as heavy as iron. Supernova explosions produce even heavier elements. In no case do we find evidence for the net amount of heavy elements being reduced by stellar evolution.

We believe that stars are formed from the gas and dust of the interstellar medium, that they convert some of this gas to heavier elements in their interiors, and that they return some of this material to the medium through mass loss. Thus the interstellar medium is a reservoir of material for star formation where heavy element abundance is continually being enriched. Hence, new stars formed from this material will begin their lives with a higher heavy element abundance than did stars formed at an earlier time. The sun is a good example of this; its central temperature is not high enough to produce much in the way of heavy elements. Even if it did, the heavy elements would probably not come to the surface. Yet, iron is observed in the spectrum of the sun and uranium is found on the earth; we believe they formed from the same material as the sun. This evidence agrees with our idea that the material from which the sun and the planets were formed *had already been inside stars* where the heavy elements were produced by nuclear processes.

A competing hypothesis might be that a cosmic mixture including heavy elements[3] was created at the beginning of the universe. But if this were true, then all stars should show similar abundances of the elements. However, this is contradicted by the observation (see Figure 10-8) that some stars are deficient in heavy elements, compared, for example, to the sun. A suitable index of heavy element content is the percentage of the total mass of a star's photosphere that is made up of heavy elements, and this quantity varies over more than a factor of 10, from roughly 0.3 percent for an old, heavy-element-poor star, to about 4 percent for a young, heavy-element-rich star. Our sun is intermediate, with a heavy element abundance of about 1.5 percent. It is estimated that the sun is a third-generation star; that is, the material from which the sun was originally formed has been inside about two previous stars. A sense of cosmic perspective may result from the knowledge that the iron in a skillet being used to prepare our food was processed from lighter atoms long ago inside some stars, perhaps even in a supernova. The same is true of many of the atoms in food and indeed in our own bodies.

Stars were divided into two *populations* by the astronomer Walter Baade.[4] Stars formed relatively recently are young in years but are made

---

[3]Elements heavier than helium are sometimes called *metals* by astronomers, as in the caption for Figure 10-8. This usage of the term is not restricted to actual metallic elements as defined by chemists.

[4]Further details of Baade's work are given in Chapter 13.

of highly processed material with a high heavy-element abundance; these are called *Population I* stars. Most stars near the sun are in Population I, as are the stars in the open clusters. *Population II* stars are found especially toward the central region of the galaxy and entirely make up the globular clusters. These old stars were formed long ago out of relatively unprocessed material that was poor in heavy elements. The significance of the classification of stars into populations will become clear as we discuss the structure of the galaxy in Chapter 12, where a more useful breakdown into three populations, depending on location, is presented.

ORIGIN OF THE SUN AND PLANETS

The evidence that we have discussed suggests rather strongly that the elements heavier than helium were formed in the stars. The very brightest stars now seen were formed within the last few million years, a very short time on the cosmic clock. We would now like to discuss where the birth of stars takes place, under what circumstances, and by what processes. The prevailing view is that planetary systems like our own solar system are formed at the same time and from the same material as their central stars. Thus, the properties of our solar system are probably relevant to the study of star formation. Recall that the orbital motions of the planets and most satellites are in the same direction as the rotation of the sun. In addition, most of the planetary and satellite orbits are roughly in the same plane, and most of the planetary orbits are nearly circular. Our birth hypothesis must explain these properties.

At present, the leading theory of the origin of the sun and planets is based on the *nebular hypothesis*, first advanced in 1755 by the German philosopher Immanuel Kant, and independently developed by the French mathematician Pierre Laplace in 1796. Since then it has been in and out of favor as new facts and ideas have emerged. According to the current version of the nebular hypothesis (Figure 11-13), the solar system formed from a large, slowly turning, nearly spherical cloud of gas and dust—the *solar nebula*. In terms of modern astrophysics, the solar nebula is a clump of interstellar matter in which a protostar formed. The theory resembles that discussed for protostar evolution. However, in this case, we are especially interested in the accretion disk, where the planets form, rather than just the core, where the star forms.

As the solar nebula contracted under its own gravitational force, it rotated faster and faster. This spin up was a consequence of the law of *conservation of angular momentum*. The angular momentum of a specific particle is the product of its mass, its distance from the axis of rotation, and its rotation speed (Figure 11-11). This definition can be generalized

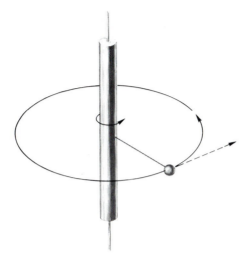

Figure 11-11. Illustration of the definition of *angular momentum* through the example of a heavy steel ball fastened to a central rod with a very light rod. Turning the central rod causes the ball to whirl about it. The ball's angular momentum is the product of the distance from the rotation axis to the ball, the mass of the ball, and the velocity of the ball.

mathematically to describe the angular momentum of any system, whether it be a person on a spinning piano stool (Figure 11-12), the rotation of the earth on its axis, or the motion of the planets around the center of mass of the solar system. The law of conservation of angular momentum states that the angular momentum of an isolated system does not change. Thus, as the nebula contracted (Figure 11-13), the average distance of the gas particles from the axis decreased, and this conservation law required that their average speed around the axis increase.

As the nebula spun faster, it became progressively more oblate and eventually became a disk. (It is a well-known property of a rotating gas or fluid to flatten out in this way;[5] it can be regarded as an effect of centrifugal force, which is greatest at the equator and smallest near the poles of a rotating object.) The disk broke up into rings of matter, in which large clumps gradually condensed to form the planets.

[5]The planet Jupiter is oblate because of its rapid rotation. The earth is also slightly oblate—we mentioned its equatorial bulge (caused by rotation) in Chapter 3.

Figure 11-12. Illustration of the conservation of angular momentum. Contraction causes the speed of rotation to increase. (From *The New College Physics* by Albert V. Baez. W. H. Freeman and Company. Copyright © 1967.)

There is still a great deal that we do not know. Even if the theory is correct in its general outline, there is very little agreement among astronomers as to exactly how the planets formed from the disk. However, it does appear that the theory is in reasonable accord with most scientists' thinking about how *stars* are formed. The central portion of the condensing nebula warmed as it contracted; gravitational potential energy of the matter contracting or falling toward the center was converted to heat, as we mentioned in the discussion of Helmholtz-Kelvin contraction (Chapter 9). When the contraction proceeded to the point that the central region

Figure 11-13. Schematic outline of the nebular hypothesis. First, a cloud (a) condenses to form a protostar (b), with contracting envelope, accretion disk, and hot core. The envelope material falls into the disk and then to the core, which evolves toward an initial main-sequence star (c). At the same time, rings of material form. While the star settles into its life as a main-sequence star (d), the rings of material condense into planets.

was hot enough for nuclear reactions to occur, the sun began to shine and a new star joined the main sequence.

A serious objection that has been raised to this theory is that the calculations of the formation of the sun in this manner show that it should have most of the angular momentum in the solar system. But when we add up the angular momenta of the various objects in the system, we find that the planets, primarily Jupiter and Saturn, have the vast majority of the angular momentum, whereas the sun (which rotates only once every 27 days) has only 2 percent of the total.

One likely answer to the above objection is that the sun may have lost much of its angular momentum since it was formed. As the ionized matter of the solar wind streams away from the sun, it is still attached to the sun in a sense by the lines of force of the solar magnetic field, which behave very much like elastic strings. As the material of the solar wind moves out from the sun, its distance from the axis increases, and hence (since angular momentum must be conserved) the sun spins less rapidly. The rate at which this effect presently occurs can be measured, and it turns out to be enough to slow the solar rotation rate down by a factor of 2 in about 2×10^9 years. As we believe the solar system is only 5×10^9 years old, this would not be enough to account for the fact that the sun has only 2 percent of the angular momentum in the solar system at the present time, and so we are obliged to study the early history of the sun to search for an additional way in which mass (and thus angular momentum) could have been shed. One possibility under study is that the solar wind was much more powerful in the early history of the solar system.

If any significant amount of solar-wind mass loss has been going on throughout the lifetime of the sun, other stars similar to the sun that have been formed more recently should be rotating faster. Younger stars of solar type (as identified by their spectra) can be found in the clusters dated by the turnoff method. Rotation speeds can be inferred from the widths of the line profiles. Suppose we are looking at a star at right angles to the axis of rotation. As the star rotates, one-half of the star is going away from us while the other half is coming toward us. The absorption lines due to atoms in the receding half are shifted toward the red (longer wavelengths), and the absorption lines in the approaching half are shifted toward the blue (shorter wavelengths) by the Doppler effect. The net effect on a line from the entire star is to broaden it; the faster a star rotates, the wider is a particular line, as illustrated in Figure 11-14. Thus, measurements of the line widths of stars tell us their rotation speeds. Of course, we do not observe all stars at right angles to their axes of rotation, and a statistical correction for foreshortening must be provided. The basic

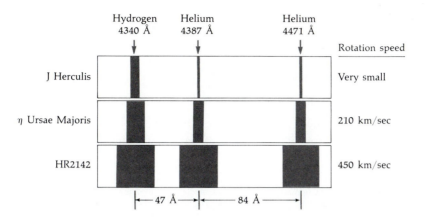

Figure 11-14. Schematic diagram of absorption line broadening caused by the different rotation rates of three stars. The widths of the hydrogen and helium lines in the slow rotator J Herculis are caused by processes in the star's atmosphere, but the larger line widths in η Ursae Majoris and HR 2142 are primarily due to the rapid rotation of these two stars. (After a drawing by I. S. Shklovskii and C. Sagan based on spectra taken by O. Struve and G. A. Shajn.)

situation, however, remains the same. Such an investigation was carried out at the Hale Observatories in 1967 by Robert Kraft. Kraft found that solar-type stars seem to rotate more slowly as they age. Thus, the nebular or Kant-Laplace hypothesis, augmented by appreciable mass and angular momentum loss from the primary star, appears reasonable. Remember, however, that we are still ignorant of the detailed processes that produce planets.

Some other hypotheses concerning the origin of the solar system, much in vogue a few decades ago, were based on a collision of the sun with another star, or at least a very close approach. These *collisional and tidal hypotheses* do not bear on star formation, but rather only on the formation of the planets. In the collision or close approach, filaments of gas could be knocked off or pulled off the sun; the filaments then cool to form the planets. The resulting planetary orbits would be highly elongated, which is contrary to the observations. In addition, the hot gas pulled out of the sun might well have dissipated, rather than condensing into planets. Further, another objection seems so fatal as to throw out all collision-type hypotheses: collisions between stars are extremely rare. The space in our galaxy is mostly exactly that—just empty space. Consider the dimensions

involved. In units of the solar radius, the earth is 215 $R_\odot$ from the sun. The nearest star is about 260,000 a.u. from the sun, or roughly 5×10^7 $R_\odot$. If we place the sun inside a cube which has a dimension equal to the distance to the nearest star, the volume of the cube is about 10^{23} times the volume of the sun. Detailed calculations indicate that the number of collisions between the 10^{11} stars in our galaxy over the past 5 billion years would be about 10 (excluding the central nucleus of the galaxy, which is obscured by dust, where the stars probably are much closer together). Perhaps the sun suffered one of these collisions, and hence the solar system is a product of a very rare event. But then we are at a loss to explain how the planets condensed and attained nearly circular orbits. The best current view holds that planets are formed as a natural by-product of star formation.

Figure 11-15 summarizes in very schematic fashion the discussions of this chapter as they bear on the evolution of a star like the sun.

POSSIBLE EXPLANATION OF BODE'S LAW— EVOLUTION OF THE SOLAR SYSTEM

Are the orbits of the planets the same today as they were in the early history of the solar system? Is Bode's Law a numerical curiosity, or is it a predictable consequence of physical laws? Many astronomers and mathematicians have studied this problem. The orbital spacings of the inner satellites of Jupiter, Saturn, and Uranus all obey mathematical relationships similar to Bode's Law. This common pattern suggests that the distances of the planets and the asteroid belt from the sun did not occur accidentally.

One of the simplest ideas yet presented to account for Bode's Law is the *theory of dynamical relaxation*. According to this theory, the orbits of the planets were originally very different from what they are today. Remember that the orbit of Halley's comet around the sun was affected by Jupiter when it passed near that planet (Chapter 7). In the same way, in

Figure 11-15. Schematic outline for the evolution of a star of 1.2 solar masses. First, a cloud (1) condenses to form a star and protoplanets (2). The main-sequence stage (3) ends with the star's expansion (4) to become a red giant, with the successive destruction of the planets (5). Later, the star may become a pulsating variable (6), and at some point it may eject a shell of gas that becomes a planetary nebula (7), which eventually dissipates into space (8). The star finally collapses into a white dwarf (9). (After *Life Outside the Solar System* by Su-Shu Huang. Copyright © 1960 by Scientific American, Inc. All rights reserved.)

the early history of the solar system, Jupiter and the other planets may have perturbed each other, changing their orbits. This process went on until the planets were set into orbits arranged in such a way that perturbations no longer seriously disturbed them. If this were not true, the planets should still be causing major perturbations in each other's orbit. Since measurements of the planetary motions nowadays show that the orbits are changing only very slowly (in fact, the changes can be ignored for most purposes), the dynamical relaxation process must be nearly complete.

In 1969, J. G. Hills (then a graduate student at the University of Michigan) did numerical experiments to test this theoretical explanation of Bode's Law. He took eleven imaginary solar systems with different sets of planetary masses and with the sizes and shapes of the elliptical orbits picked at random, and he used a computer to calculate how their orbits would change over long periods of time as the planets perturbed each other. He found that these solar systems of quite different initial properties did evolve toward stable orbital conditions, with spacings that obeyed mathematical expressions similar to Bode's Law. Furthermore, he found that this process would take place within the time available since our solar system was formed. In fact, the Jovian planets may have reached approximately their present orbits within only a million years of their formation, according to these computations. The theory of dynamical relaxation has not been fully accepted, however. Although the basic idea may be correct, it seems that additional effects, such as magnetic forces or the presence of relatively dense gas and dust left over from the solar nebula, may have played a role.

The Milky Way Galaxy

Looking through his telescope, Galileo discovered that the diffuse band of light called the Milky Way consists of a vast number of stars not individually perceived by the unaided eye. Today we know that the sun is located in a large, relatively flat, spiral-shaped system (figure 12-1a) that contains perhaps 100 billion stars, the open and globular clusters, and large amounts of gas and dust arranged in clouds and in a general medium between the stars. This system is referred to as the *Milky Way galaxy*. When we look out from the earth through the central plane of the galaxy, we see the regions where most of these stars are located—the Milky Way in its original sense. On the other hand, when we look at right angles to the Milky Way in the sky, we see relatively few stars. The Milky Way galaxy is one of a great many large star systems that are now recognized as a principal type of structure in the universe. This chapter describes how, through optical and radio astronomy observations, the size, shape, and rotational motion of the Milky Way galaxy have been determined. The appearance and properties of other galaxies in the universe are discussed in the next chapter.

It would be desirable, in trying to understand the structure of the galaxy, to be able to describe the positions of the stars in terms of their directions and distances from its center. However, finding the exact position and direction of the galactic center from the solar system has been a difficult problem. Therefore, the *galactic coordinate system* (see Figure 12-1b) that we actually use is centered for convenience on the sun. The central

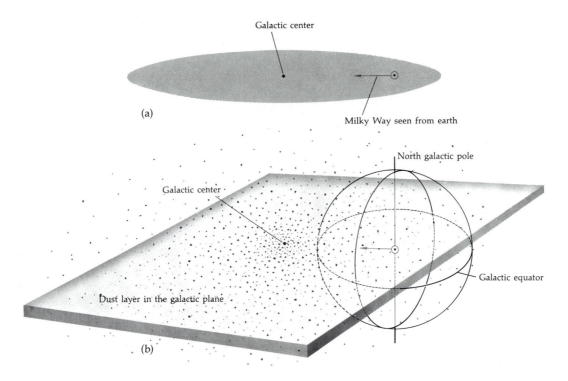

Figure 12-1.   Schematic diagram of our galaxy. (a) The sun's position (dot in small circle) with respect to the stars in the galactic disk (gray area) and the geometrical effect responsible for the appearance of the Milky Way. (b) The galactic coordinate system and the sun's relation to the dust layer and the cloud of globular clusters (dots).

plane *(galactic plane)* of the Milky Way intersects the celestial sphere along a circle called the *galactic equator. Galactic latitude* is measured in degrees north or south of this equator. *Galactic longitude* is measured in degrees to the east of the zero point, which is taken as the most likely direction to the center of the galaxy, as agreed upon by astronomers. The zero point lies in the constellation Sagittarius, at an unusual complex of radio and infrared emission known as Sagittarius A.

THE SIZE OF THE GALAXY

A monumental attempt to determine the dimensions of the galactic system was undertaken early in this century by the Dutch astronomer

Jacobus Kapteyn (1851–1922). He was a pioneer of modern statistical astronomy and attempted to compute the *star density* (number of stars in a unit volume) as a function of distance from the sun. The relative amounts of stars of different luminosities in the sample under study were assumed to be the same as the relative amounts among the nearby stars, which can be studied directly. Knowing the relative numbers of stars with different intrinsic brightnesses, Kapteyn was able to determine the distribution that they must have in space in order to produce the distribution and apparent brightnesses of the stars as seen in the sky from the earth. His galaxy model is often called the *Kapteyn universe;* it is a small, disk-shaped system with the sun at its center (Figure 12-2). In this model, the star density drops to half the value near the sun in 250 parsecs at right angles to the Milky Way and in 800 parsecs in the plane of the Milky Way. A density drop to one-sixteenth the solar neighborhood value was found at 660 parsecs in the direction at right angles, and at 3,500 parsecs in the plane.

A rather different view of the galaxy was developed by Harlow Shapley at the Harvard College Observatory. Distances were determined from observations of pulsating variable stars in the globular clusters (RR Lyrae stars). Shapley found (as shown in Figure 12-1b) that the globular clusters are distributed equally above and below the galactic plane, although they did not seem to occur near the plane. The surprise was that most of them were found in one-half of the sky, with their distribution centered about a point on the Milky Way in the direction of the constellation Sagittarius. This observation could be explained by assuming that the globular clusters were concentrated on the center of the galaxy, not on the sun. (The

Figure 12-2. The Kapteyn universe shown in cross-section perpendicular to the galactic plane. The inner and outer ellipses mark the distances at which the density of the stars (the dots) have decreased to, respectively, one-half and one-sixteenth the central density found near the sun.

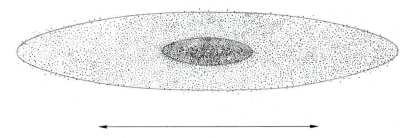

4 kiloparsecs

Kapteyn universe was centered on the sun, so if Kapteyn's model were correct, the globular clusters would be centered on a point off to the side of the galaxy.) If Shapley were correct, the sun was far out toward the edge of the galaxy, and thus the earth was not only not at the center of the solar system, but the solar system was not at the center of the galaxy. All subsequent work on the structure and dimensions of the galaxy has essentially confirmed Shapley's basic deductions, although the presently accepted distance (10,000 parsecs = 10 kiloparsecs) to the galactic center is slightly smaller than the value that Shapley deduced (14 kiloparsecs) from his globular cluster studies.

Why was Shapley right and Kapteyn wrong? Both astronomers, according to the custom of the time, had assumed (at least tacitly) that no absorbing material existed between the sun and the objects under study. If present, such material produces an additional dimming of the star besides the normal decrease in apparent brightness with distance. Distances estimated assuming no absorption are always larger than the actual distances. This discrepancy increases with the amount of absorption; that is,

Figure 12-3. The Ring nebula, a planetary nebula in the constellation Lyra. (Courtesy of the Hale Observatories.)

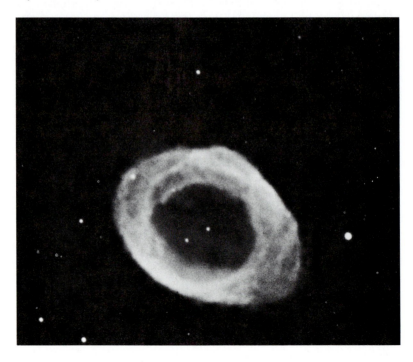

the error is larger for the more distant stars. We now know that a great deal of absorbing material is located in the plane of the galaxy. Kapteyn's work concerned objects that were embedded in the absorbing medium; he was limited by the circumstance that an observer in a fog sees only a very small "universe" around him. On the other hand, many globular clusters are found well above and below the galactic plane, and thus the light that we receive from them is relatively free from the effects of absorption. Recall Shapley's result that globular clusters were not found near the plane of the galaxy; in fact, we now know that many of them do occur there, but they are obscured by the absorbing material and were not studied by Shapley. Thus, the absorption problem was devastating to Kapteyn's studies and relatively unimportant to Shapley's results.

INTERSTELLAR MEDIUM

As we saw in the last section, an understanding of the structure of the galaxy requires an appreciation of the gas and dust between the stars. Gas and dust can be thought of as coming in two varieties—obvious and devious.

The obvious examples of interstellar material have been known for years. They include the gas masses ejected from stars, such as the planetary nebulae (Figure 12-3), obvious supernova remnants (Plate 10), not so obvious supernova remnants (Plate 9), and also the H II regions[1] (Figures 12-4 and 12-5). The term *nebula* is now reserved for the clouds of gas and dust that we find in interstellar space. However, in older books you will find it used also in describing galaxies outside our Milky Way system; the usage dates back to the time when the nature of the other galaxies as systems of both stars and gas like the Milky Way was not understood.

The spectra of bright nebulae, such as the planetaries, supernova remnants, and H II regions, show emission lines as we would expect for a hot gas seen against the dark background of space. However, some bright nebulae *lack* emission lines and have spectra rather like those of stars: absorption lines and continuous light. In addition, some H II regions exhibit these continuous spectra in addition to their emission lines. These observations are explained by the presence of dust in the nebulae, which scatters the light from adjacent stars.

The identification of the atoms responsible for some of the most intense emission lines in the spectra of bright nebulae proved to be quite

---

[1] So called because much of the gas in such a nebula is in the form of ionized hydrogen, which is abbreviated as H II to distinguish it from H I, or neutral hydrogen atoms. Neutral hydrogen in molecular form is called H_2.

Figure 12-4. The Eagle nebula, M16, a bright H II region in the constellation Serpens. (The Kitt Peak National Observatory.)

Figure 12-5. The Orion nebula, a bright H II region. (Lick Observatory photograph.)

difficult. At one time, the lines were ascribed to a hypothetical element, *nebulium*, just as some solar emission lines of unknown origin were ascribed to the then hypothetical helium. However, helium turned out to be a real element that had not yet been found on earth, whereas the nebulium lines have now been identified as the product of atoms of well-known elements in an unusual set of conditions. For example, Ira S. Bowen of the Hale Observatories showed that the intense nebulium lines near wavelength 5,000 Angstroms were due to doubly ionized[2] oxygen atoms, and that such atoms only produce light at these wavelengths when the gas density is very low (as in a nebula). Spectral lines of this type are known as *forbidden lines* because they cannot occur in normal circumstances.

The source of the energy that causes an H II region to shine is the ultraviolet light of stars that are nearby or actually located within the nebula. Photons corresponding to light of wavelengths less than 912 Angstroms have enough energy to knock the electron out of a hydrogen atom. Thus, the light of these wavelengths from a star causes the hydrogen that exists in space around the star to be ionized. The free electrons released in this way from hydrogen atoms occasionally collide with other particles, such as the doubly ionized oxygen atoms. In such a collision some of the kinetic energy of the free electron is transferred to the outer electron of the oxygen ion, raising it to a higher atomic energy level. When the latter electron later drops down to a lower level, a photon is released by the ion. This is the way in which the "nebulium lines" are produced—by *collisional excitation.* The general process by which the energy contained in an ultraviolet photon emitted by the star is eventually converted (at least partially) to light of visible wavelengths by the atoms in the nebula is called *fluorescence.* Occasionally, a free electron, produced when an ultraviolet photon ionized a hydrogen atom, meets a proton and combines with it to form a neutral hydrogen atom again. In this case, the electron may be in one of the upper energy levels of the atom, and when it drops to a lower level, one of the ordinary visible emission lines of hydrogen is emitted. This special case of fluorescence is called *recombination radiation.* In a common fluorescent light bulb, the electrical discharge (electrons) flowing through the bulb causes a vapor to emit ultraviolet light, and the ultraviolet photons strike the atoms of the bulb's inside coating, which reradiate part of the energy in the form of visible light.

It can be seen from our discussion of Planck's Law in Chapter 5 that the hotter stars will produce more ultraviolet light than the cooler stars.

---

[2]Oxygen atoms that have lost two of their eight electrons.

Therefore, the hotter stars will be the most important exciters of the H II regions. In fact, the detailed theory of this process, as worked out by the Danish astronomer Bengt Strömgren, shows that only stars of the two hottest spectral classes (O and B) emit enough ultraviolet light to produce noticeable H II regions. Several stars may be involved in the excitation of a nebula, but an idea of the effect of a given star can be gained by considering the size of the H II region it would produce by itself in a typical part of the galactic plane where the interstellar hydrogen density is about one hydrogen atom per cubic centimeter. A main-sequence O5 star with a photospheric temperature of about 56,000 K theoretically can ionize a region of about 200 parsecs in diameter, but a B1 star with a temperature of about 18,000 K will produce an H II region of less than 13 parsecs in diameter. Where the interstellar gas density is much higher (as in the central parts of the Orion nebula, where the density reaches 15,000 atoms per cubic centimeter), the size of the H II region produced by a given star will be much smaller. This effect is due to the fact that at the higher gas densities, an ultraviolet photon from the star will travel a smaller distance on the average before colliding with a hydrogen atom and ionizing it. The diameter of the Orion nebula is less than 5 parsecs.

The interstellar material appears in another obvious way in the characteristic dark lanes seen in the Milky Way (Figure 12-6) and in *dark nebulae* such as the Horsehead nebula (Figure 12-7). In this case, the light of any stars that may be adjacent to the interstellar clouds is inadequate to cause them to shine significantly by fluorescence or even by scattering. The interstellar matter is detected by virtue of its absorption of the light from background sources, whether they be stars (as in the case of the dark lanes in the Milky Way) or a bright nebula (such as the one that lies behind the Horsehead nebula). In some cases the presence of a dark nebula is not certain, and the evidence is analyzed in terms of star counts on photographs. If the counts of the number of stars per square minute of arc are much lower in the suspect region than is statistically probable (as judged from star counts in adjacent areas), the evidence suggests that a dark nebula may be present. This can be checked, for example, by studying the spectra of the few stars seen in the direction of the suspect nebula; if they show stronger indications of interstellar absorption (as discussed below) than do similar stars outside the region, the presence of the dark nebula is confirmed.

The "devious" form of interstellar matter that we referred to is a thin medium of gas and dust which pervades interstellar space and is not sufficiently concentrated to constitute a nebula. Until about 50 years ago it was generally thought that no such *interstellar medium* existed. It now ap-

350

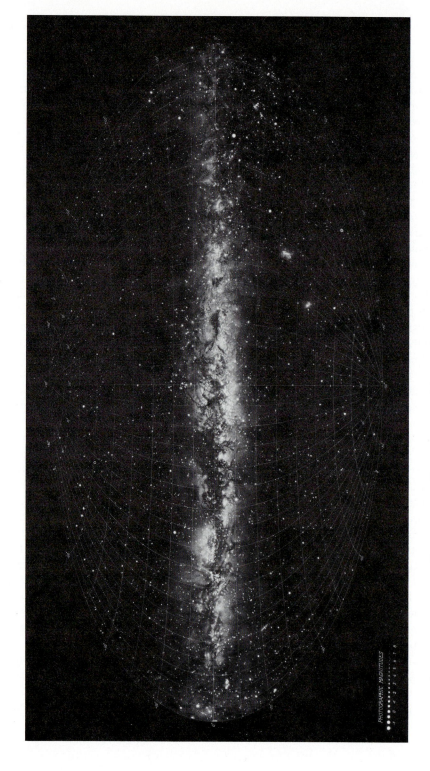

Figure 12-6. Chart-painting of the Milky Way. (Observatorium Lund, Sweden.)

Figure 12-7. The Horsehead nebula, a dark cloud seen against the bright background of an H II region. (Courtesy of the Hale Observatories.)

pears that there is an interstellar medium and that its dust component can be detected through the reddening of stars at great distances. If we study two groups of stars that are nearly identical, one group near the sun and one far away, we find that the colors of the stars in the distant group are redder than those of similar stars in the group near the sun. The effect is sometimes so great that an intrinsically blue star looks red to us because of the great amount of dust through which we view it. On the other hand, the absorption lines produced in the photospheric spectrum of a star are not affected by the dust, so we can use them to estimate the stellar temperatures and hence the *intrinsic colors* of the stars. Comparison of intrinsic and apparent colors tells how much dust is present between the observer and the star.

The reddening effect of the interstellar dust is somewhat analogous to the change in the color of the sun at sunset. The molecules in the earth's atmosphere scatter blue light better than they do red light; hence, the blue light is scattered out of the line of sight while the red comes through, and the light is reddened, as illustrated by Figure 8-11. The same thing happens to starlight, but the scattering particles in this case are not air molecules but the much larger dust particles with diameters of roughly 10^{-5} cm. We are ignorant of the composition of the interstellar dust grains, although silicate rock, iron, ice, and graphite, singly and in various combinations, have all been suggested.

Not only does the interstellar dust redden the light from a star, but it also reduces the total amount of light received. The amount is different in different parts of the galaxy, but a typical dust absorption in the galactic plane is about one magnitude per kiloparsec at the wavelengths of yellow light. The existence of absorption by a general interstellar medium (as distinct from the obvious nebulae) was finally established in 1930 on the basis of detailed studies of clusters by the American astronomer Robert Trumpler. Trumpler's work can be understood in terms of two different methods for deriving the distances of open clusters:

Method A

Adopt an average diameter in parsecs for open clusters. Measure the angular size in degrees of a cluster. Use simple geometry to calculate the distance at which a cluster of a given number of parsecs in diameter will subtend an angle equal to the measured angular diameter (Figure 12-8).

Method B

This is the *cluster parallax* technique (Chapter 11). Comparison of H-R Diagrams of the cluster under study and of another cluster of known distance tells us the absolute magnitudes of the stars in the first cluster. Comparison of the absolute magnitudes and observed apparent magnitudes then gives the distance.

Note that Method A is not directly affected by interstellar absorption, since we measure the size of the cluster, not the brightness of its stars. However, Method B involves the apparent magnitudes, and therefore is affected by absorption. In particular, absorption makes the apparent magnitude fainter, and thus makes the cluster seem farther away. In fact, Method B gave larger distances for the clusters than Method A. If Method A were correct, then interstellar matter was dimming the light of the cluster stars. The other possible conclusion was that Method B gave the correct distance, but this implied that the more distant star clusters had bigger diameters than the nearby ones. It seemed more reasonable to believe that objects of similar appearance have a similar nature, including similar

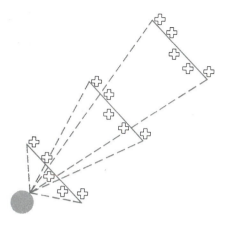

Figure 12-8. A cluster of a given linear size (represented by the solid line) subtends smaller angles at greater distances from the earth.

size, and so Trumpler concluded that interstellar absorption did indeed occur.

In 1947, W. A. Hiltner and John Hall discovered that besides dimming and absorbing the light of the stars, the interstellar dust grains also cause a slight polarization of the light.

The dust responsible for the effects described above actually constitutes only about 1 percent of the interstellar material. The other 99 percent is gas and was detected in 1904, although it was some time before the result was generally accepted. Observations of a binary star showed the usual lines that Doppler-shifted to red and to blue as the stars orbited around their center of mass. However, there were absorption lines of ionized calcium that did *not* shift as the stars moved in their orbits. It was natural to assume that these stationary absorption lines were caused by a cloud of atoms between the earth and the binary star. Subsequently, the same lines were observed in the spectra of many stars, and the presence of the interstellar gas was proven. A key point was that the same lines could be found in the spectra of both hot and cool stars; the spectra of such stars should be very different, so the lines that were common to all of them were most likely due to material outside the stars. Many substances have now been identified through their interstellar absorption features in stellar spectra, including neutral and ionized calcium, neutral sodium, neutral potassium, ionized titanium, and the molecules CN and CH. Several other strong absorption features in stellar spectra are also believed to be due to interstellar matter, but the substances that produce these absorptions have not been definitely identified.

RADIO ASTRONOMY AND THE INTERSTELLAR GAS

The gas distributed throughout the galaxy can be studied with radio telescopes. Unlike visible light, radio waves can penetrate through the interstellar dust without suffering serious absorption. Thus, we can "see" farther with radio waves. Further, the physical state of the gas between the stars is such that most of the atoms emit significant energy only in the radio and infrared wavelengths.

Radio radiation from the galaxy was detected and identified in the early 1930s by Karl Jansky in Holmdel, New Jersey (Figure 12-9), while investigating radio static for the Bell Telephone Laboratories. The source of static that he discovered was observed on a wavelength of 15 meters and crossed the meridian 4 minutes earlier each day, thus indicating its celestial nature by keeping sidereal time. Eventually, the position of the source was established with some certainty, and it was found to coincide with the center of the galaxy.

This work did not attract much attention, although in retrospect it should have. However, it did interest radio engineer Grote Reber of Wheaton, Illinois. Reber became the first radio astronomer after Jansky,

Figure 12-9. The founder of radio astronomy, Karl Jansky, and his antenna, photographed in the 1930s. (Courtesy of the Bell Telephone Laboratories.)

Figure 12-10. Grote Reber with a radio telescope that he built and used to survey the sky. (Courtesy of National Radio Astronomy Observatory, Green Bank, West Virginia.)

and for some years, beginning in the late 1930s, he constituted the entire radio astronomy profession. At his home in Wheaton he built his own radio telescope (Figure 12-10) for the observations. He was hampered by the general state of ignorance concerning his new field of study, as well as by the suspicion of his neighbors and the problem of children climbing on the structure. Reber was not only able to detect radio waves from the Milky Way, but he was able to map their distribution on the sky at a wavelength of 1.9 meters. This radio radiation has a continuous spectrum, is polarized, and has been shown to arise from cosmic ray electrons moving through magnetic fields in the galaxy *(synchrotron radiation)*. Similar emission comes from localized sources, such as the Crab nebula and other supernova remnants, and also from galaxies and quasars beyond the Milky Way.

The next advance in radio astronomy, the prediction and discovery of the 21-centimeter wavelength line of neutral hydrogen, provided students of galactic structure with a powerful tool. During World War II the normal observing facilities were not available in the Netherlands, and theoretical research was the order of the day. The Dutch astronomers were aware of the work by Reber and immediately saw the great advances that would be possible if a spectral *line* were observable with radio telescopes: Doppler shifts of the line, and thus gas motions in the galaxy, could be measured. In 1944, Hendrick van de Hulst predicted that such a line would indeed be produced by neutral hydrogen atoms in interstellar space. The atomic situation can be illustrated by thinking of the hydrogen atom as a planetary-like system with an electron orbiting a proton. In Chapter 5 we treated the proton and electron as points, but this concept is too simple. Each particle can be regarded as having a definite size and spin. Thus, there are really two ways of having the electron in the *ground level* or lowest energy level—the spins of the proton and electron can be in the same direction or in different directions, as illustrated in Figure 12-11. When the electron's spin flips from parallel to anti-parallel with respect to the proton, a very small change in energy results in the emission of a photon with a 21-centimeter wavelength (frequency 1,420 megacycles/ sec). At first glance, the problem with this line is that it is not emitted very often. In fact, a hydrogen atom on the average emits the 21-cm photon spontaneously only once in 11 million years. It may seem folly to attempt to detect photons that are emitted so infrequently, but it is not, because interstellar space is so vast and hydrogen is its most abundant constituent.

The prediction by van de Hulst was published in 1945, but, partly due to war-caused disruption of normal channels of communication, it was not widely known. Meanwhile, a similar prediction was made independently in the Soviet Union by I. S. Shklovskii. The efforts to observe this line in emission in the galaxy were rewarded in 1951 when it was detected by Harold Ewen and Edward Purcell at Harvard.

After 1951, radio astronomers studied the distribution and motion of hydrogen gas in the galaxy through measurements of the 21-cm line radiation. They found that the hydrogen apparently is confined to a fairly narrow (about 250 parsecs thick) layer in the plane of the Milky Way, and that within this layer the hydrogen is concentrated in cool *interstellar clouds,* which form a pattern of spiral arms resembling those seen in other

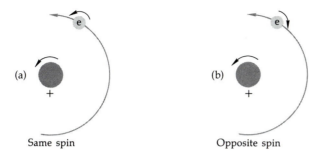

Figure 12-11. Schematic diagram describing the origin of the 21-cm emission line of neutral hydrogen. When the electron "flips its spin" from the same direction as the proton (a) to the opposite direction (b), a photon of wavelength 21 cm is emitted.

spiral galaxies (Chapter 13). These results are summarized in Figure 12-12, while the picture of the rotational motions inferred from the 21-cm line studies is given in Figure 12-18. However, newer methods have shown that the 21-cm studies give a rather incomplete picture. The problem is that only atomic hydrogen (the gas consisting of hydrogen atoms) produces 21-cm radiation. When hydrogen is in the molecular state, in which a pair of atoms join to form a hydrogen molecule, it does not produce observable radio waves, and recent studies indicate that much of the hydrogen in the Milky Way is actually in the molecular state. This includes huge regions, called *giant molecular clouds,* where there is little atomic hydrogen. The giant molecular clouds contain carbon monoxide (CO) as their second most abundant molecule after H_2. The CO is readily observed by means of its radio line radiation, and hence is used to trace out the regions of molecular hydrogen. The results of such studies, combined with the earlier 21-cm line surveys, provide the current picture of gas distribution in the Milky Way (Figure 12-13).

The striking difference about this picture, in contrast to earlier findings, is the great concentration of mass, mostly molecular hydrogen, in a ring-shaped zone around the galactic center, with inner and outer radii of roughly 4 and 7 kiloparsecs, respectively. This is where the interstellar matter of the galaxy is densest, and presumably this is where most star formation is going on at the present time.

Some of the hydrogen in the galaxy is in the ionized state. Examples are the H II regions, such as the Orion nebula. There is a form of continuous radio emission called *thermal radiation* that can be observed from

Figure 12-12.   Artist's impression of the neutral hydrogen distribution in the galaxy, based on telescope surveys of 21-cm wavelength radiation. The nature of the distribution is undetermined in the dark triangular region below the galactic center in this picture. The view is from the north galactic pole. The sun (some 10 kiloparsecs from the center) is marked by the symbol ⊙. (Courtesy of G. Westerhout, University of Maryland.)

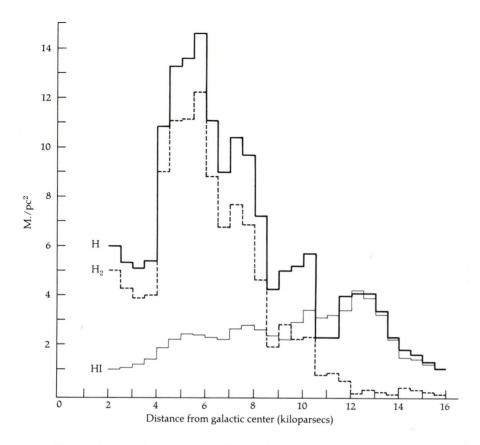

Figure 12-13. This diagram indicates the amount of gas in the galactic disk, measured in solar masses per square parsec ($M_\odot/pc^2$), as a function of distance in kiloparsecs from the center of the galaxy. The lowest curve, marked HI, indicates the amount of neutral hydrogen, which varies around an average value of about 2 $M_\odot/pc^2$. Note, however, the great concentration of gas, mostly molecular hydrogen (H_2 curve), at distances of approximately 4 to 7 kpc from the galactic center. The top curve, labeled simply H, gives the sum of atomic and molecular hydrogen. (After a diagram by M. A. Gordon and W. B. Burton.)

these objects. It can be distinguished from radio synchrotron radiation, which is also a continuum, because it may have a different variation in intensity with radio wavelength. In addition, synchrotron radiation is polarized, whereas thermal radio emission is not. There are also radio emission lines produced by neutral hydrogen atoms (and others) that have recently recombined from the ionized state in an H II region. We can observe the motions of gas in such regions by studying the Doppler effect on these *radio recombination lines*. Finally, it appears that the space between neutral hydrogen clouds in the galactic plane is occupied by a very thin, hot gas of ionized hydrogen.

In addition to CO, over 50 molecules in the interstellar medium have been found to emit radio lines. These include the hydroxyl molecule (OH), water vapor, ammonia, silicon monoxide, and formaldehyde. The intensities of some of these lines are unexpectedly strong and cannot be explained in simple terms such as the spin-flip process in neutral hydrogen. In fact, the explanation of these molecular emission lines at radio wavelengths is one of the chief problems in radio astronomy today.

Several interesting questions are raised by the discovery of the interstellar molecules. First of all, we do not understand how many of the molecules can form and persist in space. It is especially hard to explain how several atoms can come together to form the more complex molecules that have been found in the interstellar gas, such as methyl alcohol, which contains 6 atoms, and dimethyl ether, $(CH_3)_2O$, which contains 9. One would expect these molecules to be dissociated by ultraviolet light; presumably they occur in dust clouds and are shaded by the dust. A second problem, already mentioned, is why their radio emission is so strong. Finally, biochemists are wondering about the possibility that the interstellar molecules might somehow be related to processes that produce life. For example, the radio astronomers have found that both formaldehyde and ammonia molecules are present in several interstellar clouds, and laboratory experiments (similar to those described in Chapter 4) prove that when these two substances are present under suitable conditions, exposing them to heat or to ultraviolet radiation causes chemical reactions that yield amino acids.[3] This does *not* prove that living matter may somehow descend from the molecules in space, but the subject does deserve study, and such studies are under way.

STRUCTURE

Modern star-count analyses, similar to the statistical methods used by Kapteyn but with full allowance for interstellar absorption, can be used to determine the distribution of stars in the Milky Way. The results give a picture of a flattened, disk-shaped galaxy resembling that found from the

---

[3]The discovery of formaldehyde in the interstellar medium is interesting in itself, as this molecule is known to react under suitable conditions to form carbohydrates and subunits of the nucleic acids.

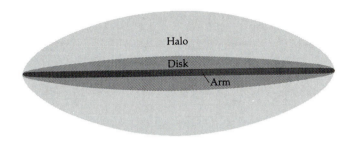

Figure 12-14.   A schematic cross-section through the galaxy, showing the increasing amount of flattening as one goes from halo to disk to arm population.

21-cm line studies. However, since our observations in visible light are limited by the absorbing effect of interstellar dust, far less of the galaxy can be mapped through observations of the stars than through the 21-cm measurements of radiation from the interstellar gas.

The picture is complicated somewhat when the shape of the galaxy, as judged from different kinds of stars, is considered. The degree of flattening varies with the type of star that is observed, as shown in Figure 12-14 (see also Table 12-1). The bright, young O and B stars have a very flat distribution concentrated close to the plane of the galaxy, as do the interstellar gas and dust. In an intermediate distribution (flattened but not as extreme) are red-giant stars and the long-period variable stars. In a more nearly spherical distribution, we find the globular clusters and the *high velocity stars.*

Looking at the flat stellar distribution in more detail, we find that the stars and interstellar matter are concentrated in *spiral arms.* This kind of structure would be expected if our galaxy were similar to others such as M31 (Plate 8), M33 (Figure 13-3), and M81 (Figure 13-5). In our above discussion of the 21-cm line, we presented the models of gas distribution and galactic rotation that are derived from the radio observations (Figures 12-12 and 12-18), but we did not indicate how the distances of the hydrogen clouds are actually determined. In fact, the method involves assuming a theoretical rotation model[4] for the galaxy and seeing if it successfully predicts the observed pattern of Doppler shifts. The distances of individual hydrogen clouds are estimated by measuring the Doppler shifts of their 21-cm lines and comparing them with the model to see at what distance from the sun

---

[4] A mathematical expression that represents the way in which the velocities of objects orbiting around the galactic center depend on the distances of the objects from the center.

Table 12-1. Types of Stars in the Galaxy

	POPULATIONS		
	Halo	*Disk*	*Arm*

Typical members	Globular clusters
	RR Lyrae stars
	W Virginis stars
	High-Velocity stars
		Bright red giants
		Novae
		Long-period
		  variables
		Sun and most
		  nearby stars
			Gas and dust
			Supergiants
			T Tauri stars
			Classical Cepheids
			Open clusters

Average distance above or below the galactic plane (parsecs)	1,000–2,000	200–500	100–150
Typical speed at right angles to the galactic plane (km/sec)	75	15–20	10
Flattening (see Figure 12-14)	Slight	Fairly strong	Extreme
Distribution	Smooth and concentrated toward galactic center	Smooth and concentrated toward galactic center	Very patchy and little or no concentration toward galactic center
Fraction of heavy elements relative to the sun	0.2	0.6–1.3	2–3
Age	Greater than about 5 billion years, or OLD	Between 1 and 5 billion years, or INTERMEDIATE	Less than 1 billion years, or YOUNG

in a given direction the observed Doppler shift can be expected. On the other hand, the distances of the stars are determined without assuming a rotational model by using the methods desribed in Chapter 10, and reasonable agreement is found between the galactic structure deduced from the stars and that deduced from the interstellar gas. As shown in Figure 12-15, the sun appears to be near the inner edge of a spiral arm.

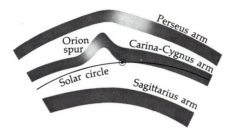

Figure 12-15.   Schematic diagram of the spiral arms near the sun. The "solar circle" has a radius of 10 kiloparsecs.

MOTIONS, ROTATION, AND MASS

The rotation of the galaxy that we observe in the 21-cm line might be expected on the basis of the flattened appearance of the Milky Way. Prior to the advent of radio astronomy it was possible to draw conclusions about the rotation, although the available data were much more limited. The Dutch astronomer Jan Oort showed in 1927 that the rotation of the galaxy could be detected in the radial velocities of the relatively nearby stars, *if* the stars at different distances from the galactic center revolved around the galactic nucleus at different speeds. On the other hand, if the galaxy rotated as a rigid body, the relative positions of the stars would remain the same, and therefore (except for small random motions) stars would not tend to move away from each other.

This is not the case if we consider a non-rigid model. For simplicity, let the orbits of the sample stars in Figure 12-16 be circles around the galactic center. The lengths of the lines represent the different orbital speeds. If we (located at the position of the sun) look exactly toward or away from the center of the galaxy, no radial motion is seen because the motions of the sun and the stars in question are at right angles to the line of sight. We do not see any Doppler shift if we look ahead or behind in our orbit at nearby stars because these stars are moving at the same speed in their orbits as we are. Only at the positions intermediate between these examples should a Doppler shift be observed; the corresponding radial velocities should show a double-wave variation (Figure 12-17). Radial velocities are termed positive (+) when the stars are moving away from the sun and negative (–) when the stars are approaching the sun. If we look in positions (b) and (f) of Figure 12-16, the sun and stars are approaching each other, and the radial velocity should be negative; in positions (d) and (h), the stars and the sun are moving farther apart, and the radial velocity

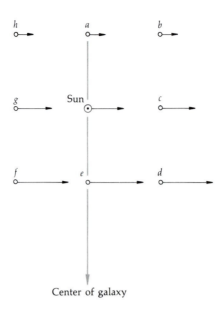

Center of galaxy

Figure 12-16. Schematic illustration of differential rotation in the galaxy. The length of the arrows represent the velocities around the center of the galaxy, with the velocity decreasing away from the center.

should be positive. In positions (a), (c), (e), and (g), the distance from the sun to the star is virtually constant, and the radial velocity should be zero. This theory, as proposed by Oort, has been verified by detailed observations. Systematic measurements of motions, along with the known distance (about 10 kiloparsecs) to the galactic center, allow a determination of the galactic rotation velocity near the sun. This speed is about 250 km (160 miles)/sec.

The determination of the galactic rotational velocity near the sun can be verified by another method that does not depend on our model for the structure and size of the Milky Way. The radial velocities of other nearby galaxies beyond the Milky Way can be measured and the results inspected for a trend, as these galaxies should not be participating in the rotational motion of the Milky Way. The problem is similar to a person on a merry-go-round looking out at the surroundings to determine which way the merry-go-round is turning. Sure enough, we find that we are approaching the galaxies in one direction and receding from those in the opposite direction. This effect could be caused by motion of the Milky Way galaxy through space, or by its rotation, or both. In fact, the direction of the galaxy's motion that we deduce from the radial velocities of other galaxies is at *right angles* to the direction of the galactic nucleus from the sun, and thus it is in the right direction to be interpreted as the result of galactic rotation.

The internal methods (radial velocities of stars and hydrogen clouds) and external methods (radial velocities of nearby galaxies) agree in assigning a rotation speed for the galaxy at the sun's distance (10 kiloparsecs) from the center of about 250 km/sec. A circle 10 kiloparsecs in radius has a circumference of about 63 kiloparsecs; at a speed of 250 km/sec (or 2.5×10^{-7} kiloparsecs per year) this means that the sun takes roughly 250 million years to complete a revolution around the galactic center. Using the distance to the galactic center and this 250 million year period of revolution, we can estimate the galactic mass using Kepler's and Newton's Laws, and we find 1.5×10^{11} solar masses. This calculation is a gross simplification, but the number is correct to within a factor of 2, and thus is not a bad approximation. A more precise treatment involves determining the variation of rotation speed with distance from the galactic nucleus, which requires both the 21-cm measurements of galactic rotation with respect to the sun and the galactic rotation velocity of the sun. The result is an observed rotational model for the galaxy, and it is represented by the *galactic rotation curve*, shown in Figure 12-18. Using the laws of motion and gravitation to interpret this curve gives an answer of about 3×10^{11} M$_\odot$ for the mass of the Milky Way.

The discovery and verification of galactic rotation helps us to understand a phenomenon of long standing, namely, the so-called *high velocity stars*. (The explanation was actually advanced by the Swedish astronomer Bertil Lindblad in 1927, before rotation of the galaxy was proven.) Most

Figure 12-17. The inferred change of radial velocities resulting from the situation shown in Figure 12-16 and described in the text. Radial velocities are positive (+), if the distance between the object and the sun is increasing, and negative (−), if the distance is decreasing.

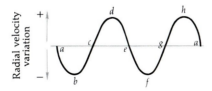

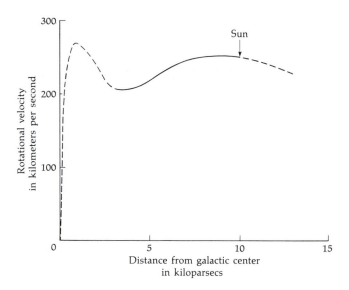

Figure 12-18. The rotation curve of the galaxy. The solid line from 3 kiloparsecs to 10 kiloparsecs is based on 21-cm line studies and is well determined. The dashed line from 0 to 3 kiloparsecs and the dashed line for distances greater than 10 kiloparsecs represent more poorly known parts of the rotation curve.

stars that have been studied near the sun have radial velocities in the range 10 to 20 km (6 to 12 miles)/sec. Some, however, have velocities of 100 km/sec or more with respect to the sun, and these are the high velocity stars. This situation results from the fact that most of the stars near the sun are in near-circular orbits close to the galactic plane; these have small radial velocities with respect to the sun. On the other hand, some stars are in highly elliptical orbits around the galactic center, as illustrated in Figure 12-19. These stars have low orbital velocities, and, when they are observed near the sun, their high radial velocity actually results from the large orbital speed of the sun. This view can be checked by seeing which way these high velocity stars seem to be moving. Their apparent motions are directed back along the sun's path in the galaxy, thus verifying that the sun's motion is really responsible for their apparent high velocity. (The effect is equivalent to the apparent backward motion of slowly walking pedestrians on the side of the road as seen by an observer in a moving car.) The high velocity stars are not closely concentrated to the galactic plane, and in addition their spectra are weak in lines of the heavy elements. Therefore, they belong to a different population than do the stars in the solar neighborhood; they are *halo stars*.

THE MILKY WAY—MODEL AND EVOLUTION

Information concerning the motions, distribution in space, and spectral properties of stars can be collected and organized into a reasonable picture. The stellar populations in the galaxy require more than the two classifications introduced earlier, and we present in Table 12-1 a simple scheme in which the stars are divided into *arm, disk,* and *halo* populations. The arm population consists of the youngest Population I objects, the disk population includes stars that we previously classified in both Population

Figure 12-19. Schematic diagram comparing the galactic orbits of the sun and a typical high-velocity star, as viewed from the north galactic pole. The apparent speed of the high-velocity star is actually due to the large orbital velocity of the sun.

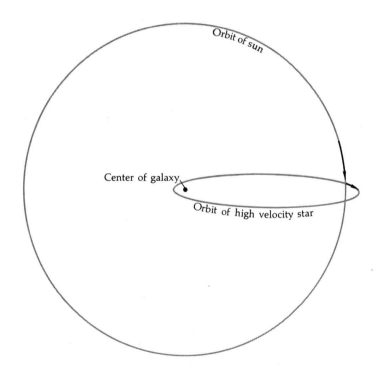

I and Population II, and the halo population consists of the older Population II stars.

Inspection of Table 12-1 suggests how the Milky Way galaxy may have evolved. The process resembles the nebular hypothesis for the formation of the solar system. The galaxy is thought to have originated as a nearly spherical rotating gas cloud, which contracted under its own gravitational attraction. As the galactic cloud *(protogalaxy)* contracted, the denser parts of the cloud condensed at a more rapid rate to form the first stars. These stars formed from the original material, which consisted of hydrogen and a small amount (probably 25 percent by mass at most) of helium. Stars formed near this time were poor in heavy elements and had a nearly spherical distribution, like the halo population, because the protogalaxy had not contracted very far.

As the protogalaxy contracted further and rotated faster, two processes continued simultaneously. First, as discussed in the case of the nebular hypothesis, the original, nearly spherical gas cloud was constantly flattening. Second, the gas cloud was partially replenished by mass ejected from the stars already formed within it by a variety of violent and nonviolent processes, as described in Chapter 11. This matter was enriched in heavy elements due to the nuclear reactions that had occurred inside the stars. Thus, as the galactic cloud continually flattened, stars (such as the sun) formed in the region that we now call the galactic disk, and they had a greater heavy element content than the halo stars.

Much of the matter in the galaxy condensed into stars, and so the amount left in the form of interstellar material decreased. On the other hand, thanks to stellar mass loss, the relative content of the heavy elements in the interstellar medium increased. At present, formation of stars with the highest heavy element content, the arm stars, is probably occurring only in isolated patches of gas and dust in the spiral arms. If the rate at which stars are formed depends on the amount of gas and dust present, this rate may have been higher in the past.

Our galaxy formed from a vast cloud of hydrogen and helium, and the first stars processed these light elements in their interiors and returned heavier elements to the still condensing cloud, from which other stars have formed more recently. These ideas are consistent with the observations of the chemical compositions, ages, and locations of the stars. However, we do not know with certainty how the galaxy assumed its pattern of spiral arms. The leading theory, as advanced by C. C. Lin of the Massachusetts Institute of Technology, is based on a concept of wavelike disturbances in the distribution of the interstellar gas, with the arms corresponding to crests (regions of enhanced gas density) of the waves.

DENSITY-WAVE THEORY OF SPIRAL ARMS

The *density-wave* theory asserts that the arms represent a spiral-shaped wave pattern that rotates around the center of the galaxy at a fixed angular speed. The individual stars in a spiral arm *do not remain in the arm indefinitely,* for the stars close to the galactic center make one revolution around the center in a much shorter time than do the stars at great distances from the center. Thus, if stars did remain in the arm for long periods of time, the inner part of the arm would soon be tightly wound around the center of the galaxy. This is contradicted by observations of the Milky Way and of other galaxies as well, which show that a spiral arm typically is wrapped at most once or twice around the center. Thus, it seems that the stars concentrated in a spiral arm at one point in time must be dispersed between the arms later on. As the wave pattern rotates, its spiral-shaped gravitational field concentrates the interstellar medium in a kind of "instantaneous" representation of the wave. A commonly used analogy is that of a freeway traffic disturbance pattern: suppose a slowly moving disturbance travels down a freeway at a constant speed (it might be the truck used for painting lane-divider lines, for example) of only 5 mph. If we view the disturbance from a safe vantage point high up in a helicopter, we see traffic approaching and receding from the disturbance at the speed limit of 55 mph. In the immediate vicinity of the lane-painting truck, the traffic moves more slowly (say, 35 mph), since one or two lanes are obstructed, and some drivers will slow down to look at the workers painting the lines. Thus, we see a greater "density" of cars (cars per unit length of freeway). The region of greatest car density moves ahead steadily at 5 mph, but the cars in this region are constantly leaving it at higher speed and being replaced by new cars that are slowing down.

Since stars are born from the interstellar clouds, we see spiral-shaped regions of gas and bright, young O and B stars in our own and other spiral galaxies, and we identify these regions as spiral arms. As the stars grow older, their respective motions carry them out of the arms, but new stars form in the region where interstellar matter is concentrated by the moving spiral disturbance. Unlike the traffic analogy, old stars don't move into the disturbance and become concentrated there; the concentration process only has a strong effect on the interstellar medium. The spiral pattern reflects a spiral-shaped gravitational field that is superimposed on (and is weaker than) the general gravitational field of the galaxy. The concentration of matter in the spiral pattern sustains this spiral gravitational field. What is not clear, however, is what *first* caused matter to adopt a spiral pattern so that it could have a spiral-shaped field. In other words, we think we know why spiral arms continue to exist, but it is less certain where the first spiral arm came from.

The density-wave theory explains why the youngest stars, namely the O and B stars, outline the spiral arms so well that they often have been called "spiral tracers," and it provides a natural explanation of why spiral arms traced out from older stars are fuzzier—the older stars have had time to move out of (or to have been left behind by) the rotating density wave, and hence they are no longer concentrated in a sharp pattern.

13

Galaxies and the Universe

NEBULAE OR ISLAND UNIVERSES?

In the mid-nineteenth century, Lord Rosse discovered spiral structure in some of the diffuse objects that were referred to collectively as nebulae. The nature of these spiral nebulae was unclear, and by 1920 opinions on this subject had crystallized into two opposing schools of thought. In April of that year the leading American proponents of the two theories debated the matter before the National Academy of Sciences in Washington, D.C.

Harlow Shapley of Harvard took the occasion to stress his conclusion that the Milky Way galaxy was much larger than was generally suspected and that the sun was not located near its center. Shapley recognized that the spirals must lie outside the galaxy, but he regarded them as much smaller than the Milky Way, and thus not very distant from it. In particular, he argued, the spirals "are truly nebulous objects," that is, clouds composed of gas and dust as is the Orion nebula, for example. Heber D. Curtis of the Lick Observatory disagreed. Although Curtis wrongly supported the concept of a rather small Kapteyn universe model for the Milky Way, he correctly argued that the spirals were at great distances, perhaps in the millions of parsecs, and that they were vast star systems like the Milky Way.

The nature of the spiral nebulae, now called *spiral galaxies*, was established largely through the efforts of Edwin Hubble at the Mount Wilson Observatory. Using the 100-inch telescope, he was able to resolve some of the brighter stars in the outer parts of M31,[1] the famous spiral galaxy in Andromeda (Plate 8). Among these stars he discovered some Cepheid variables. We explained in Chapter 10 how the measurement of the period between successive maximum brightnesses of a Cepheid can be used to determine its distance. Hubble applied this method and found the distance of M31. It was in accord with Curtis' ideas. A check on this result was provided by observing novae in M31 and, by assuming that their absolute brightnesses were similar to those of novae in the Milky Way, deriving their distances. M31 is indeed a large spiral galaxy, similar to the Milky Way, and in fact one of the nearest galaxies; the modern value for its distance is about 650 kiloparsecs. From the angular size of M31 and its distance, we can calculate its diameter and show that it is similar in size to the Milky Way galaxy and thus enormously larger than the Orion nebula.

THE SHAPES OF GALAXIES

The recognition of spirals as truly *extragalactic* objects was a great step forward. Studies of many other objects once termed *nebulae* disclosed that some of them also were distant galaxies, although they lacked spiral shape, and several types of galaxies are now known.

Spiral galaxies (Figure 13-1) are basically similar to M31 and the Milky Way. A galaxy that appears to be a spiral viewed edge on (that is, with its galactic plane parallel to our line of sight) is shown in Figure 13-2, and its similarity in appearance to the Milky Way (Figure 12-6) is obvious. By photographing the spectra of light from different parts of spiral galaxies and comparing the Doppler shifts that are measured in these spectra, it has been found that these galaxies in general are rotating, just as the Milky Way does. The rotation takes place in the sense that the spiral arms "trail." A striking property of the spirals is their variety of central concentrations. M33 (Figure 13-3) shows relatively little central concentration; the spiral arms seem to dominate its appearance. NGC 4594,[2] on the

---

[1]M stands for Charles Messier (1730–1817), the French astronomer whom Louis XV called "the ferret of comets." Since nebulae were sometimes mistaken for dim comets, Messier catalogued more than 100 nebulae (some star clusters were also included, such as M13, which is pictured in Figure 10-15) to help astronomers avoid confusion in their searches for comets. Today we remember Messier much more for this catalogue than for his comet discoveries. The "nebulae" in the catalogue included both gaseous nebulae and galaxies. M31, the Andromeda galaxy, is number 31 in the catalogue.

[2]NGC stands for the *New General Catalogue* published by J. L. E. Dreyer of the Armagh Observatory in Ireland in 1888. Together with two later supplements (called the *Index Catalogues*, abbreviated I.C.), it listed over 13,000 galaxies, clusters, and nebulae. Messier 31 is also known as NGC 224.

Figure 13-1.   The spiral galaxy NGC 4622, a galaxy similar to our own Milky Way. The view is close to face on. (The Kitt Peak National Observatory.)

other hand, is notable for the strong concentration of stars in its central region (Figure 13-4). M31 and M81 (Figure 13-5) are intermediate between these two conditions.

Barred spiral galaxies are so named because of the long bar of stars that passes through the central region of such a galaxy (Figure 13-6). Spiral arms extend from the ends of the bar. The spectroscopic observations reveal a fascinating phenomenon in the rotation of these galaxies: the bars rotate as *rigid bodies.* In other words, although a bar is composed of individual stars at different distances from the center of the galaxy, it rotates around the center as though it were a solid unit. In the solar system, the farther a planet is from the sun, the slower it moves in its orbit. As the outer planets have larger orbits, they take much longer to complete an orbit than do the inner planets. However, the stars in the outer parts of the bars in these galaxies are moving faster than the stars near the center.

Figure 13-2. NGC 891 seen edge on. (Courtesy of the Hale Observatories.)

The *irregular galaxies* have no special shape, and they are typified by the Large and Small Magellanic Clouds (Figure 13-7), which are easily visible to the naked eye in the Southern Hemisphere. The Clouds were named after the explorer Ferdinand Magellan, who saw them on his voyage through the Southern Hemisphere. (Of course, nearly everyone who lived in the Southern Hemisphere had already seen them.) Another interesting but untypical irregular galaxy is M82; photographs (Figure 13-8) taken in red light have been interpreted as suggesting that an enormous explosion in its central region may have ejected gas into space around the galaxy.

Finally, we have the *elliptical galaxies*, so called because of their smooth appearance, with an outline in the shape of an ellipse. Two such galaxies, which happen to be small companions of M31, are shown in Plate 8; another elliptical companion to M31 is shown in Figure 13-11. Elliptical galaxies are found over a great range of sizes; in fact, one of the galaxies of greatest known mass is the elliptical M87 (Figure 13-9). Among the smallest galaxies are the *dwarf ellipticals,* such as the one shown in Figure 13-10.

The masses of other spiral galaxies are determined in the same manner that we described for determining the mass of the Milky Way galaxy. The rotation curve is determined spectroscopically, and the mass can then be computed from advanced versions of Kepler's and Newton's Laws. Masses for pairs of galaxies can also be derived by a method similar in principle to the derivation of the mass of spectroscopic binary stars—the projected separation of the two galaxies is known, and their orbital velocities are found spectroscopically through the Doppler shift. Because we have only a snapshot at a given time of the galaxies in their orbits (millions of years are required to complete such an orbit), statistical methods must be applied to a large number of galaxies to obtain reliable results.

Figure 13-3.   Messier 33, a spiral galaxy with relatively little central concentration. (Courtesy of the Hale Observatories.)

Figure 13-4.   NGC 4594, a spiral galaxy with a conspicuous central concentration. (Courtesy of the Hale Observatories.)

Figure 13-5.   Messier 81, a spiral galaxy in Ursa Major. (Courtesy of the Hale Observatories.)

Figure 13-6. The barred spiral galaxy NGC 1300. (Courtesy of the Hale Observatories.)

Figure 13-7. Two irregular galaxies, the Large and Small Magellanic Clouds. Also shown is the globular cluster 47 Tucanae (lower right), which is located in our Milky Way galaxy. (Courtesy Harvard College Observatory.)

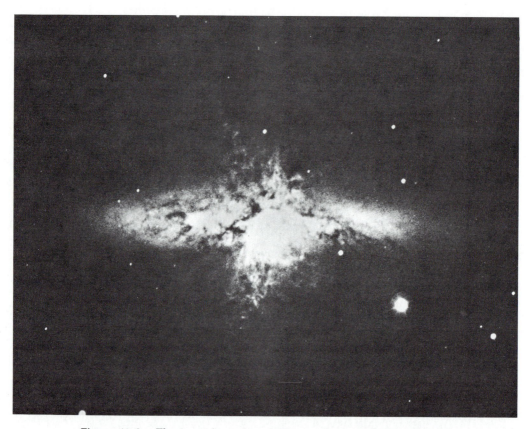

Figure 13-8.   The irregular galaxy M82, as photographed in the red light of the hydrogen alpha line. Note the filaments extending above and below the main part of the galaxy. (Courtesy of the Hale Observatories.)

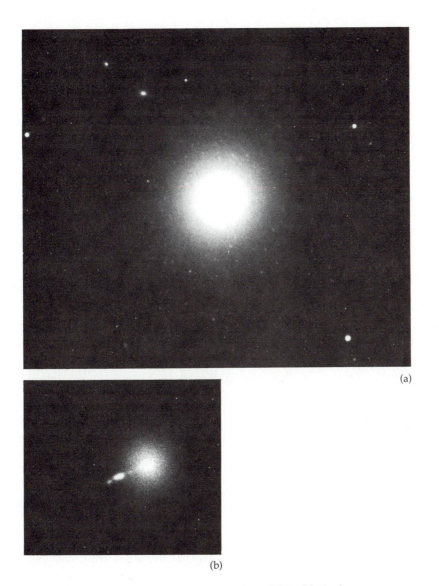

(a)

(b)

Figure 13-9.   (a) The giant elliptical galaxy M87. (b) A shorter exposure that shows the central part of the galaxy and its jet. (Lick Observatory photographs.)

Figure 13-10.   A dwarf elliptical galaxy in the constellation Sculptor. (Courtesy of Paul W. Hodge.)

The smallest known galaxies are the dwarf ellipticals, with masses of about 10^6 $M_\odot$; the largest known galaxies are the giant ellipticals with masses ranging up to about 3×10^{12} $M_\odot$. The elliptical galaxy M32 (one of the companions to M31) is intermediate, with a mass of 4×10^9 $M_\odot$. The masses of the Milky Way, M31, and M81 are all very similar, averaging about 3×10^{11} $M_\odot$. The spiral M33 has a mass of about 3×10^{10} $M_\odot$. The mass of the Large Magellanic Cloud is about 2×10^{10} $M_\odot$, and that of the Small Magellanic Cloud is about a factor of 10 less. Thus, galactic masses are roughly from 10^6 M$_\odot$ to 10^{12} M$_\odot$, with the average mass falling in the lower part of this range.

POPULATIONS REVISITED

An understanding of galaxies as star systems requires instruments and techniques capable of resolving them into individual stars for study. The outer regions of M31 were resolved by Hubble in the 1920s; the brightest stars were found to be blue supergiants. However, the central regions of M31 were not resolved until 1944.

This story revolves around the German-American astronomer Walter Baade, who was on the staff of the Mount Wilson Observatory during World War II. Technically classified as an enemy alien, Baade was restricted to the Pasadena area, where he made extensive observations with the 100-inch telescope. Because many other astronomers were assigned to wartime research, Baade had less competition for the use of the large telescope, and thus had a great deal of observing time. Furthermore, Los Angeles was blacked out at night as a military precaution, so Baade was blessed with a particularly dark sky. (In fact, such good skies at Mount Wilson are approximated today only when Los Angeles is heavily blanketed by smog.)

Baade knew the exposure required to record images of the brightest stars in the central region of M31, assuming that they were the same kind of stars as the brightest ones resolved in the outer parts of M31. Exposures with blue-sensitive plates did not show stars at the expected exposure times. Eventually, red-sensitive plates and longer exposure times yielded images of the brightest stars in the central region of M31, as well as some elliptical galaxies near M31, such as NGC 185 (Figure 13-11).

The resolution alone was a tremendous achievement which opened up the study of individual stars in the central regions of another galaxy. The importance of this was emphasized by the circumstance that the bright stars observed in the central region of M31 were surprisingly unlike those that previously had been resolved in the outer spiral arms. This is the discovery that led to the concept of stellar populations. The brightest stars in the central region were red and somewhat fainter than the brightest stars in the spiral arms, which were blue. Baade showed that the difference could be explained (Figure 13-12) if the central region stars had an H-R Diagram like that of a globular cluster and the spiral arm stars had an H-R Diagram like that of an open cluster.

Once the concept of stellar populations is explored, many facts in our own galaxy (relating to the velocities of stars, the space distribution of

Figure 13-11. NGC 185, a dwarf galaxy. This galaxy was among the first Population II galaxies to be resolved into stars by Walter Baade (note the stars resolved in this photo). This achievement was considered so important to astronomers that actual photographic prints of NGC 185 (instead of printed sheets) were bound into the scientific article that announced Baade's observation, in order that the reader could judge for himself. (Lick Observatory photograph.)

stars, and the chemical composition of stars) fall into place, as discussed in Chapter 12. Spiral galaxies such as the Milky Way have both Population I and Population II stars. Ellipticals are found to contain primarily Population II stars, like those in the center of M31. Irregular galaxies are largely composed of Population I stars, like those in the arms of M31. We know that galaxies must evolve, because the individual stars in them evolve, and it would be tempting to assign an evolutionary sequence to the different types of galaxies, but it is not clear that this would be correct. In fact, it appears likely that the overall shape of a galaxy is determined by the initial conditions of its formation, including the ratio of angular momentum to mass. The disk-shaped spiral galaxies probably have evolved from initial objects *(protogalaxies)* with high angular momentum per unit mass, while the more slowly rotating protogalaxies became elliptical galaxies. Thus the different galaxy shapes may correspond, not to evolutionary stages, but rather to the array of different initial conditions.

DISTANCES TO GALAXIES

The recognition of galaxies such as M31 as island universes basically similar to the Milky Way involved the determination of their distances, as it was necessary to demonstrate that they were comparable in size to the Milky Way. As mentioned before, the method involved the observation of Cepheid variable stars. This is a suitable technique for the few dozen closest galaxies, in which individual bright stars like the Cepheids can be resolved. (The Magellanic Clouds are the closest definitely known galaxies, at a distance of 50 kiloparsecs. The spirals M31 and M33 are each at a distance of about 650 kiloparsecs.) Individual Cepheids cannot be resolved in the more remote galaxies, and thus other techniques must be employed to measure the distances.

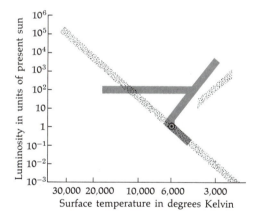

Figure 13-12. Schematic H-R Diagram showing regions occupied by the two basic populations of stars as originally classified by Baade. Population I stars, dotted area; Population II stars, shaded area. (Adapted from *Stellar Populations* by Geoffrey and Margaret Burbidge. Copyright © 1958 by Scientific American, Inc. All rights reserved.)

The methods used to extend the distance scale include the determination of the absolute magnitude of the brightest star in a galaxy of a particular type. Studies show that the absolute brightnesses of these supergiant stars in different galaxies of the same type are remarkably similar, and therefore such stars can be used as *distance indicators*. This method has the advantage that the supergiants are among the easiest stars to observe.

Observations also show that the average size of the largest H II regions in a galaxy of a particular type is nearly constant. For example, in a loosely wound spiral such as M33 (Figure 13-3), the average diameter of the five largest H II regions is about 175 parsecs. If similar nebulae are observed in a more remote galaxy of similar shape to M33, their angular diameters can be measured and the distance determined.

CLUSTERS OF GALAXIES

It appears that most galaxies occur in distinct groups called *clusters of galaxies*. A sample large cluster of galaxies, found in the constellation Hercules, is shown in Figure 13-13, where a variety of galaxy types can be seen. The distance-measuring techniques that involve the brightest stars and the largest H II regions can be used out as far as the Virgo cluster of galaxies at a distance of about 11 megaparsecs (millions of parsecs) and the Fornax cluster at 13 megaparsecs. However, beyond the distances of the Virgo and Fornax clusters of galaxies, individual features such as stars or bright nebulae cannot be resolved or even recognized (except for supernovae). At these distances we must use properties of the galaxy clusters themselves as distance indicators. The absolute magnitude of the brightest galaxy in a cluster can be determined in the nearby clusters (such as Virgo) and then used to calculate the distances of those more remote. The most distant known clusters are some 1,000 megaparsecs away, but some quasars (Chapter 16) are much farther. This process is simple in principle but complicated in practice. For example, the brightnesses of distant galaxies must be corrected for the Doppler effect, as they are found to have large red shifts (see next section). Thus, when we measure the V magnitude of a nearby galaxy of negligible red shift, we are measuring the light that the galaxy *emitted* in the wavelength range of the V filter. But when we measure the light of a distant galaxy that has a large red shift with respect to the earth, the light that passes through the V filter actually was *emitted at shorter wavelengths* and has been red shifted. Thus, we do not get a true measure of the intrinsic V magnitude of the galaxy unless the red shift is found from the spectrum and a suitable correction is applied to the V photometer observation. The interpreta-

Figure 13-13. Part of the Hercules cluster of galaxies. The "crosses" around the bright stars are due to an optical effect in the telescope. (Courtesy of the Hale Observatories.)

tion of our observations of distant galaxies is also complicated by the circumstances that *we are looking backward in time*. The farther away a galaxy is, the longer the light has been on its way. The light from a galaxy 1,000 megaparsecs away has been en route to us for 3.3 billion years. Hence, the observations of distant galaxies are observations of younger galaxies, and because of evolutionary changes, the stars and nebulae seen in the distant galaxies may differ systematically from those in nearby and thus "older" galaxies.

TWO BASIC PROPERTIES OF THE UNIVERSE

In this section we discuss two basic observed properties of the universe: (1) the galaxies have red shifts, and the more distant the galaxy, the greater is its red shift; (2) radio waves, with a continuous spectrum corresponding to a black body of temperature 3 K, are received by us in equal amounts from all directions in space. Any attempt to explain the nature and origin of the universe must account for these two phenomena.

The Red Shift

The distances to clusters of galaxies can be estimated using the methods described in the preceding section. Spectra of the galaxies can be taken and analyzed by the Doppler effect for radial motions. It turns out that the absorption lines in the spectra of most galaxies are shifted to the red, as first noted at Lowell Observatory by V. M. Slipher in 1912. If these red shifts are interpreted as a result of motion, the galaxies must be moving away from us in a systematic fashion. Exceptions to this are a small number of relatively nearby galaxies.

Figure 13-14. The spectra of galaxies located in increasingly distant (top to bottom) clusters, with arrows showing the red shifts of the H and K lines of singly ionized calcium. The distances are estimated from the brightest galaxies in the clusters. The red shifts (compared to the distances) do not exactly agree with the adopted Hubble constant of 100 km/sec/megaparsec, because each spectrum shown corresponds to a specific galaxy in the cluster, whereas the Hubble constant is determined from the average red shifts for each of many clusters. If the Hubble constant is 50 km/sec/megaparsec, as some evidence now indicates, then the correct distances are twice as large as those listed in the figure. (Adapted from an illustration supplied courtesy of the Hale Observatories.)

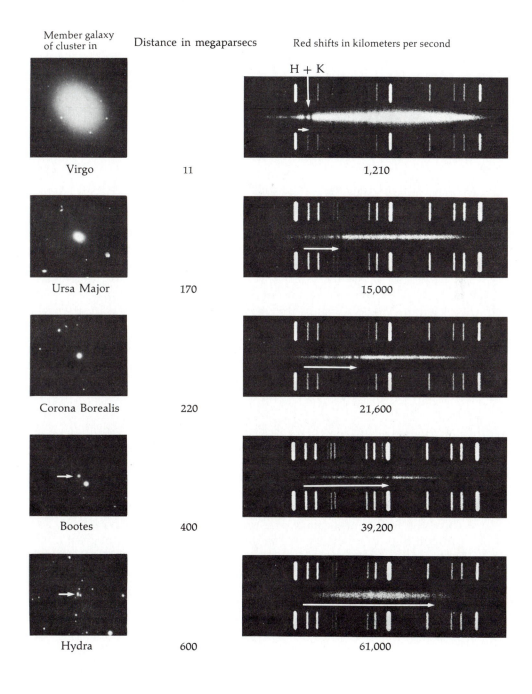

Member galaxy of cluster in	Distance in megaparsecs	Red shifts in kilometers per second
Virgo	11	1,210
Ursa Major	170	15,000
Corona Borealis	220	21,600
Bootes	400	39,200
Hydra	600	61,000

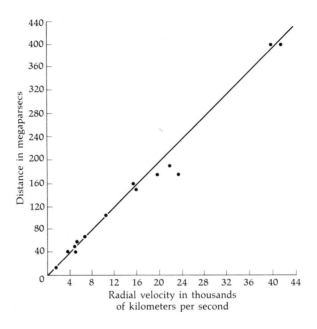

Figure 13-15.   Graph showing the distance versus radial velocity relationship for the better studied clusters of galaxies. Each plotted point
represents a cluster. The straight line corresponds to a Hubble constant of 100 km/sec/megaparsec.

In 1929, Hubble turned his attention to this problem and found that
clusters of galaxies all exhibited systematic red shifts; most important, the
amount of the red shift depended on the distance, as shown in Figures
13-14 and 13-15. The straight-line form of the dependence of radial velocity on distance indicates that the velocity is proportional to distance.
Thus, the farther the galaxy, the greater the velocity of recession. The
constant of proportionality, or *Hubble constant,* is a quantity of great interest to astronomers. The value currently in use is about 100 kilometers (62
miles)/sec/megaparsec.[3] Thus, an object 1 megaparsec distant has a recession velocity of 100 km/sec (62 miles/sec); an object 3 megaparsecs distant
is receding at 300 km/sec; and the Virgo cluster galaxies have an average
radial velocity of about 1,100 km/sec and a distance of about 11 megaparsecs.

_____

[3]However, many astronomers now believe the correct value may be as low as 50 km/sec
megaparsec.

For galaxies close to the Milky Way, the velocity of recession calculated from the distance and the Hubble constant is very small, and the effect is masked by the ordinary random motions of these galaxies. This explains why some nearby galaxies are approaching us rather than receding.

Cosmic Background Radiation

One of the most remarkable discoveries about the general properties of the universe is that of the cosmic background radiation. The first evidence of this radiation was obtained in 1965 at the Bell Telephone Laboratories by Arno Penzias and Robert Wilson, who were investigating some unexplained receiver noise while planning a study of the radio emission from the Milky Way. In fact, the noise was actually due to radio emission from the sky, and further observations by many astronomers have established that this radiation is *isotropic:* we receive it in equal amounts from all directions. (A slight departure from isotropy has recently been reported.) Over the wavelength range of 0.2 to 21 cm, where it has been well measured, the radiation approximates thermal emission (that is, follows a Planck Law) at about 3 K. Studies of absorption lines due to CN molecules in the cold interstellar clouds of the Milky Way show that the molecules are heated by the background radiation, and thus confirm the existence of a 3-K radiation that apparently pervades space. This radiation was anticipated on theoretical grounds some years ago by the physicist George Gamow.

WHY IS THE SKY DARK AT NIGHT?

Olbers' Paradox is a problem posed by the German scientist Heinrich Olbers in 1826.[4] He asked a deceptively simple question: "Why is the sky dark at night?" The point of this question can be illustrated by considering a distribution of stars that *does not* produce a dark sky.

Suppose that the number of stars in a given volume of space is constant and that it remains constant for similar volumes at all distances from the earth. If this were true, and all stars were roughly like the sun, one could show mathematically that the entire sky should be as bright as the disk of the sun. The reasoning is illustrated in Figure 13-16. If the stars

---

[4]The problem was discussed earlier, in 1743, by the Swiss astronomer J. P. Löys de Chéseaux. However, he published his conclusions in an appendix to a book on comet studies, and they were soon forgotten.

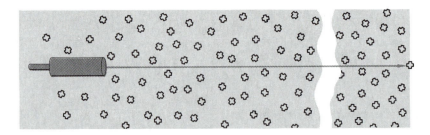

Figure 13-16. Olbers' Paradox. If the stars are uniformly distributed throughout space, a line of sight ultimately intersects the disk of a star. Even if this has not happened by a given distance from the observer, one has only to continue to ever greater distances and it will eventually occur.

are distributed with a constant density throughout space, and if space extends without limit, it is impossible to draw a straight line from the earth in any direction without intersecting the disk of a star. The situation is somewhat akin to constructing a mosaic of the sky; as the stars at greater and greater distances are included, more and more of the sky mosaic is filled in by pieces with the same disk brightness as the sun. If the distribution of stars extends as far as one pleases, all vacant places would be filled in, and the entire sky would be as bright as the solar disk.

The situation discussed in the preceding paragraph can be visualized in a different way, with a geometrical argument that involves two hypothetical regions in space having the form of thin spherical shells (Figure 13-17). We take the simple case that (1) the two shells are of equal thickness; (2) all of the stars in the two shells are alike (have equal intrinsic brightness); (3) the star density (number of stars per cubic parsec) is equal in the two shells; and (4) the outer shell is twice as far from the sun as the inner shell. We then ask the question: Which shell furnishes more light, as seen at the sun? Since the stars are all of equal brightness, a given star in the inner shell will appear to be four times brighter than an outer-shell star as seen from the sun, due to the inverse square law. On the other hand, although the thicknesses of the two shells are equal, the volume of the outer shell is greater; in fact, it is four times greater, as one can easily calculate, and so it must contain four times as many stars. The result is that *each shell* contributes an *equal amount* of light, as seen from the sun. Thus, if space had the simple and uniform properties assumed in this example, every shell-like region—*no matter how far away*—would contribute an equal amount of light, and the contributions from the many shells would cause the sky to be dazzlingly bright.

This situation can be checked by stepping outside on a clear night. Clearly, the above discussion does not hold in practice—the sky is fairly dark. The simplest way to avoid the paradox is to postulate a limited or finite distribution of stars. The problem was regarded as solved when it was realized that the individual stars seen in the sky were, in fact, part of the Milky Way galaxy, just an isolated system of stars in space. However, the discovery that space is filled with galaxies reopens the question. Simply substitute galaxies for stars, and Olbers' Paradox again asks: "Why is the sky dark at night?" We shall answer this question in the next section.

THE BIG BANG

Cosmology is the field of science that seeks to determine the nature and origin of the universe. The conclusions are still uncertain, but the *Big Bang* or *Primeval Fireball* theory presented here represents the current consensus among astronomers. It was first suggested by the Belgian astronomer Abbé Georges Lemaître in 1927, and like many other accepted ideas this theory may change or fall when further evidence is obtained.

The Big Bang idea is based on a somewhat literal interpretation of the observed red shifts in the spectra of galaxies as Doppler shifts caused by receding motions. The fact that the motions are systematically away from us reopens the problem of the center of the universe. Does this mean that

Figure 13-17.   Sketch showing that a shell of stars at a larger distance has more stars than a nearer shell with the same thickness and star density.

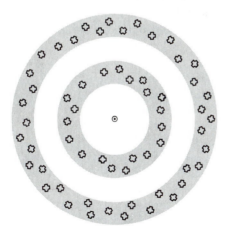

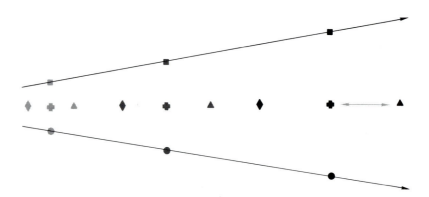

Figure 13-18. The universal red shift does not imply that the Milky Way galaxy is at the center of the universe. A small group of galaxies in the expanding universe is shown schematically at three different times. An observer in either of the two galaxies connected by an arrow would observe the other to be red shifted. In fact, the distances between all galaxies increase on this diagram.

the Milky Way galaxy is at the center? No. Exactly the same picture would emerge if the red shifts were measured by astronomers located in another galaxy, as illustrated in Figure 13-18. The situation is like that of a rubber balloon with black dots on it that is being inflated. All other dots on the surface are moving away from *any* given dot on the surface.

An age of the universe can be estimated by considering the past history of these motions. If the motions indicated by the red shifts were run backward, all galaxies would reach approximately the same point in space. Since objects 1 megaparsec apart are separating from each other at 100 km/sec, we can estimate the age of the universe (*Hubble time*) by assuming that this velocity has remained constant. A parsec is 3×10^{13} km, and a megaparsec is 3×10^{19} km. The time required to traverse the latter distance at 100 km/sec is 3×10^{17} seconds, or about 10^{10} years. This value is close to the age of the oldest stars in the Milky Way. The exact age computed from the Hubble constant depends on several detailed effects, such as the possibility that the expansion rate has slowed down since the universe was born. In addition, the exact value of the Hubble constant is in dispute. Ages from 10 billion to 20 billion years are commonly quoted.

Thus, in the Big Bang theory, the universe originated in the explosion of a single concentration of matter some 10^{10} years ago, which began the continuous expansion that we see now in the red shifts of galaxies. Some of the helium found in stars today originated by nuclear reactions in the dense cosmic fireball. Gamow studied this model for the origin of

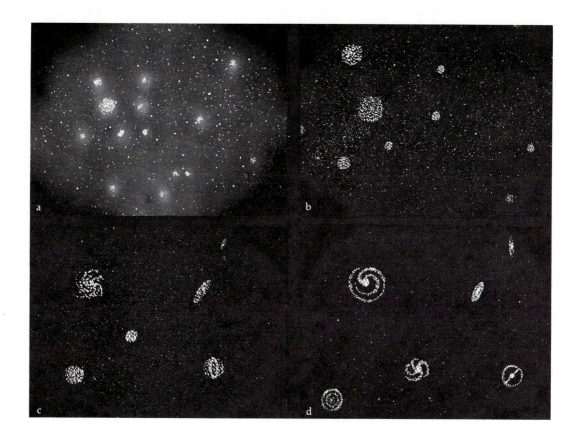

Figure 13-19. Artist's impression of the formation of galaxies in a universe created by the Big Bang. According to the theory, the events shown here took place in a time period bridging the gap between stages (b) and (c) of the evolution of the universe, as depicted in Figure 1-2. (a) The expanding universe was filled with ionized matter (white dots) and radiation (white glow) from the primeval fireball. It was sufficiently dense so that photons could travel only short distances before striking matter (that is, the universe was opaque). The radiation pressure kept the clouds of matter from contracting. However, the universe was cooling, and after 100,000 years had elapsed, the temperature had dropped to about 3,000 K, and the gas began to recombine to the neutral state. (b) The neutral gas absorbed radiation less effectively than it had done in the ionized state. Hence, photons could travel long distances without being absorbed, and so the universe was transparent. They escaped from the clouds, and the radiation pressure dropped, permitting the clouds to contract gravitationally. Thus, the protogalaxies formed (c) and evolved into galaxies (d), which we still observe today. (After *The Origin of Galaxies* by Martin Rees and Joseph Silk. Copyright © 1970 by Scientific American, Inc. All rights reserved.)

the universe and calculated that the explosion would have produced an enormous burst of photons. He predicted that these photons, red shifted by the expansion of the universe, would be observed in present times as radio wave photons, and this is the best current explanation of the 3-K background radiation discovered by Penzias and Wilson.

Formerly, it was believed that the expansion of the universe was the key to the modern resolution of Olbers' Paradox. However, E. R. Harrison of the University of Massachusetts has shown that the paradox would not occur in a static but otherwise realistic universe. We need only consider that the universe was created at a definite time in the past and that stars are luminous for a limited length of time. Then, an observer generally receives light only from a shell of stars. The light from stars beyond the shell has not had time to reach the observer, and stars closer than the shell have finished their luminous lifetimes and no longer emit light. Of course, the example is artificial, but it shows that expansion is not crucial to resolving Olbers' Paradox.

OTHER SCHEMES

The Big Bang theory seems at present to be the most likely hypothesis for the origin of the universe. However, other theories have also been advanced. The principal one in recent years was the *Steady State* theory proposed in 1948 by H. Bondi, T. Gold, and F. Hoyle at Cambridge University. According to this theory, there was no beginning to the universe and there will be no end. The universe has always looked more or less the way it does now and it will always look this way. Matter is continually coming into existence in the form of hydrogen atoms in space. These eventually make up new galaxies, which replace the old ones that are moving away from us in the general expansion.

In the Steady State scheme, there was no cosmic fireball, and hence there should be no 3-K background radiation. If the identification of this radiation is correct, then the Steady State hypothesis is wrong. During the 1960s, evidence from radio astronomy suggested that the space density (number per cubic parsec) of radio-emitting galaxies was greater in the distant past than it is now. This also contradicts the idea that the universe has always been the same, as assumed in the Steady State theory.

The *oscillating universe* theory supposes that there is no beginning and no end. However, it proposes that, rather than being constant, the universe is currently in the course of an expansion that began with a Big Bang and is gradually slowing down. At some future time, gravitation will overcome the expansion effect, and the universe will begin to collapse, finally reaching a point of ultimate coalescence in which the high temperature and pressure break down all matter into elementary particles; a new Big Bang occurs, and the expansion begins again.

The universe may have begun in a Big Bang, or it may be in a steady or oscillating state. In any case, it is characterized now by the process of expansion and is filled with radiation resembling that expected from a Big Bang. The great distances between galaxies make it a nearly empty universe.

DECELERATION AND DENSITY OF THE UNIVERSE

In a Big Bang universe, the expansion is continually slowed down because of the mutual gravitational attraction of all the galaxies in the universe. This gravitational attraction itself decreases as the expansion proceeds, because the galaxies get farther apart. If the attraction is sufficient to slow the expansion to zero speed and then reverse it (leading to an oscillating universe), the universe is described as *closed*. If the gravitation is inadequate to reverse the expansion, the galaxies fly apart forever and the universe is *open*.

The situation is comparable to that of a bullet shot up from the surface of the earth. The earth's gravity exerts a force that slows the bullet's upward motion. If the bullet is shot with a speed less than the earth's escape speed of 11 km/sec, it eventually slows to zero speed and falls back to the ground. However, if it is shot at the escape speed or a greater velocity, it escapes from the earth, even though the earth continues to exert a retarding force. The gravitational attraction of the earth gets smaller as the bullet gets farther away.

In cosmology, an open universe corresponds to the situation of "escape"—the galaxies would be receding from each other so rapidly that the gravitational deceleration would never slow them down to zero speed. To determine whether the universe is open or closed, we estimate the average density of matter. If the density is high enough, *averaged over all space*, then the gravitational attraction is strong enough to "close the universe." The *critical density* above which this would occur is about 5×10^{-30} grams per cubic centimeter, which corresponds to about three hydrogen atoms per cubic meter. Unfortunately, different teams of qualified astronomers have made very different estimates of the average density of the universe. Some of these estimates are higher than the critical density, and some are lower. The problem is that the masses of objects in extragalactic space, and even the distance scale, are poorly known. Further, there is probably a considerable amount of unobservable mass, such as mass in the form of black holes or extremely thin and hot intergalactic gas. It is obviously very hard to estimate this unobserved mass. Recently, a diffuse background of x-rays mapped by the HEAO-1 satellite has been attributed to an intergalactic gas of this type. This conclusion may allow better estimates to be made. However, the interpretation of the HEAO-1 observations is itself new and unconfirmed. Thus, another method of determining whether the universe is open must be tried.

The second method for investigating whether the universe is open or closed consists of actually determining the manner in which the expansion has been decelerating since early times. As we look farther out in space, we observe the red shifts of galaxies at greater distances, and hence at correspondingly more ancient times. At these more ancient times, the galaxies were closer together. Thus, their mutual gravitational attraction was stronger, and the deceleration was greater. In other words, as time passes, the deceleration of the expanding universe decreases. By comparing red shifts of galaxies at different distances, we can see how the deceleration has changed with time.

The situation is depicted schematically in Figure 13-20. Although this method is fine in principle, it has several defects in practice. One defect is that the red shifts, and hence the recession velocities, of galaxies at different distances are roughly the same for the various universe models— except at extremely great distances. But we see very few galaxies at such great distances, because the galaxies are simply too faint. Another fundamental problem is that it is very difficult to estimate the distance of a galaxy. The usual method is to compare the apparent brightness of the galaxy with the absolute magnitude and use the inverse square law to determine the distance. But what is the absolute magnitude of the galaxy? As we have seen, galaxies come in different shapes, sizes, and masses, and there are corresponding differences in their absolute magnitudes. The absolute magnitudes of very distant galaxies are very poorly known, since they are just too faint and too small in angular size for detailed study. Thus, in effect we are limited to simply measuring their apparent magnitudes. Since galaxies come in clusters, the usual practice is to assume that a well-defined property of the cluster can be used as a "standard candle," that is, an object whose absolute magnitude is believed equal to that of similar nearby objects. A typical standard candle would be the brightest galaxy in a cluster, or the tenth-brightest galaxy in a cluster, for example.

However, the results of studies of this type indicate that the standard candles are far from perfect. Thus, when diagrams of the type illustrated by Figure 13-20 are prepared with actual measured data, there is scatter

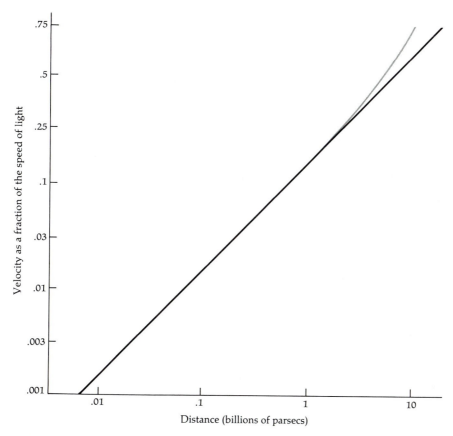

Figure 13-20. Deceleration of the universe could be detected by departures from Hubble's law (dark line). If there were no deceleration, the relation between velocity and distance would be the straight line. Deceleration would produce the departure shown (gray line), but note that it is apparent only for the most distant galaxies. (After *Will the Universe Expand Forever?* by J. Richard Gott III, James E. Gunn, David N. Schramm, and Beatrice M. Tinsley. Copyright © 1976 by Scientific American, Inc. All rights reserved.)

in the plotted points because the assumed absolute magnitudes are not perfectly correct. The measured apparent magnitudes from which the distances are estimated are also uncertain due to errors in the observations. As a result, different studies of the deceleration of the universe according to the method of Figure 13-20 yield different results, and even the results of the best-regarded studies are quite uncertain.[5] We could extend Figure 13-20 to much greater distances if we were able to use quasars in addition to (or instead of) galaxies, since quasars, being very bright, are observed at much greater distances. Since the curves in Figure 13-20 diverge by increasing amounts at greater distances, it would then be easier to tell whether the universe is open or closed. However, the evidence available until now indicates that individual quasars have very different absolute magnitudes. There is no certain way of judging the absolute magnitude of a quasar from its observed properties, although some astronomers are building a case for their idea that the absolute magnitude can be estimated from a feature due to ionized carbon emission in the ultraviolet spectrum. If this idea is correct, then quasar distances can be estimated, quasars can be used in place of galaxies, and it should be possible to determine whether the universe is open or closed.

---

[5] Another problem is that the absolute magnitude of a galaxy may change with time. Thus, as we study galaxies further back in time (younger galaxies), we have to allow for this kind of unknown evolutionary effect.

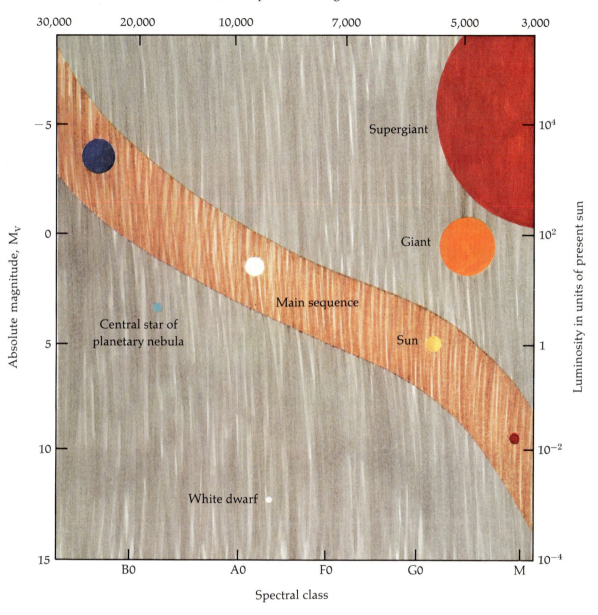

Surface temperature in degrees Kelvin

Absolute magnitude, M_V

Luminosity in units of present sun

Spectral class

Plate 1 In this artist's version of the Hertzprung-Russell Diagram, a graph of star brightnesses (vertical scale) versus temperatures (lower scale, higher temperatures at the left), the positions of typical kinds of stars are indicated by colored disks whose sizes indicate the wide range in stellar diameters. (Not to scale.) Shown are the sun and other stars of the main sequence, as well as a red giant, a red supergiant, a white dwarf star, and the central star of a planetary nebula.

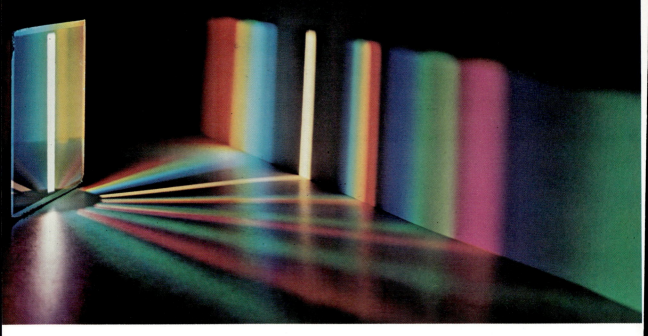

Plate 2 A beam of white light strikes a diffraction grating and produces a spectrum. (Courtesy of Bausch and Lomb.)

Plate 3 A loop-shaped prominence in the corona, as photographed from SKYLAB on December 19, 1973. The loop was moving rapidly away from the sun and had grown to forty times the size of the earth. The original photograph was taken in the light of ionized helium at 304 Å. (National Aeronautics and Space Administration and Naval Research Laboratory.)

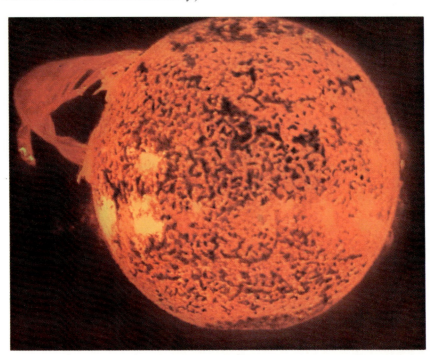

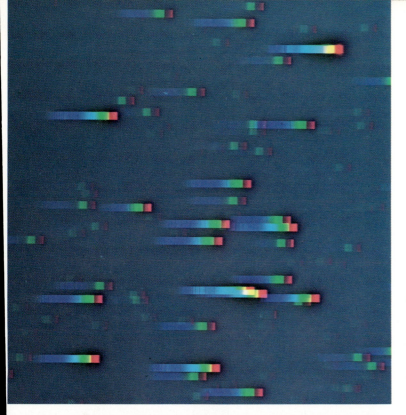

Plate 4 Photograph of the Hyades star cluster, taken with a large prism mounted on the front end of a Schmidt telescope. The spectrum of each of the brighter stars is seen. (Courtesy of the University of Michigan.)

Plate 5 Comet Bennett photographed on April 16, 1970. (J.C. Brandt, R.G. Roosen, and S.B. Modali, Goddard Space Flight Center.)

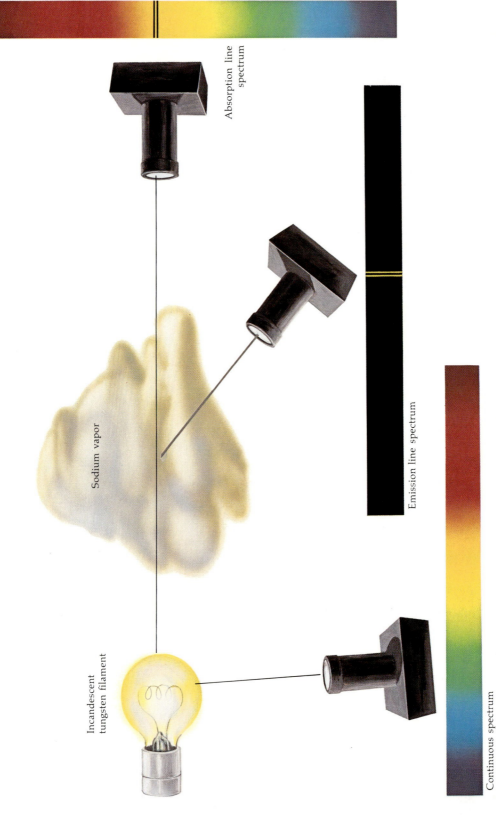

Absorption line spectrum

Sodium vapor

Emission line spectrum

Incandescent tungsten filament

Continuous spectrum

Plate 6 Sketch illustrating Kirchhoff's laws.

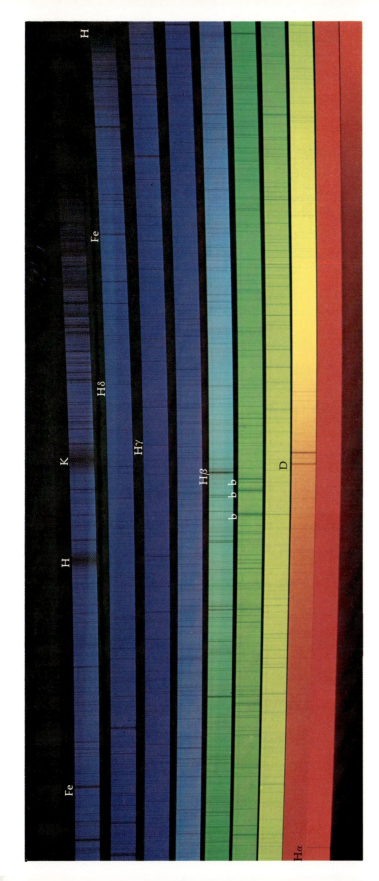

Plate 7 The visible light spectrum of the sun. Some of the principal absorption lines, the elements producing them, and the wavelengths are given in the table. (Courtesy of the Sacramento Peak Observatory, Air Force Cambridge Research Laboratories.)

IDENTIFYING SYMBOL		WAVELENGTH IN ANGSTROMS			ELEMENT
FRAUNHOFER	OTHER				
C	Hα	6,563			Hydrogen
D		5,890	5,896		Sodium
	b	5,167	5,173	5,184	Magnesium
F	Hβ	4,861			Hydrogen
	Hγ	4,340			Hydrogen
	Hδ	4,102			Hydrogen
	Fe	4,046			Iron
H		3,968			Calcium
K		3,934			Calcium

Plate 8 The spiral galaxy in Andromeda, Messier 31. (Courtesy of the Hale Observatories. Copyright © by the California Institute of Technology, and the Carnegie Institution of Washington.)

Plate 9 The Veil nebula.

Plate 10 The Crab nebula.

Plate 11 View of Mars from the Viking 1 spacecraft in June 1976. Note the Grand Canyon of Mars at left. (National Aeronautics and Space Administration.)

Plate 12 Close-up picture of the planet Jupiter, telemetered back to earth from a from a Pioneer interplanetary spaceprobe. The large oval feature at upper left is the great red spot, a Jovian atmospheric disturbance that has been studied by astronomers for more than three centuries. (National Aeronautics and Space Administration, Ames Research Center.)

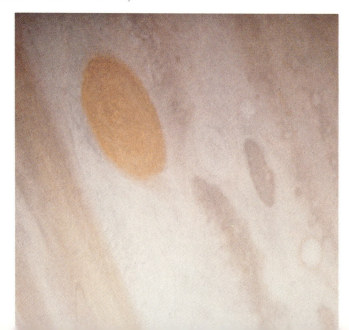

Space Age Astronomy

ATMOSPHERIC LIMITATIONS OF
GROUND-BASED OBSERVATORIES

In the discussion of observatories (Chapter 6), we mentioned some factors that affect the selection of an observatory location. For some astronomical research, however, there is *no suitable location on earth*—good weather and remoteness from city lights are not sufficient. This circumstance arises from four basic properties of the atmosphere: *seeing, scattering, absorption,* and *airglow.*

Seeing refers to the smeared appearance of the image of a star as seen through a telescope. It arises in turbulent air masses, which bend starlight by different amounts. The atmospheric seeing effect limits the sharpness of astronomical photographs and reduces our ability to study fine details, such as the filaments and dark globules in nebulae, the narrow dark lanes between the bright granular elements in the solar photosphere, and the surface markings of the planets. With the atmosphere above one's telescope, it is rarely possible to photograph detail much smaller than one second of arc. Seeing even affects the limiting magnitude of a telescope, since the image of a very faint star may be spread out so much that it is no longer detectable.

The *scattering* of sunlight by air molecules, which causes the sky to appear blue, hinders our ability to study the faint solar corona. Starlight is also scattered, and this decreases the apparent brightnesses of the stars, especially at the shorter wavelengths, where the effect is most pronounced.

Absorption of light by atoms and molecules in the atmosphere is a much more severe effect. Ozone absorbs ultraviolet radiation (below 3,000 Angstroms), and other gases contribute to the absorption at even shorter wavelengths, so ultraviolet, x-rays, and gamma rays from the sun and stars do not penetrate to the earth's surface (see Figure 14-1). Water vapor, oxygen, and carbon dioxide are among the chief absorbers at the infrared wavelengths between 1 micron (10,000 Angstroms) and 1 millimeter wavelength. However, there are a number of "windows" in the infrared spectrum, corresponding to limited wavelength regions where the absorption is small. Infrared observations can thus be made in these wavelength windows, especially from mountaintops that are above much of the earth's water vapor. Beyond the infrared, in the radio portion of the spectrum, there are a few windows at millimeter wavelengths, and the atmosphere is essentially transparent between 1 centimeter and 10 meters. Beyond about 10 meters, the waves are reflected back into space by the electrons in the earth's ionosphere. Thus, cosmic radiation with very long wavelengths cannot be received at the surface of the earth.

In addition to atmospheric scattering and absorption, which reduce the light that reaches us, there are the *airglow* emissions by atoms and molecules in the upper atmosphere. This dim light in the sky interferes

Figure 14-1. Radiation from space is heavily absorbed by the atmosphere (dark shading), and only certain wavelengths (light shading) penetrate to the earth's surface. The situation is improved at balloon altitudes, and the full electromagnetic spectrum is accessible at satellite altitudes. (From *Observations in Space* by Arthur I. Berman. Copyright © 1963 by Scientific American, Inc. All rights reserved.)

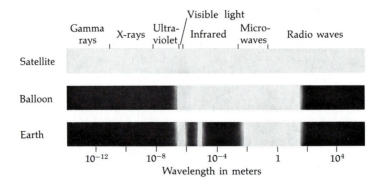

with observations of the faintest stars and galaxies. Further, since the airglow is emitted preferentially at certain wavelengths, it can obscure the spectra of astronomical objects at those wavelengths.

Space astronomy (observations from high-altitude aircraft, balloons, rockets, and satellites) became necessary because of these atmospheric limitations. It was principally stimulated, however, by the possibility of observing in previously inaccessible wavelength regions. *The radiation in different parts of the spectrum appears to be dominated by different physical processes, or to arise under different physical conditions, or to come from different sorts of celestial objects.* There are exceptions to this statement, but generally speaking, it is valid. Take the sun, for example: in the visible region of the spectrum most light comes from the photosphere; in the ultraviolet, most of the radiation comes from the chromosphere and corona; at x-ray wavelengths, most of the solar emission is due to flares and to active regions in the corona; and in the long radio wavelengths that do not penetrate to the surface of the earth, bursts of radiation are detected from particles (ejected by flares) that travel out beyond the limit of the visible light corona. Ultraviolet astronomy concentrated initially on the hot O and B stars, which emit most of their radiation below 3,000 Angstroms, and on the interstellar gas, which has its strongest absorption lines in this part of the spectrum. Infrared astronomy has involved the search for new types of cool stars and for emission from dust clouds. X-ray astronomy includes the study of high-temperature flares on the sun and of phenomena that occur when matter streams toward a highly condensed object such as a neutron star or black hole. Gamma ray astronomy includes the measurement of radiation from processes in the nuclei of atoms (rather than the electron shells). Radio waves and infrared radiation are able to penetrate interstellar dust that is opaque to optical light; at these longer wavelengths it is possible, therefore, to observe previously unexplored regions such as the center of the galaxy.

TRADITIONAL REMEDIES

High-Altitude Observatories

For some astronomical investigations there is considerable improvement in the atmospheric limitations when one observes from a mountaintop. The decrease in scattered light is sufficient to enable limited observations of the outer solar atmosphere with coronagraphs (Figure 14-2), which has encouraged the establishment of solar observatories on mountains in the southwestern states. The decrease in water vapor at high altitude makes a

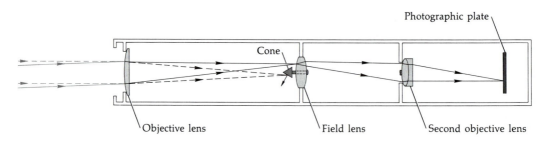

Figure 14-2. Optical sketch of a coronagraph showing the blocking devices, lenses, and baffles. Most of the photospheric light (dashed lines) is removed by the metal cone. Coronal light (solid lines) passes through to the plate. Stray light is caught by the baffles. A tiny mirror on the front of the second objective catches a spurious image of the sun. Finally, every optical element, but particularly the first objective, must be kept dust free, to avoid scattering of light by the dust particles. Even when the sky is clear, and the optics clean, the corona cannot be photographed to as great a distance from the solar limb as during a total eclipse—when no coronagraph is needed.

significant difference in the transmission of infrared radiation by the atmosphere. An observatory at 4,105 meters (13,600 feet) on Mauna Kea, Hawaii, has been built for infrared astronomy. The altitude is good for astronomy, but hard on astronomers, who experience considerable fatigue while working at the facility. Observations of the faint zodiacal light (Figure 2-1) are especially sensitive to scattered light. This has led to expeditions to quite high mountaintops, such as Chacaltaya in Bolivia at 5,307 meters (17,630 feet). The seeing is also better on some mountains; this depends on the local weather conditions as well as the altitude. At the Pic-du-Midi (2,807 meters or 9,400 feet) in the Pyrenees, the excellent seeing enabled French astronomers to observe relatively fine details on planetary surfaces prior to the availability of even better close-up views from spaceprobes.

High-Altitude Aircraft

Aircraft are especially valuable for infrared astronomy studies, as the improvement in transmission of the atmosphere at altitudes of 11,000 to 15,000 meters (35,000 to 50,000 feet) makes a considerable difference to infrared observations. One of the most remarkable contributions came from observations of the center of the galaxy. University of Arizona astronomer Frank Low, flying in a Lear jet near an altitude of 15,000 meters, used a telescope (Figure 14-3) mounted in a port on the side of the airplane to measure infrared emission from the galactic center at wavelengths between 40 and 350 microns. This region of the galaxy cannot be seen in visible light because of the large amount of dust in the galactic plane between the earth and the center. The airplane observations indicated that in a region of less than 10 parsecs diameter at the galactic center, infrared energy is being released at a rate of about 8 million times the total bolometric energy emission of the sun. The explanation for this phenomenon is not yet clear.

A more advanced facility is the Kuiper Airborne Observatory, consisting of a computer-controlled, 90-cm (36-inch) telescope mounted in a C141 jet plane (Figure 14-4). This aircraft is based at the NASA-Ames

Figure 14-3. Astronomer Frank Low at the observer's position in a Lear jet used as an infrared observatory. Face mask provides oxygen as he operates the telescope. (NASA.)

Figure 14-4. The Kuiper Airborne Observatory. (NASA-Ames Research Center.)

Research Center, Moffet Field, California, and accommodates teams of astronomers who study the infrared spectra of celestial sources. An aircraft can also be flown along the path of totality to extend the effective duration of a solar eclipse. In 1973, a supersonic Concorde gave astronomers an unprecedented *74 minutes* of total-eclipse viewing time!

Balloons

Balloons are useful at much higher altitudes than aircraft. Many balloon experiments are conducted at altitudes as great as 46,000 meters (150,000 feet). After a few attempts at manned experiments, nearly all of the astronomy in this field has been done with unmanned balloons. The telescopes can be controlled by radio command from the ground. In one case, superb photographs of the photospheric granulation and of filamentary structure in the penumbrae of sunspots (see Figures 9-5 and 9-6) were obtained with the Project Stratoscope 30-cm balloon telescope. Another Stratoscope experiment recorded an extremely compact nucleus in a peculiar galaxy—a nucleus so small that it is smeared out by atmospheric seeing when the galaxy is observed from the ground. A sketch of a balloon telescope is given in Figure 14-5.

Figure 14-5. The Balloon-Borne Ultraviolet Stellar Spectrometer (BUSS). Key elements of this payload are: (1) pointed section; (2) gondola; (3) azimuth shaft; (4) electronics box; (5) elevation shaft; (6) telescope; (7) star-tracker; (8) spectrograph; (9) detector container; (10) roll cage structure; (11) battery; (12) electronic systems; (13) balast box; and (14) magnetometer. (NASA.)

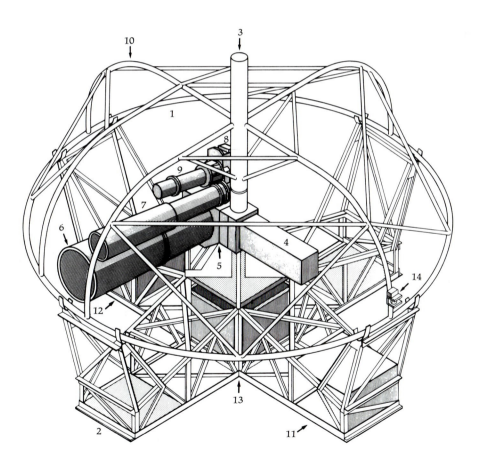

Balloons have been used extensively for "far-infrared" astronomy—for example, to map the Milky Way at wavelength 300 microns—and for gamma-ray, cosmic-ray, and "near-ultraviolet" (2,000–3,000 Angstroms) observations. In 1976, for example, Dutch and American balloon experimenters recorded previously unseen ultraviolet emission lines that originate in the chromospheres of cool, giant stars.

Rockets

X-rays as well as those ultraviolet waves at 100 to nearly 2,000 Angstroms are absorbed in the thin atmosphere above balloon altitudes. Therefore, in order to observe at these wavelengths, the greater heights reached by rockets and satellites are necessary. In the study of objects beyond the solar system, rocket experiments opened up a previously unsuspected field of astronomy. In fact, the discovery of the strongest galactic x-ray source, Scorpius X-1 (Sco X-1), was a happy accident in a rocket flight that failed in its intended purpose to detect x-rays from the moon!

The study of x-ray sources[1] depended primarily on the use of rockets from 1962, when the first source was discovered, until UHURU, the first x-ray observatory satellite, was launched in 1970. About 50 x-ray sources were discovered during this time, and it was found that most of them belonged to two basic classes: supernova remnants and *x-ray binaries*. The supernova remnants were objects, such as the Crab nebula, that were already well known to astronomers. However, the x-ray binaries were associated with relatively inconspicuous visible-light stars that were little studied prior to the discovery of their x-rays. Since only a small fraction of known binary stars were observed to produce detectable x-rays (and these were *not* simply those binaries that happen to be closest to the earth), it was clear that the x-ray binaries must be basically different from ordinary binary stars. However, studies with rockets were hampered by the basic disadvantages of this technique: the observing time per rocket flight is very short (typically only a few minutes in those days), and, due to the high cost of such experiments, only a modest number of rocket flights can be conducted each year. Thus, it is not feasible to make an intensive study of a binary star that has an orbital period of, say, a few days, with rockets, nor is it easy to make a systematic survey of the whole sky. Such research had to await the advent of UHURU and other appropriate satellites.

---

[1]This excludes x-ray sources on the sun, as special satellites to monitor solar x-rays were already operating during this interval.

A third and apparently much smaller class of x-ray sources was discovered with the rockets of the late 1960s. These objects, called *transient x-ray sources,* were distinguished by short lifetimes (for example, a few months), during which they attained great brilliance in x-rays. The first of this class, called Centaurus X-2, was discovered in 1967, while the next source was not found until 1969. We later learned, with satellites, that these objects flare briefly and brightly, sometimes becoming more intense than the Crab nebula and Sco X-1, and are in fact very common. However, satellites must continuously monitor the sky with x-ray detectors to find them before they fade away.

The advantages of using a rocket for astronomical research are that one can reach altitudes of several hundred kilometers, where the limitations of the atmosphere no longer apply, and that it can be done at a much lower total cost than is involved in launching a satellite into earth orbit. On the other hand, the observing time above the atmosphere is very limited. In fact, the great contributions of rocket experiments have been possible because of the very limited number of satellites available for astronomical research until quite recently. With the advent of orbiting observatory satellites, the role of the rocket experiment is increasingly becoming that of a test bed for important new ideas and instruments before the scarce satellite accommodations are assigned to them.

The best known rocket vehicles used by American astronomers belong to the Aerobee series (Figure 14-6). These liquid-fuel rockets lift typical payloads of 15 to 270 kg (33 to 600 pounds) to altitudes ranging from 240 to 480 km (150 to 300 miles), with exposure times of two to six minutes above the level where the atmospheric effects interfere. Astronomers are currently switching over to the newer, solid-fuel rocket, the Astrobee, which has approximately 25 percent greater payload capacity than the Aerobee. For most work in astronomy, the key component of a rocket vehicle is the guidance system. In the early history of x-ray astronomy it was not possible to point continuously at a specific target during the flight. While spinning and tumbling, the payload instruments scanned large regions of the sky. This resulted in the discovery of many celestial x-ray sources, but provided only meager data on the properties of each one. Modern rocket guidance systems allow the astronomer to select a specific target before launch and to use the full duration of the flight above the atmosphere to study that particular target.

ORBITING OBSERVATORIES

Mountaintop observatories and telescopes flown in aircraft and balloons are still limited by the atmosphere, especially by its low transmission of

Figure 14-6.   The launch of an Aerobee rocket. (NASA.)

stellar radiation at the very short and very long wavelengths. Rockets can carry intruments above the atmosphere, but the short duration of a flight limits the observation time available. The obvious solution is to have a telescope permanently (or at least for an extended period of time) located outside the offending region of the atmosphere. Two methods of achieving this were considered by space scientists and engineers. One method was that of a permanent observatory on the moon (*lunar base*), either operated by astronauts or occasionally visited and serviced by them. In terms of astronomical research, this method has not been adopted, although the Apollo astronauts did make some astronomical observations from the moon, and they did install some automated geophysical instruments that continued to telemeter information on the occurrence of moonquakes and other local phenomena for several years. The other approach to a semi-permanent space telescope is that of the *orbiting observatory*, an artificial earth satellite carrying a telescope and other instruments operating via electronic commands from a control room (Figure 14-7) on the earth.

Figure 14-7. Eclipse of the sun, March 7, 1970, as observed in ultraviolet light by the Harvard College Observatory's instrument on satellite OSO-6. The different shadings represent different brightnesses; observations made at the different locations on the solar disk by scanning the instrument across the sun are combined to produce the image shown here on a computer display device at the OSO Control Center. (Goddard Space Flight Center.)

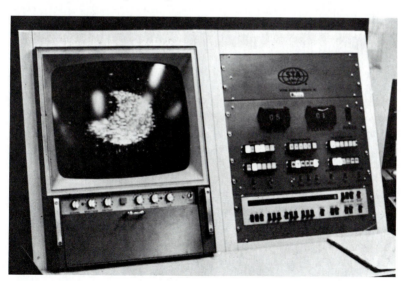

Solar Research Satellites

The most extensive work with orbiting telescopes has been the program of solar research from space. The first *Orbiting Solar Observatories* or OSOs (Figure 14-8) were built and launched before the other astronomy satellites because, as a practical matter, it was easier to develop apparatus to stabilize a spinning satellite to point its telescopes at a target as bright as the sun. In addition, the sun is unique to astronomical research in that it is the only star close enough for detailed studies of its atmospheric structure and changing features. Obviously it is also of much greater practical interest to us on earth than any other star. Eight OSOs were put into orbit between 1962 and 1975. Together with the important solar investigations carried out by SKYLAB astronauts in 1973–1974, they illustrate the way that space research in a given technical discipline usually proceeds. The

Figure 14-8. Photograph of the OSO-5 satellite that was used to study solar flares. Principal parts are labeled; additional experiments carried in the wheel section are not indicated. The wheel rotated about 30 times per minute to stabilize the satellite as the sail pointed to the sun. Gas jets controlled the rate of rotation. (Courtesy of Ball Brothers Research Corporation, Boulder, Colorado.)

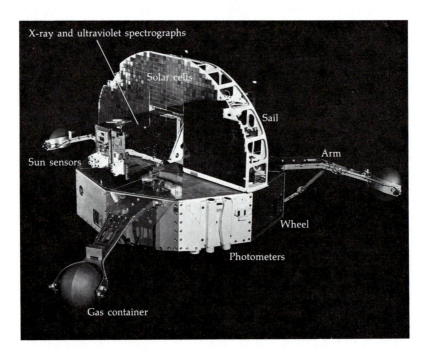

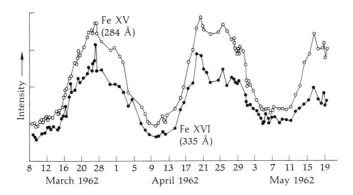

Figure 14-9. Variable solar emissions of highly ionized iron lines as determined from the first Orbiting Solar Observatory. (Goddard Space Flight Center.)

overall trend is from spacecraft of limited capabilities, bearing instruments with rather crude resolution in space, wavelength, and time, to more complex and capable satellites, with instruments of higher sensitivity and resolution that are designed for detailed investigations of the phenomena discovered by the previous satellites.

The instruments on OSO-1, in 1962, simply measured the ultraviolet, x-rays, and gamma rays from the whole sun—technology did not yet allow the mapping of the sun's detailed appearance. No gamma rays were detected, but the observations showed that the intensity of x-ray emission was correlated with the presence of active regions on the sun, and that sometimes rapid bursts of x-rays were emitted. A similar result was found in the ultraviolet spectrum, where emission lines produced in the solar corona underwent substantial variations. Some of these changes took place fairly gradually; a tendency of the peaks (Figure 14-9) in the intensity of the emission to occur at intervals of about 27 days suggested that much of the line radiation was coming from active regions in the corona that were passing across the solar disk as the sun rotated.

With the early solar satellites, solar x-rays and ultraviolet light were studied with instruments that actually could not resolve separate portions of the solar disk. This was not a serious drawback for some x-ray research, since the intensity of x-rays from a solar flare or other active event is sometimes much greater than the total emission at the same wavelength from the rest of the sun. This provides a sort of *de facto* resolution of the active event on the sun, and indeed from such studies we learned that temperatures in a solar flare sometimes briefly reach 50 million K, hotter even than the center of the sun! However, in the ultraviolet, the emission arises in many parts of the sun, and often the two or three brightest regions have comparable intensity. Thus, the ability to resolve small parts of the sun for separate study was needed. By 1969, an OSO-6 instrument was operating with the capability of mapping the whole sun in ultraviolet light every eight minutes, with a spatial resolution of 30 arc seconds (about one-sixtieth of the solar diameter). Studying such maps,

Figure 14-10. The corona in x-rays as photographed from SKYLAB on June 12, 1973. Numerous x-ray bright points are clearly visible, as is the conspicuous dark coronal hole. A small flare is in progress in the coronal hole. (Courtesy Harvard-Smithsonian Center for Astrophysics.)

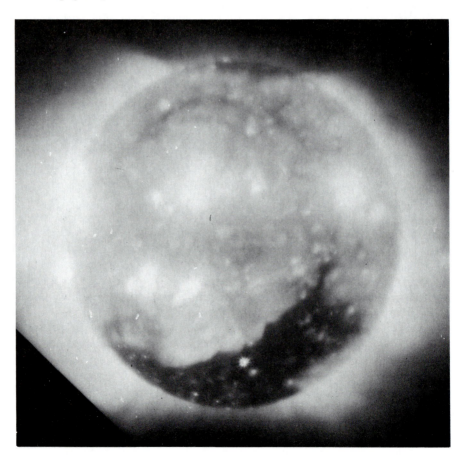

the experimenters were able to identify interesting regions and command the satellite instrument to concentrate on them to determine their physical properties. Research of this kind led to the discovery of an *isothermal plateau*, a narrow zone of height in the chromosphere where the temperature rises much less rapidly with altitude than elsewhere above or below. The final OSO (OSO-8) carried an instrument capable of resolving ultraviolet features only a few arc seconds in size. (At the distance of the sun, one arc second corresponds to a length of 725 km or 450 miles.) Among other studies, this allowed the experimenters to monitor small oscillating regions and to scan the limb of the sun to determine the precise altitudes at which various spectral lines are most intensely emitted. These studies provided the raw data for improved physical theories of the structure of the solar atmosphere.

In 1973–1974, a battery of large solar telescopes was carried on the manned spacecraft SKYLAB. The astronauts monitored viewing screens in order to locate interesting features and events on the sun and direct the instruments to them. Photographs of exceptional quality were obtained, since the astronauts were able to change film canisters and bring the exposed film back to earth. (Unmanned satellites transmit cruder pictures by means of radio telemetry.) Among the many phenomena discovered was that of the *x-ray bright points* (Figure 14-10). These are small, hot structures in the corona that last only about eight hours. As many as 1,500 of them may occur on the sun each day, and, surprisingly, they occur about equally all over the sun, whereas flares, sunspots, and most other previously known manifestations of solar activity occur mostly in restricted regions of the disk. By comparing the SKYLAB x-ray photos with solar magnetograms made at ground-based observatories, the investigators learned that the x-ray bright points occur at locations where new magnetic field lines were emerging from below the visible surface of the sun. Longer-term studies then showed that within 30 degrees of the solar equator, occasional longer-lived bright points, which may survive for more than a day, seem to grow into the active regions of sunspots and flares that are the most familiar components of solar activity.

A white-light coronagraph, which produced artificial eclipses, also was launched on SKYLAB. It obtained far better photographs than possible with coronagraphs based on earth, thanks to the absence of atmospheric scattering at satellite altitudes. Already, from a more primitive television coronagraph operating in 1971–1972 on OSO-7, we knew that the corona occasionally ejects huge "bubbles" of solar plasma at high speed. The SKYLAB instrument recorded many series of time-lapse photos of such events (Figure 14-11).

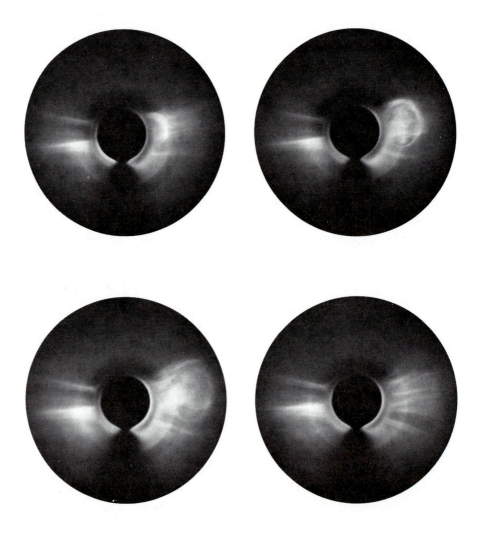

Figure 14-11. The ejection of a solar "bubble" as photographed from SKYLAB on August 10, 1973. The Universal Times of the different photographs are: upper left, 13:43; upper right, 14:24; lower left, 14:48; lower right, 16:37. The central, dark circle corresponds to the occulting disk, which blocks light from the photosphere. The disk's projected diameter, equal to 3 $R_\odot$, reveals by comparison the immense size and rapid motion of the bubble. (Courtesy High Altitude Observatory.)

Ultraviolet Astronomy Satellites

An *Orbiting Astonomical Observatory* (OAO-2) was launched in 1968. Much larger than an OSO, the 10-foot-long OAO carried eleven instruments into orbit for ultraviolet studies of stars, galaxies, and other objects. At one end of the spacecraft, four 32-cm (12.5-inch) telescopes focused ultraviolet light through filters onto television cameras. This equipment (called the *Celescope* experiment) was intended to answer the basic question: What does the sky look like in the ultraviolet? Its purpose was to send back pictures of as much of the sky as possible. These comprised space astronomy's first attempts to produce an equivalent of the Palomar Sky Survey, which mapped the sky at visible wavelengths. Each of the four Celescope instruments obtained its pictures in light of a different ultraviolet wavelength region, depending on the properties of the filters and television tubes. These wavelengths ranged from 1,050 to 3,200 Angstroms. Two Celescope pictures of the Pleiades open cluster are shown in Figure 14-12; visible-light photos were given in Figure 6-19.

At the other end of OAO-2 was a 40-cm (16-inch) telescope and six smaller instruments. Instead of taking survey pictures as in the Celescope experiment, these instruments were used to measure the ultraviolet brightnesses and spectra of individual preselected targets. For example, this equipment was used by Arthur Code of the University of Wisconsin

Figure 14-12. The Pleiades as seen by the Celescope experiment. Compare with the ordinary photographs given in Figure 6-19. (Courtesy of Project Celescope, Smithsonian Astrophysical Observatory.)

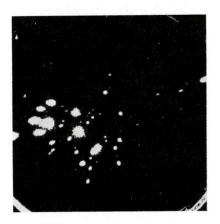 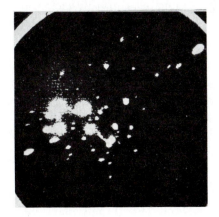

to study two bright comets during the early months of 1970. Observing Comet Tago-Sato-Kosaka on January 14, he detected free hydrogen atoms (that is, hydrogen atoms not contained in molecules). This was done by observing radiation in the form of an emission line at wavelength 1,216 Angstroms. This line, called *Lyman alpha,* is known to be produced by hydrogen atoms. The most exciting part of the discovery was that the hydrogen light did not come from one of the three then-known parts of a comet (nucleus, coma, and tail). Instead, it came from an enormous region, slightly larger than the sun, which surrounded the comet. This showed that there is a fourth basic constituent of comets—the hydrogen cloud that shines in ultraviolet light. As described in Chapter 7, the nucleus of a comet consists of frozen water (H_2O) and other icy substances. When the comet is near the sun, the outer skin of ice evaporates, producing water vapor. The sun's ultraviolet light then breaks up the H_2O molecules in the vapor, liberating the hydrogen (H) gas, which flows outward and constitutes the hydrogen cloud.

Many of the spectral lines that are most useful for studying the interstellar gas are located in the ultraviolet part of the spectrum. Therefore, a second ultraviolet satellite, Copernicus, was launched in 1972 to study the ultraviolet gas by means of spectroscopy. One of its most important findings concerned the "missing matter." Since the stars are made from the interstellar gas and eventually return much of their material to the interstellar medium by means of stellar winds, planetary nebula ejection, and supernova explosions, you might expect the interstellar gas to have roughly the same chemical composition as the atmospheres of stars. However, Copernicus measurements of interstellar absorption lines showed that many of the heavier elements, such as iron, were deficient in the interstellar gas, by factors of up to 100. In other words, compared to hydrogen, there is perhaps 100 times as much iron in the atmosphere of a star such as the sun as there is in the interstellar gas. The leading theory to account for the missing matter proposes that it *is* present in space, but in *solid* form. In other words, the iron (and other heavy elements) has condensed out of the gas in the form of tiny dust grains, much like soot particles in the effluent gas from a powerplant smokestack. (Presumably the condensation occurs as the hot gas leaving a star cools down in space.) Unlike gas atoms, the dust grains do not produce spectral absorption lines, although they do cause a general dimming and reddening of the light of background stars.

The Princeton University astronomers under Lyman Spitzer, Jr., who operated Copernicus, also studied the nature of clouds in the interstellar medium. They found that a typical interstellar cloud consists largely of atomic hydrogen, but with molecular hydrogen (H_2) at its center, where dust grains prevent the ultraviolet light of stars from penetrating in sufficient amount to break up the molecules into atomic hydrogen (H). It had previously been believed that the interstellar gas consists of these cold, relatively dense clouds (with typical temperatures of 100 K, or –173° C), surrounded by a warmer (1,000 K), much thinner gas. However, the Princeton scientists found no evidence for this warm gas. Instead, Copernicus revealed the presence of a very hot (100,000 K or more), even lower-density gas that is ionized (each hydrogen atom has lost its electron). A major question, which is stimulating many different theories, is: What causes this gas to be so hot and to be fully ionized? Some astronomers believe that energy from supernova explosions is responsible.

X-Ray Observatory Satellites

In 1970, the launch of the UHURU[2] satellite gave us the first orbiting celestial x-ray observatory. It was a spinning spacecraft, making one revolution on its own axis every 12 minutes, designed to scan the sky and map the locations of x-ray sources. Among the many scientific reports that resulted from the UHURU research was a catalog of 161 sources (Figure 14-13) observed during the first five and one-half months of operation. The strongest source was the well-known Scorpius X-1, over 10,000 times brighter in x-rays (wavelengths 2–6 Angstroms) than the weakest of the sources, which is known by its catalog number, 3U1237-07. In a lecture to the American Astronomical Society in 1974, Edwin M. Kellogg, one of the UHURU scientists, estimated that about 100 of the x-ray sources were objects in our own galaxy, including binary stars, globular clusters, and supernova remnants; about 30 sources are definitely extragalactic in nature. Some of the extragalactic sources are associated with galaxies known to have undergone eruptions or other unusual activity, and with quasars. However, most of these sources were *not* associated with individual galaxies, but rather with galaxy clusters. It is believed that the x-rays from the clusters are produced in a hot gas or "intracluster medium" at a temperature of around 100 million K.

The remaining UHURU sources are not yet fully understood; in many cases we have no idea whether or not they are even located in our galaxy. Later x-ray satellites, such as SAS-3, a spacecraft developed at the Massachusetts Institute of Technology, and the first High Energy Astrophysical Observatory (HEAO-1), studied the detailed spectral and temporal properties of the x-ray sources. In particular, these investigations

---

[2]From the Swahili word for freedom; the satellite was launched from a platform off the coast of Kenya on the seventh anniversary of that country's independence.

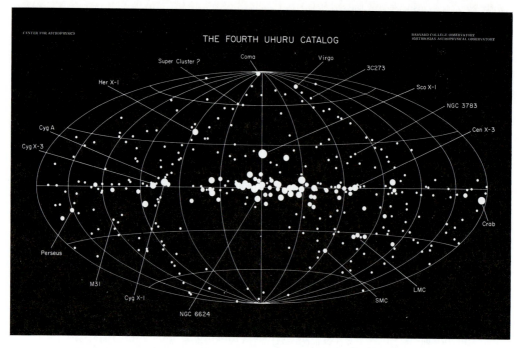

Figure 14-13. Map of the x-ray sky as determined from UHURU. The central, horizontal line corresponds to the galactic equator. (Courtesy R. Giacconi, High Energy Astrophysics Division, Harvard-Smithsonian Center for Astrophysics.)

revealed that many of the sources associated with stars in our galaxy are highly variable in intensity, with x-ray light curves revealing the effects of binary motion (including eclipses), stellar pulsations, and other phenomena. HEAO-1 also measured the diffuse background of x-rays from extragalactic space and found evidence that it may be produced in an extremely hot (500 million K), thin gas of unknown origin. As mentioned in Chapter 13, there may be enough of this gas to close the universe.

There are other astronomical research satellites besides those we have mentioned; they include the Radio Astronomy Explorer, which mapped the long-wavelength radio waves that cannot penetrate the earth's ionosphere, the Vela satellites, which discovered the as-yet-unexplained bursts of gamma rays from the sky, and many spacecraft built by European and Asian nations. As space astronomy advances from its initial stage of determining the basic appearance of the universe at wavelengths that we cannot see from the ground, specially designed rocket and satellite experiments are being launched to test theories based on the initial discoveries.

15

Exploration of Space

We hope someday, having solved the problems we face, to join a community of galactic civilizations.

JIMMY CARTER
President of the United States of America
(Recorded message carried on the Voyager
spacecraft, launched in August 1977
toward the Jovian planets and
interstellar space.)

In the preceding chapters we outlined many of the techniques and results of historical and modern astronomy. This work was carried out almost exclusively with purely *observational* data; that is, the only aspect of an astronomical object that was directly sampled was its photon radiation. Thus, unlike other scientists, astronomers who did this research did not make laboratory or field experiments. Except for a few planetary discoveries mentioned in Chapter 8 and an occasional meteorite, the astronomical material was beyond our reach. Rocket technology was required to send instruments or people to explore the moon and planets and the interplanetary medium. Although the first artificial satellite, Sputnik 1, was launched into earth orbit by the USSR on October 4, 1957, the development of rocketry had been under way for quite some time.

ROCKETRY

The basic principles of rocketry and satellite orbits were known to Isaac Newton. He considered the effects of firing a projectile horizontally from a high mountain. If the speed is low, the projectile travels some distance and falls to the earth's surface. If the speed is increased to about 8 km/sec, the projectile does not fall to earth but travels around the earth in a low, circular orbit. The various situations for the ideal case, which ignore the effect of air resistance (friction), are shown in Figure 15-1. This drawing is based on one by Newton, with some additional ellipses added to show the situation for higher altitude satellites. The period of a hypothetical artificial satellite was easily calculated by Newton, for the earth's circumference of 40,000 km (25,000 miles) is traversed at 8 km/sec in 5,000 seconds, or about 83 minutes. Actually, satellites are usually placed in orbits at a distance of more than 200 km above the surface of the earth to reduce the effect of air resistance. Sputnik 1 had an orbital period of 90 minutes.

When the speed is increased to 11 km/sec (7 miles/sec), a rocket has reached the *escape speed* necessary to leave a closed trajectory or orbit and pass into space beyond the earth. Sometimes one hears this situation described as "escaping from the earth's gravitational field," but this is incorrect. At escape speed the rocket is going fast enough so that it does not follow a closed orbit around the earth, but there is always an attractive force between earth and rocket in accordance with Newton's law of grav-

Figure 15-1. Drawing adapted from a sketch by Newton showing the possibility of launching artificial satellites. (From *The New College Physics* by A. V. Baez. W. H. Freeman and Company. Copyright © 1967.)

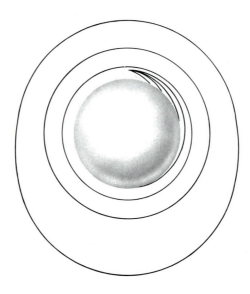

ity. When astronauts walk on the moon under the influence of the lunar gravity, they are still clearly affected by the earth's gravity, because the moon itself is in orbit because of the gravitational force exerted by the earth. In planning a deep space probe, such as a mission to Mars, the gravitational attraction of several objects, certainly including the earth, moon, sun, and Mars, must be taken into account. The necessary computations of the orbit are complex, and there are technical difficulties in launching and inserting the probe into the proper trajectory, so that mid-course corrections (firing of small rockets at times during the flight) usually are necessary to keep the probe on the desired track.

As we have seen, the theory involved in sending a rocket to Mars was known to Newton. But how can the speeds needed for orbit and escape be attained in practice? In the long run it is better to achieve the necessary speed relatively slowly than to acquire it suddenly. Too rapid accelerations tend to squash astronauts and damage the instruments. Thus, Jules Verne's (*From the Earth to the Moon*, 1865) old idea of shooting a space ship from a gun with muzzle velocity equal to the earth's escape speed is unsatisfactory. Moreover, a projectile shot from a gun, carrying no rockets, has no provision for additional maneuvers. All of these problems are solved, at least in principle, by the use of rockets. A gradual acceleration is possible, and fuel reserves can be used for mid-course corrections.

The scientific principle of rocket flight is contained in Newton's Third Law, that for every action there is an equal and opposite reaction. If we produce hot gases in a rocket chamber with a nozzle at one end, the rocket and gases both push on each other, with the result that the gas goes out the nozzle and the rocket goes the other way. This is the principle of the jet engine and the rocket engine. In the jet engine, air from the surroundings is combined with fuel to produce the hot gas. In the rocket, all the needed substances are carried along in the form of gaseous, liquid, or solid propellants. Rockets do not require air from their surroundings to function, and therefore can travel through the near vacuum of space.

The application of these principles to develop useful rockets for space travel was pioneered by three men. The first was a Russian, Konstantin Tsiolkovsky (1853–1935). He was an "idea man" who originated many important concepts. For example, he realized that the use of a single rocket to place a payload in orbit was inefficient and that the efficiency could be greatly increased by the use of multiple stages in which the required final high speed was imparted only to a relatively small upper stage and payload. The second pioneer was Robert H. Goddard (1882–1945) of Clark University. His talent combined theory and experiment. His test flights in Massachusetts and near Roswell, New Mexico led to fundamental innovations in rocketry, including the development of practical liquid propellant rocket motors. He was awarded over 200 patents for his technological developments. In 1959, NASA named the Goddard Space Flight Center, nucleus of its unmanned scientific satellite program, for him. The third pioneer was Hermann Oberth (1894–   ). Although he was associated with a number of practical and theoretical developments, Oberth's contributions were substantially through the inspiration of others by his books on rockets and space travel. Somewhat in contrast to Goddard, Oberth sought to publicize rocketry; for example, he served as a technical advisor to an early science fiction film. His efforts in rocketry and its popularization were important to the rapid development of rocket technology in Germany.

The role of science fiction in these men's careers is worth noting. Tsiolkovsky, Goddard, and Oberth were all initially inspired by the descriptions of voyages to the moon by Jules Verne. Goddard also acknowledged a debt to the writings of H. G. Wells.

The Chinese in the thirteenth century are generally given credit for the use of the first rockets. A great deal of research and technical development, mostly in the twentieth century, has led to advanced rocket vehicles such as the Saturn 5 shown in Figure 15-2. This rocket with payload measures about 110 meters (360 feet) in height, and the weight at liftoff is about 2.9×10^6 kg (6.4 million pounds).

SPACE NEAR THE EARTH

Since the launch of Sputnik 1, a great many artificial satellites and space probes have been used to study the properties of space around the earth. This work has resulted in a detailed, mostly unexpected picture of the remarkable phenomena that surround our planet. The thin upper atmosphere of the earth is strongly affected by the ultraviolet and x-radiation of the sun. This radiation is absorbed in the ionosphere and other regions of the upper atmosphere and constitutes a heat source that affects the density and temperature of the atmosphere at these levels. Since the amount of ultraviolet and x-ray emission varies with solar activity, the upper atmosphere is directly affected by variable phenomena on the sun. For example, when a major solar flare occurs, the absorption of the enhanced ultraviolet radiation at heights near 160 km (100 miles) heats the thin air at those levels, causing it to expand. Satellites orbiting at higher altitudes encounter denser air than usual, because gas from below has expanded up to their altitudes. This in turn causes the satellite to encounter in-

Figure 15-2.   The lift-off of Apollo 11 on July 16, 1969. Aboard were the first men to land on the moon. (NASA.)

creased air resistance, or "drag." in fact, these solar activity effects on the upper atmosphere were discovered when scientists noted that unexpected changes in the orbital motions of satellites were correlated with the occurrence of solar flares.

Trapped Particles

Above the ionosphere there exists a region (*magnetosphere*) in which fast-moving electrons and protons are trapped in the earth's magnetic field. Spiraling along the magnetic lines of force, they bounce back and forth between *mirror points* located in opposite hemispheres. These trapped particles were discovered in 1958 by James Van Allen of the State University of Iowa, using a geiger counter carried on the first American satellite, Explorer 1. Two regions of trapped particles (Figure 15-3) were soon mapped; they were named the *Van Allen Belts*. From additional measurements by subsequent satellites, we now recognize that the entire

Figure 15-3. The Van Allen Belts. The dots indicate schematically the number of trapped particles per unit volume in the earth's magnetic field. (After *Radiation Belts around the Earth* by J. A. Van Allen. Copyright © 1959 by Scientific American, Inc. All rights reserved.)

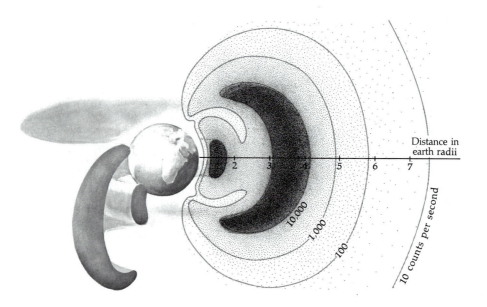

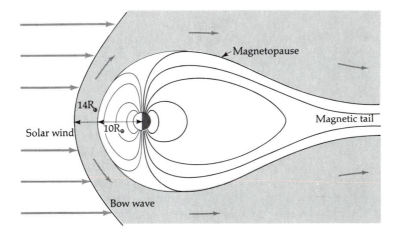

Figure 15-4. The magnetosphere as shaped by the solar wind. The outer boundary of the magnetosphere, called the *magnetopause*, is not generally penetrated by the solar wind plasma. The direction of the solar wind flow is indicated by the arrows. The half-shaded circle represents the earth.

magnetosphere is filled with the trapped particles, and thus the idea of two distinct belts was an oversimplification. On the inner side, the magnetosphere is limited by the collision of the trapped electrons and protons with atoms and molecules of the atmosphere. On the outer side, it ends at a surface called the *magnetopause*. The shape of the magnetopause is shown in Figure 15-4. This surface lies at a distance of about 10 earth radii in the direction toward the sun, and it is the place where we can regard the earth's magnetic field as ending due to the pressure of the solar wind. The solar wind streams around the earth, as shown in Figure 15-4, and in the anti-solar direction a *magnetic tail* extends far out into space. The magnetic tail has been studied with many spacecraft, including several of the Explorer, Interplanetary Monitoring Platform (IMP), and Pioneer series. Note that the tail extends for millions of kilometers in the direction opposite the sun. In the sunward direction, the *bow wave* indicated in Figure 15-4 is formed as the rapidly moving solar wind adjusts itself to stream smoothly past the obstacle of the earth's magnetosphere. It is analogous to the bow waves formed around supersonic aircraft and projectiles in the atmosphere.

For many years before the advent of space satellites, it was known that the earth has a magnetic field, with a pattern resembling that assumed by iron filings sprinkled around a bar magnet (Figure 3-15). From the satellite measurements, we now find that although this pattern, called a *dipole field*, holds true near the earth's surface, it becomes distorted at high altitudes due to the effects of the solar wind. Sometimes a major solar flare will eject protons from the sun—in small amounts, but at speeds greatly exceeding those of the solar wind particles. These flare protons are deflected by the earth's magnetic field, and they strike the ionosphere above the polar caps, producing increased ionization that blacks out many radio communication channels over the poles. In addition, we often have striking auroral displays in the temperate or even tropical zones of the earth after solar disturbances occur. The *aurora borealis* (northern lights, Figure 15-5) and *aurora australis* (same effect when seen in the Southern Hemisphere) are produced by electrons *(auroral electrons)* that collide with the atmospheric gases and cause them to glow. A third type of particle, besides trapped particles and auroral electrons, should be mentioned: the *cosmic rays* are electrons, protons, alpha particles, and nuclei of most types of atoms. They come from somewhere in our galaxy, possibly from supernovae and pulsars, and perhaps also from beyond the galaxy. A fraction of them have enough energy to penetrate the outer parts of the earth's magnetic field and strike the atmosphere, or even the surface, of the earth.

When the Van Allen Belts were discovered, many physicists believed that they were the source of the auroral electrons. The idea was that solar activity produced changes in the solar wind, which, in striking the magnetosphere, was thought to cause some of the trapped electrons to rain down on the earth and produce aurora. Unfortunately, this beautiful idea is not supported by evidence from the satellites. Among other problems, the auroral electrons have slower velocities (typically less than 2,000 km/sec) than the Van Allen electrons (typically more than 4,000 km/sec). In the early days of space research, it was also thought that the Van Allen particles were trapped in the belts from either or both of two sources, namely, the solar wind and the upward moving products of collisions between cosmic rays and atmospheric gas molecules. Once again, more detailed studies show that, in fact, these sources cannot be responsible for most of the trapped particles. It appears that both the trapped particles and the auroral electrons come from the region of the magnetic tail, and presumably the tail particles somehow originate in the solar wind, but the problems are far from solved. Thus, we know that the earth is surrounded by a region of fast-moving electrons and protons trapped in the magnetic field, and we know that aurorae are caused by somewhat slower moving electrons that strike the atmospheric atoms and molecules, but

Figure 15-5.   A bright *aurora borealis* as seen from the University of
Alaska campus in the fall of 1958. (V. P. Hessler.)

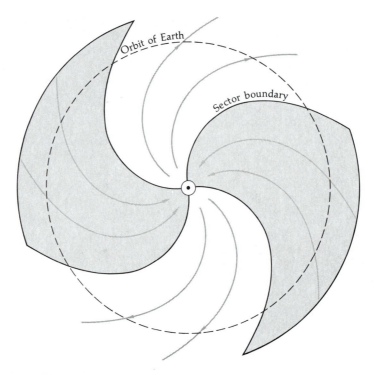

Figure 15-6. The interplanetary magnetic field as seen in the plane of the ecliptic, showing the sector structure. The magnetic field changes from north-seeking to south-seeking in adjacent sectors, as illustrated by the arrows. The magnetic field is drawn into the spiral shape shown by the combined effects of (1) the ejection of matter that carries the field out from the sun, and (2) the sun's rotation. An analogy is the spiral pattern produced in the streams of water from a rotating lawn sprinkler.

we still are not sure exactly how either the trapped particles or the auroral electrons occur.

Instruments carried on satellites and space probes have measured the strength, shape, and variations of the magnetic field surrounding the earth, and have determined the distribution, composition, and velocities of the charged particles in and beyond the magnetosphere. In this latter regard they have established the properties of the trapped particles, the solar wind particles, and the cosmic rays. The properties of the magnetic field and the various sorts of particles are found to respond to variations in the solar activity. In the case of the cosmic rays (which arise outside the solar system), the solar magnetic field, as it is carried outward into interplanetary space by the solar wind, affects the numbers and energies of the cosmic particles that are able to penetrate the solar system to the vicinity of the earth. The field in interplanetary space, arising at the sun, is weaker at sunspot minimum, and so cosmic particles are less strongly deflected by it at that time, and we then detect more of them in our part of the solar system. The measurements show that the interplanetary magnetic field does not have the simple bar magnet pattern that the earth's field assumes near the surface of our planet; rather, there is a spiral *sector structure* (Figure 15-6) that is caused by the combined effects of rotation of the sun and the outward flow of the solar wind. This structure is present at low heliographic latitudes. However, at the high latitudes above the polar regions of the sun, the spiral shape vanishes, and the field lines probably extend radially out into space. A proposed pair of future spacecraft, called the Solar Polar Mission, is intended to explore this region.

Dust in the Solar System

Satellites and space probes also carry instruments to determine the distribution of dust particles or micrometeoroids in circumterrestrial and interplanetary space. The aim is to see whether the dust particles come from the dissolution of comet tails, from debris in the asteroid belt, or from a more general medium that pervades space, and to determine whether there are concentrations of the dust at particular locations in space, such as the *libration points* shown in Figure 15-7.

In the early years of the space age, satellites detected unexpectedly large numbers of the dust particles, and some astronomers thought that the earth and moon each might have a dust cloud. Later work showed that the early experiments had given erroneous overestimates of the amount of dust particles. One of the chief techniques originally used was the method of *acoustic detection,* in which vibrations caused by the impact of an interplanetary dust particle on a microphone-like device mounted on a satellite were converted to electrical signals that were measured and telemetered back to earth. Now it appears that some form of noise in the instruments masked the true impacts of dust particles. More reliable measurements were subsequently obtained with *time-of-flight detectors* (Figure 15-8) carried by the space probes Pioneer 8 and 9. The heart of such a detector is two very thin, electrically sensitive films, placed about 5 cm (2 inches) apart and parallel to each other. As the space probe travels

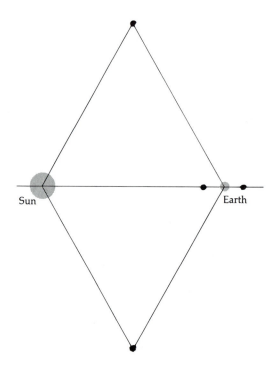

Sun

Earth

Figure 15-7.   Libration points in the sun-earth system. In theory, interplanetary dust could concentrate at the locations marked, but no large concentrations are found. Two of these points are on the earth-sun line at a distance of 0.01 a.u. on either side of the earth, and two more are located at the equilateral triangle points equidistant from the earth and the sun.

through the solar system, a fast-moving interplanetary dust particle will occasionally penetrate first one film, and then the other. The instrument measures the position on each film and the brief time interval between the two penetrations. From this information we can deduce the speeds and directions of motion of the micrometeoroids, as well as count the rate at which their impacts occur.

The results of interplanetary dust particle measurements are still in a preliminary state, since much of the older data are now in dispute. Although the dust has been detected, there is no convincing evidence that it is concentrated around either the earth or the moon. At the average distance of the earth from the sun, and in the plane of the earth's orbit, the most numerous dust particles are those with a diameter of about 1 micron (10^{-6} meter or about 0.00004 inches), and the scarcity of smaller par-

5 cm

Figure 15-8. Interplanetary dust detector carried by Pioneer 8. (Courtesy of O. E. Berg and U. Gerloff, Goddard Space Flight Center.)

ticles may be due to their being blown away by the pressure of sunlight. On the average, there are about 20 particles in the micron size class in every cubic kilometer of space (85 particles/cubic mile) in the vicinity of the earth's orbit. It seems, based on the orbits of the dust particles, that those found in the vicinity of the earth's orbit probably come from comets. There also have been ambitious attempts to actually collect dust particles with rocket-borne traps, and these experiments have brought back some particles from the upper atmosphere. Whether these are truly of interplanetary or earth origin has been argued by scientists. It appears that at least some of them probably are genuine interplanetary dust. Later spacecraft of the Pioneer series carried instruments to study dust particles in the asteroid belt.

THE MOON

The moon's surface has long captured the imagination of amateur astronomers. A good pair of binoculars or a small telescope reveals the wealth of interesting geological or *selenological*[1] features on the lunar surface.

The moon is at a distance of about 60 earth radii, or about 384,000 km (240,000 miles), and has a diameter of about 3,475 km (2,160 miles). The mass was not so easy to determine in the days before spaceprobes were

---

[1]From *selene,* Greek word for the moon.

sent to the moon. At that time, the lunar mass (it is about $1/81$ of the earth's mass) had to be inferred from a wobble in the earth's orbit caused by the earth executing a small orbit around the center of mass of the earth-moon system. (This wobble was detected because it produces a small monthly parallax in the apparent positions of Venus and Mars.) Using the moon's mass and diameter, we can compute its density: it is 3.3 times as dense as water—a value similar to that of the rocks in the earth's crust.

This simple comparison of densities at once tells us that the interior of the moon is unlike that of the earth. The average density of the earth is 5.5 times that of water, much more than the moon's value of 3.3. This occurs because the earth's core is made of iron and nickel, materials denser than crustal rocks, and the material is compressed to densities even greater than those of ordinary nickel and iron, due to the weight of the outer layers. Thus, we would not expect the moon to have an earth-type core.

The moon's gravitational attraction at its surface is about one-sixth of the surface gravity on earth; thus, an astronaut who weighs 180 pounds on earth would weigh only 30 pounds on the moon (the astronaut's mass does not change; the weight of an object is the force exerted by gravity on the object). The accuracy with which we know the mass of the moon has been improved through study of the orbits of circumlunar spacecraft such as the Lunar Orbiters, which made close-up moon photos prior to the manned Apollo missions. Small irregularities in the moon's gravitational force were detected in this way and attributed to concentrations of mass below but near the lunar surface; they are called *mascons*. One theory of the mascons is that they are large and dense meteoroids that struck the moon in the past.

Surface Features

Examining a picture of the moon (Figures 2-2 and 15-9), one notes at once the craters, the mountainous regions, and the *maria*. We now know that there is no water on the moon, but the maria (from the Latin word for seas) were so named on the basis of their appearance as large, relatively dark areas. The pattern of these dark areas, typically several hundred kilometers in extent, with the brighter *highland areas* produces the familiar man-in-the-moon effect. The heights of features on the moon's surface were originally estimated from the lengths of their shadows as seen through earth-based telescopes (note the shadow of the central peak in Figure 15-10). It was found in this way that the maria are generally low-lying and relatively flat. Radar echo timings were subsequently used to study the heights and slopes of the maria. Maria occupy about 25 percent

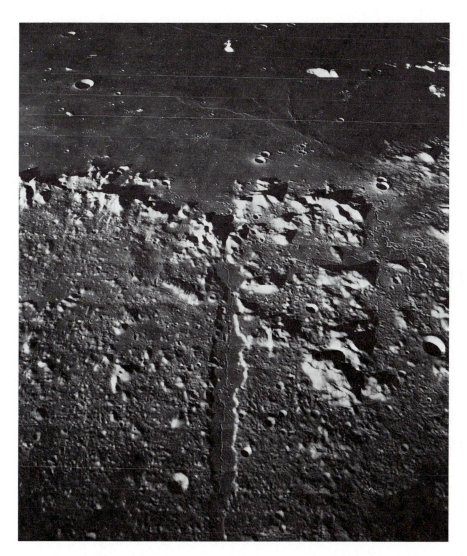

Figure 15-9. A lunar landscape, the region of the Alpine Valley, as seen from Lunar Orbiter 5. The Valley extends through the mountainous region in the foreground to the Mare Imbrium. There are craters in the mountainous area and in the floor of the mare, and there is a rill in the valley floor. A peak is located on the mare plain in line with the Alpine Valley, and another is to the right of the first; these may be volcanic in origin. "Wrinkle ridges" (probably associated with lava flows) are visible on the mare. (NASA.)

Figure 15-10.   The crater Tycho as photographed by Lunar Orbiter 5. Notice the collapsed outer wall and the central peak (with shadow). Tycho is about 100 km (60 miles) in diameter. (NASA.)

of the lunar surface. The other 75 percent is covered by the relatively bright and heavily cratered highlands.

The *craters* are found essentially all over the moon, both in the highland areas and in the maria. A cross-sectional diagram is shown in Figure 15-11. The crater floor lies below the surrounding terrain, and the crater rim rises above the surroundings. Geometrical calculations based on the depths, heights, and lateral extent of crater features show that the volume of material missing from the pit is approximately the same as the volume of material piled up in the crater rim. Many craters have central mountain peaks (Figures 15-10, 15-18, 15-22, and 15-24). There are craters over 150 km (95 miles) in diameter, and, as the Surveyors (unmanned spacecraft that landed on the moon and sent back close-up photos of the surface) revealed, they range down in size to the smallest pits that can be discerned amidst the rocky soil. The ubiquity of craters cannot be overstressed. They are everywhere on the moon. There are even craters within craters within craters . . . (Figure 15-12).

A variety of other structures decorate the surface of the moon. Presumably each type of geological feature provides a clue to the processes that have operated on the lunar surface in the past, just as the geological features of the earth reveal the erosion, glaciation, mountain building, and volcanism that have occurred here. There are linear depressions in the surface, called *rills* (Figure 15-13), and there are the conspicuous *rays*, the bright streaks that radiate from the larger craters such as Tycho (Figure 15-14). The rays extend for hundreds of kilometers, suggesting that material was ejected from the crater region by an explosion or an impact. Small *domes* (Figure 15-15) are also seen on the lunar surface; their appearance is reminiscent of small mountains of volcanic origin that are found on earth.

The principal questions about the lunar surface through the years have concerned the origin of the craters and the nature of the rocks and soil. The controversy over the cause of the craters has concerned the alternative theories of *meteoroid impact* and *volcanic origin*. The fact that craters are mostly located at random on the moon supports the impact hypothesis, since volcanoes on earth usually occur in chains.

The appearance of the rays suggests an impact to most astronomers, although a volcanic origin for them might be possible. The approximate equality in volume of the material in a crater rim and the material missing

Figure 15-11. Profile of a typical lunar crater. (Not drawn to scale.)

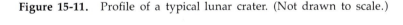
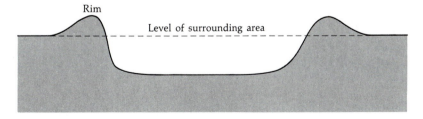

Rim

Level of surrounding area

(a)

Figure 15-12.   Lunar craters. (a) The region of the craters Stöfler and Maurolycus as seen from earth. (Lick Observatory photograph).

(b)

(b) Panorama photographed from the Apollo 11 Lunar Module showing numerous pits and small craters. (NASA.)

Figure 15-13. The region of crater Julius Caesar showing the Ariadaeus Rill. (Lick Observatory photograph.)

from the pit strongly suggests an origin by meteoroid impact. (On earth, the volume of a volcanic mountain far exceeds the capacity of the crater at its top, since the lava that built the mountain came up from deep below the crust.) Since we know that erosive processes on the moon are much less effective than on earth (because there is no water and virtually no air), at least *some* of the craters that we see must have resulted from impact by the smaller bodies in the solar system. In fact, the prevailing body of evidence suggests that most of the lunar craters originated by impact.

On the other hand, there certainly is evidence for some volcanism on the moon. The domes are most likely volcanic; they hardly could be due to impact. Some craters are not randomly distributed but found in *crater chains,* which may have a volcanic origin (Figure 15-16). In addition, the phenomenon of filled-in or *drowned craters* (Figure 15-17) argues for volcanism in the form of lava flows. Finally, there have been occasional reports of gas clouds or reddish glows observed on the lunar surface. These have been discounted many times in the past, but an early record (1783) of such an occurrence was made by a truly expert observer, William Herschel. Several astronomers, notably N. A. Kozyrev in the Soviet

Union, have reported seeing gas clouds in the crater Aristarchus (Figure 15-18). These observations, if correct, could be the result of local subsurface heating.

Rocks and Soil

Now what about the surface material? This can be studied in a variety of ways. The total light reflected from the lunar surface is relatively low. The ratio of the amount of light that is reflected in all directions to the total

Figure 15-14. The region around crater Tycho, showing the rays. (Courtesy Lunar and Planetary Laboratory, University of Arizona.)

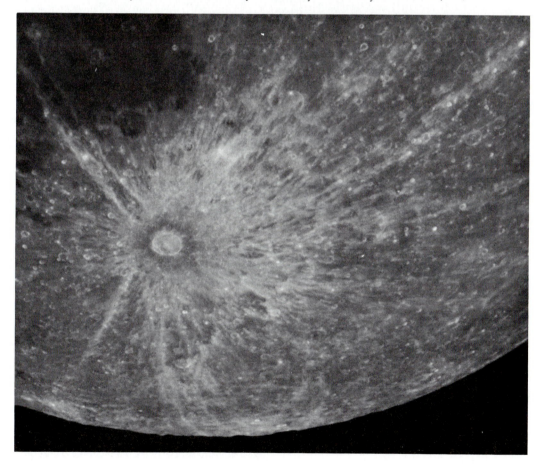

(a)

(b)

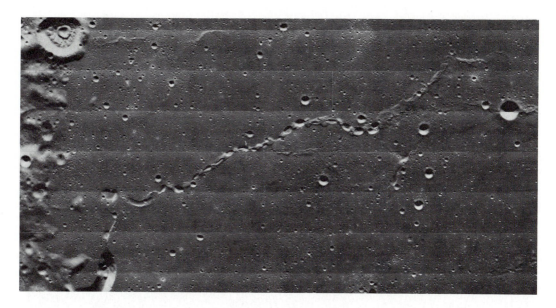

Figure 15-16.   A lunar crater chain (about 35 km or 22 miles in length) as observed by Lunar Orbiter 5. (NASA.)

Figure 15-15.   Lunar domes. (a) The region of the Marius Hills, showing several dome features thought to be volcanic in origin; also note the many wrinkle ridges. The crater Marius is in the background. (b) Close-up view of the region north of crater Gruithuisen. The large symmetrical dome on the left is about 20 km (12 miles) in diameter. (NASA.)

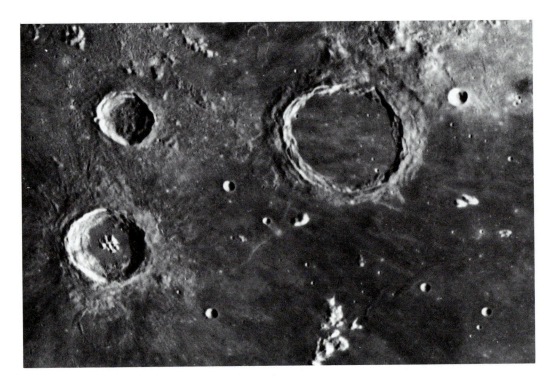

Figure 15-17. The drowned crater Archimedes (right) is seen in this view, which also shows the craters Aristillus (bottom left) and Auto-lycus (top left). (Lick Observatory photograph.)

light that falls on a moon or planet is called the *albedo,* and this fraction is only 0.073 for the moon; individual regions do reflect much more (high-lands) or much less (maria). The albedo may indicate a dark rock, such as basalt, or a lighter rock that has been darkened by solar wind bombard-ment, a process that has been checked and simulated in the laboratory. The brightness changes dramatically with the relative orientation of the earth, moon, and sun, as explained in Figure 15-19; this implies a rough lunar surface. When we observe the full moon, we are looking with the sun behind us, and we see no areas in shadow. At other lunar phases, many areas are shadowed by local height irregularities, from mountain ranges to tiny craterlets and rock particles, and so the lunar surface seems less bright. Observing the way in which radio waves of different wavelengths, beamed to the moon by powerful radar transmitters, are

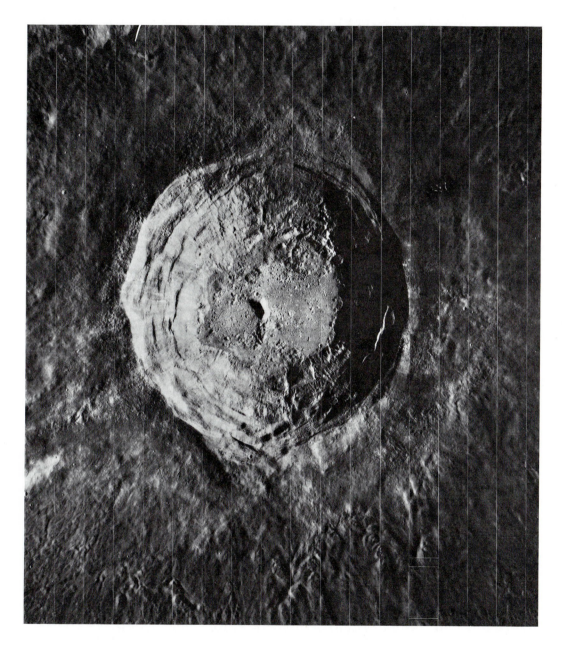

Figure 15-18.   The crater Aristarchus as recorded by Lunar Orbiter 5. (NASA.)

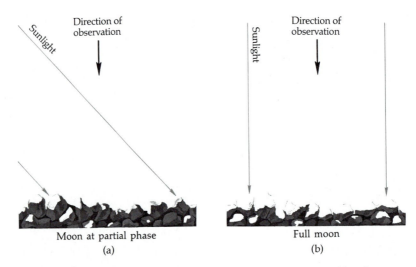

Figure 15-19. Schematic diagram showing the increase of brightness of the lunar surface at full moon. (a) At partial phase, the observer sees *both* illuminated surface and surface in shadow. (b) At full moon, *only* illuminated surface is visible.

reflected by the lunar surface, we can determine that the surface is very rough, on the scale of 1 cm (0.4 inch) and less. Well before men walked on the moon, the radar echoes showed that it must be largely covered with small rock particles and other irregularities (perhaps tiny crater pits) of these sizes.

Visible light from the illuminated lunar surface is, of course, reflected sunlight. But at longer wavelengths, electromagnetic radiation received from the moon is dominated by the thermal emission of the surface. Thus, further information about the lunar surface material came from measurements of its infrared and short-wavelength (millimeter and centimeter waves) radio emissions. Measurements of this radiation give the temperature of the emitting region, regarded as approximately an ideal radiator. The different wavelengths are emitted at different average depths,[2] so we can probe the first 30 cm (12 inches) or so of the lunar surface. These studies show that the surface temperature changes from about 373 K to 123 K as the lunar day goes from noon to shortly after sunset.[3] On the other hand, at depths of several centimeters, the change

---

[2]The longer the wavelength, the greater the depth.

[3]Recall that the boiling and freezing points of water at sea level are 373 K and 273 K, respectively.

in temperature is already much less, and it amounts to only a few degrees K below 30 cm. This provided strong evidence that the moon was covered with a very effective insulating layer, composed of some kind of rock dust. However, it left several questions begging: How deep is the dust layer? How compact is the dust? Could a spacecraft land on it safely, or would it settle in unstable sands, shifting due to the impact of landing?

The advent of the space age enabled scientists to test previous deductions about the moon by directly sampling the surface. Further, we have seen the surface close up (Figure 15-20), and from photographs of the

Figure 15-20.   Impressions on the lunar surface. (a) Astronaut's footprint from the Apollo 11 mission; (b) the padprint of the Surveyor spacecraft that landed on April 20, 1967, as photographed by the Apollo 12 astronauts on November 20, 1969. (NASA.)

(a)   (b)

Figure 15-21. The Oriental Basin as seen from lunar orbit. This region is on the moon's western limb and is not clearly visible from earth. (NASA.)

Figure 15-22. View of the moon's far side showing the crater Tsiol-kovsky. (NASA.)

Figure 15-23. Rubble-strewn lunar landscape as photographed by Surveyor 7. The rock in the foreground is about 0.6 meters (2 feet) across and cast a 1.2-meter (4-foot) long shadow at the time the picture was taken. (NASA.)

hidden side of the moon we have found that although both sides have the same type of features, there are noticeably more highlands and fewer maria on the far side. A Soviet spacecraft, the Automatic Interplanetary Station, took the first pictures of the far side in October, 1959; Figures 15-21 and 15-22 show portions of the lunar surface not visible from earth, as later photographed by the Lunar Orbiter spacecraft.

Figure 15-24. Oblique view of crater Theophilus—about 100 km (60 miles) in diameter and about 6.5 km (4 miles) deep—as seen by Lunar Orbiter 3. (NASA.)

The Ranger spacecraft (in 1964 and 1965) transmitted photographs of the lunar surface en route to hard landings on the surface. In February, 1966, a Soviet probe, Luna 9, achieved the first soft landing on the moon and sent clear views of its surroundings back to earth. This showed that the dust coating of the lunar surface did not prevent it from providing a stable platform for a spacecraft. The first of the Surveyors had a soft landing in June, 1966; they obtained many photos (Figure 15-23). Later, several spacecraft were put into lunar orbit.

What did we learn from these activities? The soft landings of several spaceprobes established that the lunar surface can support the weight of these craft, an important consideration in determining not only the nature of the material but also the risk in sending a manned space vehicle. Close-up views of the landing pads and their impressions (or padprints) showed that the lunar surface material near the spacecraft was soil-like and compactible. The panoramic views telemetered back to the earth were of rock-strewn, cratered landscapes, and established that the presence of lunar craters extends down to pits of very small dimensions (Figure 15-23). The photographs from orbiters showed oblique views (Figures 15-24, 15-25) and vistas that increased our comprehension of the lunar surface. It was clearly very rough, and manned landing sites would have to be carefully chosen.

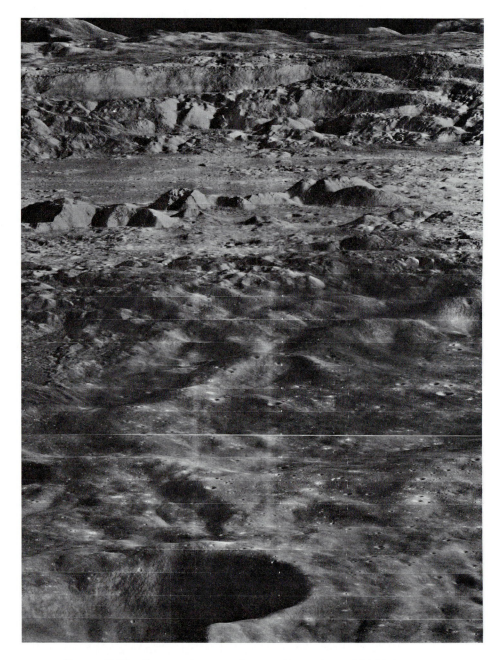

Figure 15-25.   Panoramic view of crater Copernicus. (NASA.)

Figure 15-26. The solar-wind-collecting foil placed on the lunar surface by the Apollo 11 astronauts. (NASA.)

Manned exploration of the moon was achieved in July, 1969, by the Apollo 11 crew of Neil Armstrong, Edwin Aldrin, and Michael Collins.[4] Armstrong and Aldrin gathered rocks from the surface and took core tube samples of the soil. They used a foil to collect solar wind particles (Figure 15-26), set up a reflecting device that was used for laser beam measurements by earth-based astronomers, and installed a lunar seismic station (Figure 15-27) to study moonquakes. The solar wind foil was rolled up and brought back to earth along with the rock and soil samples. They also took close-up stereo photographs of the appearance of the lunar soil.

The Apollo 11 astronauts found that there is indeed dust on the moon; it seemed to be everywhere and to cling to everything, including their spacesuits and equipment. From the analysis of the rock and soil samples, and by study of visual and photographic observations on the

---

[4]Collins, the Command Module Pilot, remained in lunar orbit while the other two astronauts explored the region of Tranquility Base.

Figure 15-27.   The seismometer placed on the moon by the Apollo 11 astronauts. (NASA.)

moon, we know that the lunar surface in the vicinity of the landing spot in the Mare Tranquillitatis consists of a few centimeters of sandy soil lying atop a layer of larger fragments of broken rock that is several meters deep (the *regolith*). Beneath the regolith is the *lunar bedrock,* which is exposed in the bottoms of craters. Many rocks lie on the surface, and are of two basic types: *igneous rocks* (including basalt) that were probably formed by the cooling of lava on or close below the lunar surface, and *breccia,* or rocks formed by the pressing together of small fragments of older rocks. In addition, there are iron meteoroid fragments and glass particles in the soil.

The igneous rocks, soil, and breccia were dated by radioactive methods similar to those described in Chapter 3. The soils and breccia gave ages up to 4.6 billion years.[5] The igneous rocks gave smaller ages, ranging up to 4.0 billion years. (Deductions from the rock and soil dating are summarized in the next section.)

---

[5] The oldest rocks found on earth have an age of about 3.5 billion years.

The shapes of the lunar rocks are especially interesting; exposed surfaces of rocks lying on the moon tend to have more rounded contours than the unexposed bottom sides. This suggests that the tops have been eroded or weathered by the impacts of tiny meteorites. Some of the rocks are partially covered by glassy crusts, and Armstrong discovered that in several craters of about a meter in diameter, there were droplet-shaped blobs of glass, larger than the tiny glass particles found in the soil sam-

Figure 15-28.   Lunar Orbiter 5 view of Mare Imbrium. The area in the upper part of this photograph is at a lower elevation than the region in the lower part, which is apparently covered by a lava flow. (NASA.)

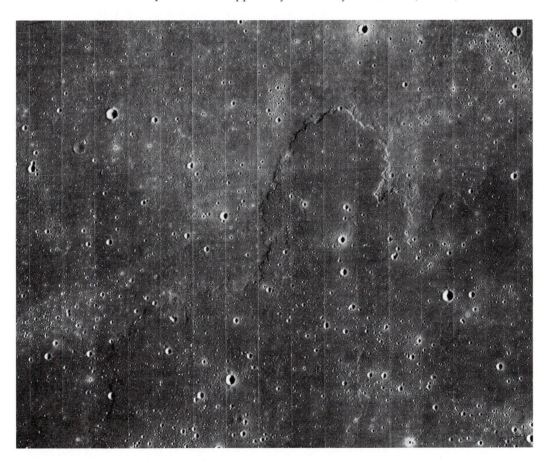

ples. Similar droplets and glass crusts appear in the centers of smaller craterlets that were photographed with the stereo camera.

Several theories were proposed to explain the lunar glass. Thomas Gold ascribed the glass that is specifically found in small craters to melting of rock by radiation from extremely large solar flares. The point is that reflection from the crater walls would have concentrated the heating effect at the crater bottoms, where Armstrong found this glass. However, there is no proof that such giant solar outbursts actually occur. Another idea is that glass is formed by rock-melting due to the escape of hot gases from the interior of the moon. A third suggestion, which seems most reasonable to us, is that the various kinds of glass, including the soil particles and the rock crusts, originated when meteoroids fell on the moon, heating the material they struck and splattering the molten rock over surrounding areas.

The rocks have the same bulk composition as the soil, consisting of minerals similar to those formed from magma within the earth. There are some differences from earth rocks; for example, the titanium content is higher in the lunar rocks, and they have many glassy inclusions. The lunar lavas (Figure 15-28), as compared to terrestrial lavas, are found to have been very hot and very dry (no minerals with water bonds).[6] Such lavas would flow more easily than terrestrial ones; this may explain the flat appearance of the maria. There was no evidence of living, dead, or fossilized organisms in the lunar samples.

Moonquakes and the Lunar Interior

During their lunar missions, the Apollo astronauts placed seismometers at a variety of locations on the surface of the moon. The purpose of measuring moonquakes was to probe the lunar interior, just as earthquake waves are used to study the interior of the earth. The lunar seismometers were much more sensitive than the usual terrestrial ones, whose ability to detect weak quakes is limited by interfering noise from large storms and human activities. This was fortunate, since the moonquakes turn out to be very weak. The Apollo instruments continued to function for years, recording thousands of moonquakes, but even the strongest of these rate only about 2 on the Richter scale of earthquake magnitude. A Richter 2 earthquake is one that people do not notice, although it can be detected by seismometers. Likewise, the total seismic energy (counting all moonquakes that occur in a year) released in the

---

[6]Recall the discussion of molecular bonds in Chapter 5.

moon was found to be less than one-billionth of the corresponding quantity on earth.

The locations or *foci* of the rock motions that cause moonquakes were found to lie deep within the moon. Furthermore, about 70 "repeating foci" have been found, where moonquakes occur again and again. There is a strong tendency for moonquakes generated at a particular focus to recur with a 27-day periodicity, equal to the period of the moon's orbit around the earth. This has helped to convince scientists that the moonquakes are triggered by the earth's tidal force. On earth, the most prominent tides are those raised in the oceans by the moon. On the moon, the most prominent tides seem to be rock motions in the deep interior.

The Apollo seismic instruments also recorded lunar vibrations from the impact of meteoroids and the crashing of space vehicles onto the moon's surface. For example, after the Apollo 12 explorers left the moon and returned to their command module in lunar orbit, their lunar ascent stage was sent crashing back to the surface, striking about 75 km (47 miles) from the Apollo 12 seismometer. On that occasion, the moon was found to "ring," vibrating at about 1 cycle/sec for fully 40 minutes. Similar behavior was observed when the Apollo 14 crew conducted a similar experiment. The ringing phenomenon is very rare on earth; it occurs only after an exceptionally strong quake.

Several theories to explain these lunar vibrations were proposed. The spacecraft impact could have kicked up lunar material into high trajectories, from which it rained down on the surface over many minutes. Or perhaps the impact triggered landslides on crater rim slopes, which persisted for the observed duration of the ringing. Neither of these theories seemed to fit the facts sufficiently well. A more plausible suggestion to account for the ringing interprets it in terms of inhomogeneous material or *rubble*. As we know from studies of the earth, seismic waves are reflected or bent by changes in the material through which they are traveling. If the lunar subsurface material were very inhomogeneous, some waves might propagate more or less directly through it and reach the seismometer quickly, while other waves might be reflected many times, bouncing around and taking a long time to reach the seismometer. Thus, multiple reflections of seismic waves and a thick layer of rubble may explain the seismic events that last so long on the moon.

Explosions were also used to produce vibrations in the lunar material for structural research. For example, during the final Apollo mission to the moon (Apollo 17, December, 1972), the astronauts set up a network of eight explosive devices at locations hundreds of yards apart surrounding the seismic detectors they had installed. After the crew returned to earth, the devices were triggered by radio command. The resulting seismic disturbances were analyzed to determine the subsurface structure (using a technique much like that used by oil-exploration geophysicists on earth). Some of the Apollo 17 instruments and many of those from the earlier lunar landings were still operating in September, 1977, when NASA terminated the program of monitoring them for further scientific data. They exceeded their design lifetimes by so much that the unplanned expenditures of a few million dollars per year for continued operations could no longer be paid.

Combining all of the information from the Apollo seismometers and from related studies, the following basic model (Figure 15-29) has been formulated for the interior of the moon: the crust extends to a depth of about 60 km (37 miles); below it is the lithosphere, also a region of solid rock, which reaches to 1,000 km (620 miles) below the surface. The vast majority of moonquakes originate in the lower 400 km (250 miles) of the lithosphere, at foci where rock slippage is induced by the earth's tidal force. Below the lithosphere is the lunar asthenosphere, whose rock may be partially molten near its upper boundary, but which is still relatively cool (not much hotter than 1,000° C) by comparison to the deep interior of the earth. The asthenosphere extends to the center of the moon, 1,738 km (1,080 miles) below the surface. There is no evidence so far of an iron core within the asthenosphere, or a molten core of any kind.

Supporting evidence for the lack of a hot, molten core at the center of the moon comes from the moon's *shape*. The moon has three unequal radii, as measured (1) along its rotational axis, (2) in the direction of the earth, and (3) in the direction along the moon's orbit. The radii directed toward the earth and along the moon's orbit are, respectively, 1.1 km (0.7 mile) and 0.7 km (0.4 mile) longer than the polar radius. These differences imply large stresses in the interior that could not be supported if it were very hot. In other words, a fluid or near-fluid would have adjusted itself to a spherical shape.

History of the Lunar Surface

The six Apollo moon-landing crews (Apollo 11, 12, 14, 15, 16, and 17) brought back a total of 380 kg (840 pounds) of rock and soil. Three automated Soviet sample-return spacecraft, Luna 16, 20, and 24, retrieved small quantities of lunar material, and two small, automated moon rovers, the Lunokhods, also explored the surface. The samples originated in a variety of geological settings on the moon. From radioactive dating, mineralogical analysis, and related studies of this "hard evidence," a

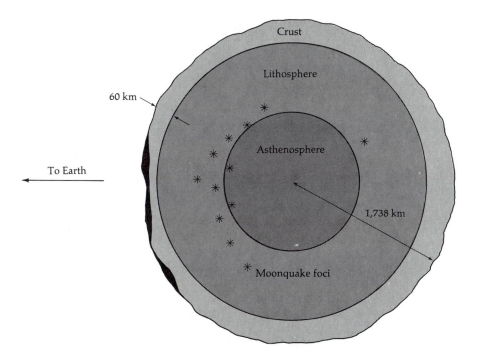

Figure 15-29. Schematic diagram of the lunar interior. The thickness of the crust is exaggerated for clarity. Basalt filling under the maria is shown in black. Note that nearly all moonquake foci are located in the half of the lithosphere that faces the earth. (After a diagram by M. N. Toksöz.)

rough overall history of the moon's surface has been deduced. Researchers disagree on some of the details, but the broad picture may well be correct.

To begin, it appears that there was an era, 4.6 billion years ago, when the whole surface of the moon was in a molten state, probably as a result of intense bombardment by meteoroids in the final stage of the formation of the moon. This hot layer cooled*[7] rapidly, with the lighter minerals floating upwards, forming the 60-km-thick crust that has been detected by the Apollo seismic studies. Subsequent impacts occurred at a lesser

---

[7] Each event or process marked with an asterisk is among those that have been deduced from the age, nature, and specific place of origin of particular rocks brought back from the moon.

rate, but included some individual great collisions, with asteroid-sized bodies scooping out huge basins hundreds of kilometers in diameter. The impacts melted the surface rock in these basins, and so new rock formed as it cooled,* about 4 billion years ago. Following this stage and enduring for more than a half billion years (ending about 3.3 billion years ago), lava erupted into the basins from deep under the surface, where heat generated by radioactivity in the lunar interior had produced temporary zones of molten rock. The great impact basins were what we now call the maria; the upswelling lava filled them and froze,* giving the maria their present dark, smooth appearance. As the radioactivity inside the moon died down in accordance with the half-lives of the active isotopes of uranium and a few other elements, the hot interior regions cooled and solidified, producing the present cool lithosphere. The lithosphere must be still getting thicker, extending downward as the softer asthenosphere beneath it continues to cool.

The lunar highlands represent portions of the surface that have been scarred by meteoroid impacts ever since the crust cooled,* but which have never been erased by huge impacts and subsequent lava flooding, as were the regions of the maria. Note that the earth's lithosphere is only about 70 km thick on the average, while that of the moon is about 1,000 km thick. The moon is smaller and cooler, and its lithosphere is so thick and strong that it cannot break up into the thin plates that float on the partially molten rock below, as on earth. This explains why the earth has continental drift, but there is no corresponding process on the moon. Since continental drift is responsible for the formation of mountain ranges (recall Chapter 3), this also explains why mountain ranges are not found on the moon. What passes for mountains on the moon are simply the central peaks and rims of impact craters or occasional low volcanic formations.

EVOLUTION OF THE EARTH-MOON SYSTEM

There are three distinct traditional ideas on the origin of the moon. (1) It could have been formed elsewhere and captured by the protoearth long ago. (2) It could have formed near the earth as a by-product of the same process that formed the earth. (3) It could have once been a part of the earth and separated from it by fission. Each of these ideas is open to objection when examined in detail, and recently this led to a fourth idea—that the moon was formed by the conglomeration of a considerable number of smaller satellites that were previously captured by the earth. This may be the leading theory at present. The exploration of the moon has yielded much information on its interior, its surface properties, and its

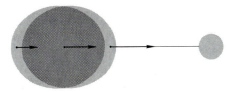

Figure 15-30. Schematic diagram of the moon's tide-producing forces (shown by the heavy arrows). The *differences* in the forces exerted at different points in the earth produce the tides on both sides of the earth. View is from the plane of the moon's orbit around the earth. (Not drawn to scale.)

history since formation, but the origin of the moon is still an unsolved problem.

The earth-moon system is changing. This fact is intimately tied up with the tides, as shown in Figure 15-30. The moon's gravitational attraction pulls the water on the side of the earth facing the moon a little away from the earth, and it pulls the earth a little away from the water on the side opposite the moon. The general phenomena of tides are well known to many people, but they are complicated by the local topography so that tidal effects are more dramatic in some locations than in others. The sun is also a tide producer. When the sun and the moon are on the same or opposite sides of the earth, the highest tides (called *spring tides*) are produced. When the moon and the sun are 90 degrees apart, as seen from the earth, the lowest *(neap)* tides occur. One generally thinks of tides in the oceans, since those are the ones that we notice, but the same forces also cause tides in the solid part of the earth.

If the earth were not rotating on its axis, its lunar tidal bulges would occur on the line directed toward the moon. However, it is rotating, and this causes the earth's tidal bulge to precede[8] (in the direction of the earth's rotation) the earth-moon line (Figure 15-31). The pull of the moon on this tidal bulge is slowing down the earth's rotation at the rate of about 0.002 seconds per century. This means that the day was shorter in the past. In Devonian times (some 350 million years ago), the day must have been about 22 hours long, and this value can actually be checked. Some reef corals show yearly bands that are composed of about 365 finer

---

[8]This happens because it takes a finite amount of time for the tidal bulge to form or to recede, and during this time, the earth's rotation carries the bulge away from the earth-moon line.

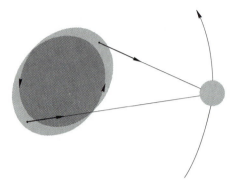

Figure 15-31. Schematic diagram showing the earth's tidal bulges. The tidal forces are shown by the long arrows. The force on the part of the bulge nearer the moon is larger than the force on the more distant bulge, and hence the earth's rotation is slowed. The view is drawn as seen from above the North Pole of the earth. (Not drawn to scale.)

rings, which biologists have shown correspond to a daily growth. Fossil corals of a similar type from Devonian time show 400 fine rings per yearly band. Since there is no evidence for (or reason to suspect) any real change in the length of the year, each day must have been shorter, and we can calculate the length of the day at that time as about 22 hours.

Previously, the earth probably exerted a tidal force on a faster-rotating moon, causing it to slow down to its present rotation rate of once per month. Tides produced in the moon by the earth are the reason that the moon keeps the same face toward the earth.

The earth also has an *equatorial* bulge, which is a feature of its shape and not a changing effect of tidal forces. The pull of the moon on this bulge is largely responsible for the phenomenon of precession. The earth's rotation axis is tilted by 23.5 degrees from the direction at right angles to the plane of the ecliptic. Hence, the equatorial bulge is not in this plane, although the pull of the moon and other solar system bodies acts to bring the bulge toward the plane. The bulge is not pulled into the plane of the ecliptic any more than the earth's pull brings the moon down from orbit. Instead, the result is the precession of the earth's rotation axis around an axis perpendicular to the plane of the ecliptic, with a period of 26,000 years (Figure 2-10).

The slowing down of the earth as revealed by the coral studies leads to an interesting situation. As mentioned in Chapter 11, in our discussion of the origin of the solar system, the quantity of *angular momentum* is conserved in any isolated system. The angular momentum can correspond to an object's rotation on its axis, or its revolution in orbit, or both. If the earth's rotation is being braked by the tides, its angular momentum is decreasing and must reappear somewhere else. Presently, it is taken up in the moon's orbit around the earth: the moon's orbital angular momentum is being gradually increased, and the moon is slowly spiraling outward, away from the earth. As it moves away, the orbital period increases in length according to Kepler's Laws. Thus, the day and the month (in its original sense) are both lengthening, but the day at a faster rate. If the earth-moon system survives the sun's evolution to a red giant, the day and the month will be equal at approximately 43 of our present 24-hour days some 5 to 10 billion years from now. At that time, the earth will keep the same face toward the moon, and the lunar tides will no longer change. Beyond this point, the further evolution of the system is a much more speculative subject. Even when the moon's tidal influence on the earth stops, tides will still be produced by the sun. This will tend to slow the earth's rotation even more and make the day longer than the month. When this happens, the lunar effects must start up again but in the opposite direction, shortening the day and bringing the moon closer to the earth. The situation is complex, because the solar-produced tides in the earth slow down the rotation while the lunar-produced tides will speed it up. The detailed calculations indicate that the moon will spiral in toward the earth, continually coming closer. Eventually, it will get so close that the difference in the earth's gravitational pulls on the near and far halves of the moon will literally tear the moon apart. When this happens, the lunar remains might take the form, at least in part, of a ring of particles around the earth, perhaps like the rings of Saturn.

STRATEGIES OF PLANETARY EXPLORATION

There are a variety of ways to design an automated mission to a terrestrial planet.[9] Simplest of these is the *flyby* of a spaceprobe that takes pictures and measurements for a brief interval as it passes the planet. It has the advantage of simplicity and relatively low cost, since it does not have to be equipped or controlled for orbital and landing operations. Another advantage is that it may be launched into a trajectory that will allow it to gain energy as it falls through the gravitational field of the planet, giving

---

[9]Somewhat different techniques may be involved for a Jovian planet, where the solid surface, if any, is inaccessible.

it a "gravity-assisted" boost toward a second planet. Thus, two or more planets can be visited for the cost of one launch rocket and spacecraft. The obvious disadvantage of a flyby is that the time spent in the immediate vicinity of a planet is brief, so detailed and extensive studies are impossible.

Much more extensive studies are possible with an *orbiter* spacecraft—one that literally becomes an artificial satellite of another world. It can map and study large areas of the planet to provide a general overview of planetary properties. But, from the height of orbit, highly detailed studies are impossible. Such studies can be made with a *soft-lander* spacecraft, which provides close-up photos of its immediate surroundings, direct sampling of the surface material, weather and seismic measurements, and so forth. Biological experiments can also be done with a lander. The obvious disadvantage of a lander, however, is that it lands in one place; the surface at that location may not be representative of other parts of the planet. *Lander-rovers*, spacecraft that land and then move around the surface (or unload a smaller surface vehicle), are a major step beyond the capabilities of stationary landers. They have not yet been used except in a very limited manner on the moon. Actually, a rover need not be a surface vehicle, but could possibly take the form of a pilotless aircraft, cruise missile, or balloon launched either from a lander or an orbiter.

THE EXPLORATION OF MARS

Studies of Mars as a planet began with telescopic observations, then continued with each of the space mission strategies (up through the soft-lander) described in the preceding section. The most famous of the problems posed by telescopic observation of Mars was the report in 1877 by Giovanni Schiaparelli that he had seen canals. Actually, his word was *canali*, Italian for "channels." Channels might well be of natural origin, but canals (as the word was usually mistranslated) are by definition artificial waterways, such as those at Suez and Panama. Percival Lowell in Arizona became the leading exponent of this interpretation. Planetary photography was in an unsatisfactory state (the pictures were just too fuzzy), and so astronomers relied on drawings made at the telescope. Unfortunately, no two astronomers drew exactly the same thing, and some drawings were completely different from others made by equally reputable observers. Lowell drew an almost rectilinear system of canals (Figure 15-32) resembling a vast engineering project, which some believed the enterprising Martians had constructed to bring water down from the polar caps to irrigate the temperate and equatorial regions! Although their

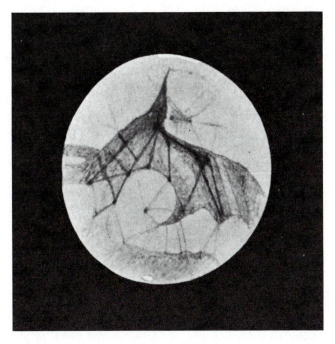

Figure 15-32.   A drawing of Mars by Percival Lowell showing canals. (Lowell Observatory photograph.)

drawings differed, some astronomers reported that groups of canals tended to radiate from (or meet at) certain spots, which were compared to oases in a desert.

Mars has seasons for the same reason that the earth has them; its axis is tilted about 24 degrees to the plane of its orbit. As the seasons change, so do the polar caps, as the early observers noted. When it is summer in the northern hemisphere of Mars, the north polar cap is small, but in the southern hemisphere winter prevails, and the south polar cap is large. Soon, as spring comes to the southern hemisphere, the south polar cap is seen to decrease in size, retreating toward the south pole. At the same time, however, the appearance of the southern hemisphere changes; the dark areas in the region just outside the polar cap get darker, and the effect seems to spread gradually toward the equator. This progressive enhancement of the contrast of the dark areas was called the *wave of darkening*. (The same phenomenon happens half an orbital period later in the Martian northern hemisphere, when spring occurs there.)

The observation that the wave of darkening proceeded from polar latitudes toward the equator, and that it coincided with the decrease in the size of the polar cap, led to the idea that the two effects were causally related. It was natural to believe that the polar caps were formed in whole or in part of frozen water, and that as the cap receded in the spring, water vapor was carried equatorward by the wind, or liquid water spread down over the planet, thanks to the convenient network of canals. The water might cause the darkening by producing physical or chemical changes in the rock and soil of the Martian surface, or it might be triggering a much more interesting phenomenon. *Perhaps the wave of darkening indicated the presence of life on Mars;* that is, perhaps the water was rejuvenating vegetation that had lain dormant or dead throughout the winter season. A difficulty was that the spectra of Mars showed little evidence for water; some astronomers believed the polar caps might actually be dry ice, frozen carbon dioxide.

Thus, before the advent of the spaceprobe, astronomers studying Mars were faced with three basic questions: Was there life (at least vegetation) on Mars? What were the canals—natural rock channels, artificial canals produced by an extraterrestrial civilization, or figments of the zealous observers' imaginations? And what were the polar caps? Now, as a result of several close-up studies of Mars by Mariner and Viking spacecraft, we have some answers for these questions and also a quite different view of Mars.

Discoveries of the Mariner Spacecraft

The modern exploration of Mars began with three Mariner flybys. The first was Mariner 4, which came within 9,900 km (6,200 miles) of the planet in July, 1965. It carried a television camera and transmitted about 20 images of Mars with higher resolution than had ever been attained with earth-based telescopes. These pictures showed a cratered surface (Figure 15-33) resembling that of the moon much more than that of the earth. An astronomer had actually predicted this, but the prediction had been mostly overlooked and forgotten, and so the discovery of craters came as great surprise. It also was a great disappointment, for it implied that Mars was a geologically and probably biologically dead world, like the moon. Mariner 4 also discovered that the Martian atmosphere was much thinner than expected, with 100 or 200 times less surface pressure than that of the earth's atmosphere at sea level. Earlier observers had thought that the Martian atmosphere might be as dense as that of the earth in the high reaches of the Andes, where some life exists and hu-

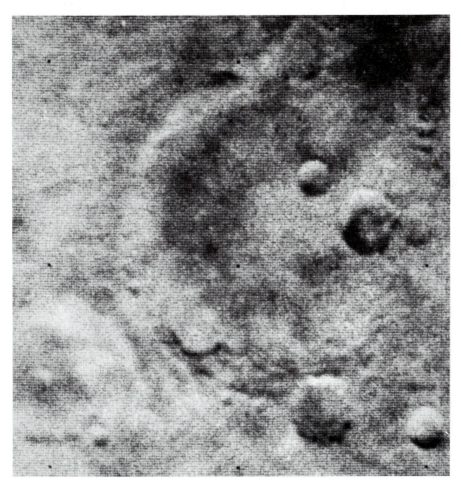

Figure 15-33.   A close-up view of Martian impact craters obtained by Mariner 4. (NASA.)

mans can breathe. It turned out that the Mariner 4 pictures gave an unduly pessimistic idea of Mars. As the spacecraft passed the planet, it photographed only one percent of it, which by chance was an ancient, geologically undisturbed, and most unrepresentative terrain!

The next two spacecraft, Mariners 6 and 7, flew by Mars in the summer of 1969, obtaining about 200 photos plus infrared and ultraviolet measurements. The ultraviolet data gave details about the atomic compo-

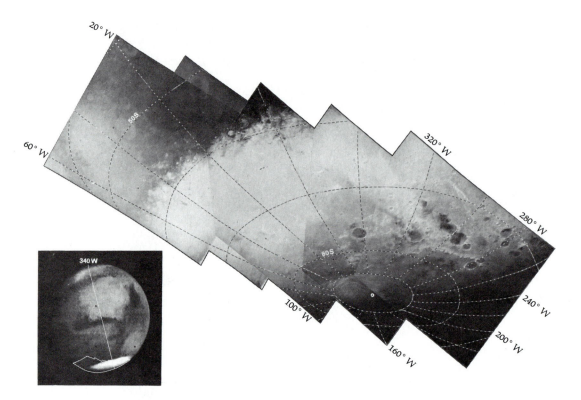

Figure 15-34. Mosaic of Mariner pictures of the Martian south polar cap, showing craters and "snow"; Martian latitude and longitude are marked. (NASA.)

sition of the Martian atmosphere; the infrared instrumentation allowed us to measure temperatures on the surface. The area photographed by these two Mariners was ten times as large as that surveyed by their predecessor and included some interesting views of the south polar cap (Figure 15-34). Again, however, the pictures showed a mostly cratered surface. However, the pictures were much sharper than those obtained by Mariner 4, and they revealed finer details about the craters (Figure 15-35). It was seen that many craters had been subject to erosion; their rims had been worn down a bit, and their interiors had filled with debris, possibly wind-blown sediment. Mars still appeared to be relatively less interesting than astronomers had hoped, but the next stage in exploration of the red planet was to revise our ideas drastically again.

The early history of the Mariner 9 mission was a striking demonstration of the advantage of an orbiter spacecraft. As Mariner 9 approached Mars in November, 1971, the planet was enveloped in a vast dust storm which totally obscured its surface. Had Mariner 9 been on a flyby trajectory, its photographs would have been almost worthless, for there was hardly anything to be seen. At the same time, two Soviet spacecraft, Mars 2 and 3, were en route to Mars. They were programmed to eject capsules to land on the surface. Neither Soviet Mars lander sent back useful pictures, and radio contact was lost shortly after they landed. The same storm that was defeating Mariner 9's photography may have attained wind speeds of almost 200 km/hour (125 mph), and possibly damaged the Soviet landers.

Mariner 9 was maneuvered into an elliptical orbit, revolving about Mars once every 12 hours. As it continued to circle the planet, a few dark spots emerged and grew amidst the bright background of the dust storm. Over the next few weeks, as the dust settled, it became clear that the spots were the summits of a group of enormous volcanoes in the Tharsis region of Mars. The largest, named Olympus Mons (Mt. Olympus; Figure 15-36), is more than twice as high as Mt. Everest and as big as the state of Missouri. Volcanic craters are present at the top of each of the Martian

Figure 15-35. A cratered portion of the Martian surface (103 km wide by 77 km high) as transmitted by Mariner 6. (Courtesy of the National Aeronautics and Space Administration, Jet Propulsion Laboratory.)

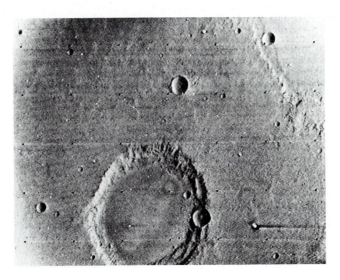

Figure 15-36. The huge Martian volcano Olympus Mons as photographed by Mariner 9. (NASA.)

volcanoes, and there are hardened lava flows on their slopes. There is no evidence that the volcanoes are alive today, but their presence at least shows that Mars underwent an era of intense vulcanism sometime in the past. These enormous mountains had been missed by the earlier flybys.

Mariner 9, the first manmade satellite of another planet, sent back data and photos from Mars orbit for almost one year, during which it satisfactorily executed more than 37,000 radioed commands. It mapped the whole planet, observed the local and global responses to the changing seasons, and proved that the ancient, cratered terrain discovered by Mariner 4 corresponded to just one aspect of a diverse geological history. The Mariner 9 observations showed that Mars is dominated by two surface processes at present: the airborne transport of large quantities of dust or sand and the seasonal transfer of carbon dioxide between the two polar caps via the atmosphere. It solved two of the old controversies about Mars: the wave of darkening was seen to be caused by seasonal wind patterns, which change the arrangement of sand deposits on the Martian surface. The canals of Mars, with the exception of one or two of the largest ones, simply do not exist.[10] They seem to have been the products of the observers' imaginations.

The geological features discovered by Mariner 9 included the Tharsis region of huge volcanoes and some smaller volcanoes elsewhere on the planet; an enormous canyon (Figure 15-37) much longer, wider, and deeper than the Grand Canyon of Arizona; numerous dry canyons (Figure 15-38) and riverbeds, many resembling the arroyos of the southwestern United States; fields of sand dunes (Figure 15-39); and layered geological structures near the poles (Figure 15-40).

Mariner 9 also found that at each of the poles of Mars there are two kinds of polar caps, one that appears to be permanent, and one that comes and goes with the seasons.

The evidence accumulated by Mariner 9 showed that, unlike our moon, Mars has undergone volcanic activity of the type that formed the very large volcanic mountains found in Hawaii and Japan. Another significant clue to the physical history of the planet is the great canyon photographed by Mariner 9. This canyon, now called Vallis Marineris (Mariner Valley), is almost 4,000 km (2,500 miles) long; in places it is wider than our Grand Canyon is long. It appears to be a fracture in the crust of Mars much like the Rift Valley of East Africa, and hence it is taken as evidence that some fragmentation has occurred in the Martian lithosphere. This tectonic activity is much less than that present on the earth, for there seem to be no mountain ranges of the ordinary folded type (such as the Rockies and the Himalayas) on Mars. Further, even if the Martian

---

[10] A large canyon was found at the location of Coprates, one of the widest canals indicated on the early telescopic drawings. It is a perfectly normal canyon, with no evidence of artificial construction.

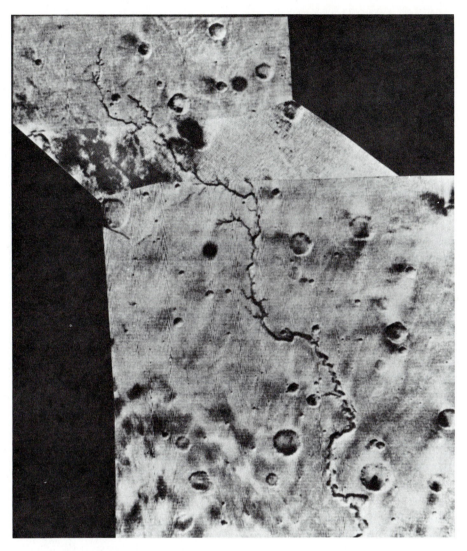

Figure 15-38. A possible river channel on Mars, as shown in a mosaic of Mariner 9 photographs. The form of this channel is convincing evidence that water once flowed on the planet's surface. (NASA.)

Figure 15-37. The Grand Canyon of Mars, as shown by a mosaic of Mariner 9 photographs. The canyon is more than 5,000 km long and at least 6 km deep. The immensity of the canyon is illustrated by the inset, which shows the Grand Canyon of Arizona to the same scale. (NASA.)

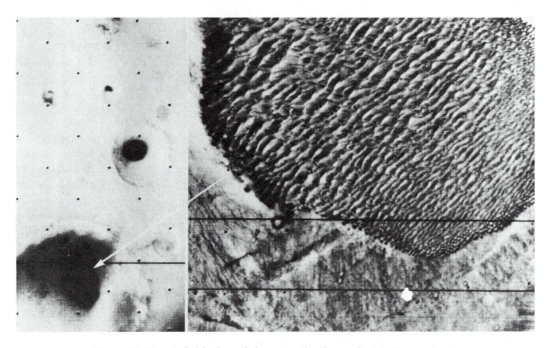

Figure 15-39.   A field of sand dunes in the floor of a Martian crater, as photographed by Mariner 9. The dune field is approximately 130 by 65 km in size. (NASA.)

lithosphere is slightly fractured, the huge size of Olympus Mons and the other Tharsis volcanoes is evidence that the plates do not drift. These volcanoes have built up to so great a height and volume because the surface layers at their locations have not drifted across the hot plumes of magma welling up from below, but have remained stationary. Putting the various clues together, we would estimate that the lithosphere of Mars is thicker than that of the earth, but much thinner than that of the moon.

The discovery of what appear to be canyons cut by ancient rivers is potentially the greatest contribution of Mariner 9. The implication is that much liquid water was once present on Mars, and where there is water, we would think there might be life. There is a modest amount of water visible on Mars now, in the form of ice in the permanent polar caps. In the present climatic conditions on Mars, it is always too cold at the poles for water ice to melt. On the other hand, the *seasonal* polar caps are made of dry ice (frozen carbon dioxide). In the Martian spring hemisphere, the

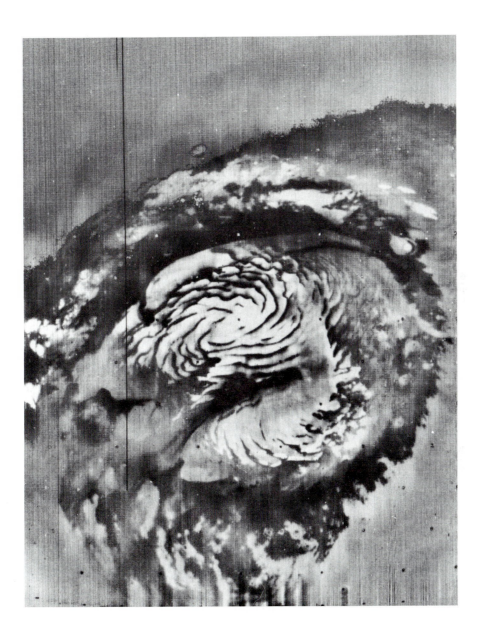

Figure 15-40. The north polar frost cap is approximately 1,000 km across in this Mariner 9 photograph. Notice the layered structure. These are surface features, *not* clouds. (NASA.)

temperature is high enough for the dry ice to sublimate to the vapor state, and the seasonal cap shrinks as its material reenters the atmosphere. Meanwhile, in the opposite hemisphere, carbon dioxide freezes out from the atmosphere, building up the seasonal polar cap there.

However, it is suspected that the climate was once very different. The layered structure of the Martian terrain near the polar caps, discovered by Mariner 9, has been interpreted as having been caused by the deposition of sediments (probably by the wind) during times of great climatic changes, perhaps comparable to the variations between the ice ages and the warmer interglacial periods on earth. The presence of apparent dry river beds on Mars, with its implication of copious liquid water in the past, is also evidence of major climatic change. Indeed, Mars would have had to be warmer with a significantly denser atmosphere in the past for liquid water to persist on the surface.

It is fascinating to speculate (although this is a minority opinion) that the past climatic changes on Mars were caused by the same phenomenon that produced the earth's ice ages. If so, one would have to ascribe the process to something external to Mars and the earth affecting both of them. The obvious candidate would be the sun.

Another unsolved question is, what happened to all of the water on Mars? There is not enough of it in the present atmosphere or permanent polar caps to have produced all the canyons photographed by Mariner 9, with their forked tributaries and curved meanders (features characteristic of rivers on the earth). One theory is that the water might be preserved in thick layers of permafrost just below the surface. If Mars had a regolith made of rock rubble, as the moon does, water might have settled below the surface amidst the rubble and then frozen when the climate became colder. Another suggestion (which more scientists seem to favor) is that most of the liquid water evaporated and that the water vapor was then broken up into oxygen and hydroxyl (OH) gases by the action of ultraviolet light from the sun. These two gases are very active chemically. They would have reacted and combined with minerals in the crust of Mars, so they would have gradually been removed from the atmosphere. On earth, there is always much oxygen in the air, because plant life continues to produce it by photosynthesis. Also, the ozone layer formed from the oxygen in the earth's atmosphere shields the water vapor below it from the effects of the sun's ultraviolet light.

Viking and the Search for Life

In June, 1976, the first of two Viking spacecraft entered a closed orbit around Mars. Each Viking consisted of an orbiter spacecraft, equipped to

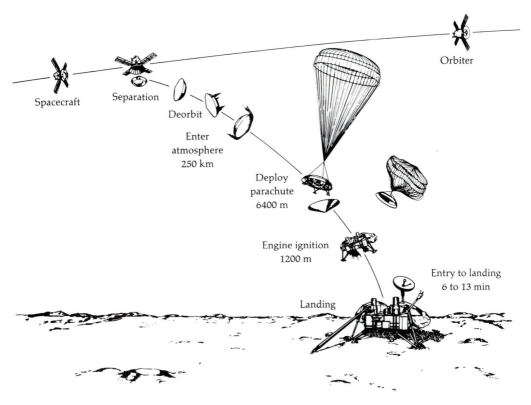

Figure 15-41. Schematic diagram of the Viking landing sequence. (NASA.)

make a variety of measurements and to map the planet below at even higher resolution than was accomplished by Mariner 9, and a lander section designed to explore the properties of the Martian air and soil.

Each Viking, larger and better equipped than Mariner 9, weighed 3,500 kg (7,700 pounds) when launched from Cape Canaveral. Upon reaching Mars and entering their elliptical orbits, the Vikings began transmitting pictures of the surface; these were examined by scientists at the Jet Propulsion Laboratory in Pasadena in order to make the final selection of the safest and most interesting landing sites. On July 20, 1976, the Viking 1 lander separated from its mother craft and dropped toward the surface (Figure 15-41), where retrorockets and a parachute helped it to land gently on its three sturdy aluminum legs.

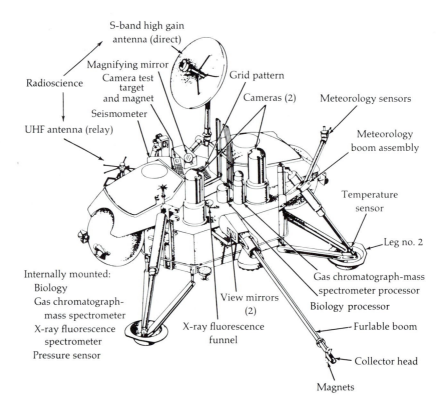

Figure 15-42.   Schematic diagram of the Viking lander's external appearance. (NASA.)

The lander weighed 605 kg (1,330 pounds) and contained instruments, communications equipment, and a computer (Figure 15-42), as well as electronic systems for controlling its temperature and distributing its nuclear-generated electrical power. The instruments included two electronic cameras for stereoscopic imaging, three biological laboratory experiments to test for the presence of microscopic life, a "gas chromatograph-mass spectrometer" to search for organic molecules in the soil, another device to determine the chemical composition of the soil minerals, a small weather station to record the wind speed, air pressure, and temperature, and a seismometer to monitor marsquakes.

The cameras on the Viking 1 lander gave us a panoramic view of a rock-strewn desert landscape covered with fine, red dust (Figure 15-43). Sand dunes were within easy walking distance, and the rim of an impact

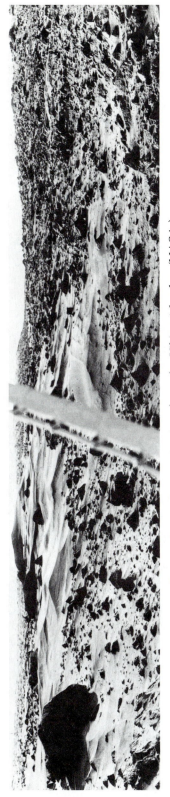

Figure 15-43. Panoramic view from the Viking 1 lander. (NASA.)

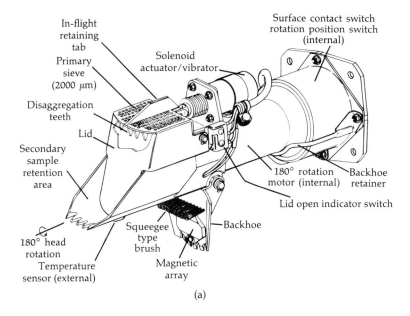

(a)

(b)

(c)

Figure 15-44. The scoop on the end of the Viking sampler boom. (a) Schematic diagram. (b) As photographed during tests on earth. (c) As photographed on Mars by the Viking cameras. (NASA.)

crater was visible on the horizon a few kilometers away. The largest rock near the spacecraft was 3 meters (10 feet) high. Semicircular depressions in the soil at the base of the rocks were evidence of "wind scouring," while on the opposite sides of the rocks were piles of soil, showing that this was the "lee" side, protected from the prevailing wind.

Viking 1 detected nitrogen in the Martian atmosphere, providing the first proof that this gas, which is a necessary component of organic molecules and hence of life as we know it, is present on Mars. The pictures sent back from the lander showed no signs of "macrobes" (large visible life forms such as a tree, bush, or rabbit). However, only the most optimistic scientists and science fiction fans anticipated finding macrobes.[11] The main thrust of the biological studies was to search for the most primitive kind of life, microorganisms, since it is hard to imagine more advanced life forms evolving in their absence.

The soil samples analyzed by the laboratory instruments aboard the Viking landers were selected from photographs of the spacecrafts' im-

---

[11] Indeed, the cameras were designed to work so slowly (for technical reasons) that if a rabbit had hopped through the field of view, it might not have been recorded—at most a blurring of the background rocks would have been seen.

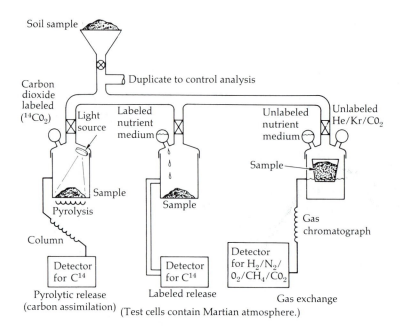

Figure 15-45. Simplified schematic diagram of the Viking biology experiments. Martian soil samples are provided by the scoop. Pyrolytic release (left): Photosynthesis utilizes carbon dioxide (CO_2), and this experiment attempts to produce biological synthesis with so-called labeled gases. The labeling uses the isotope carbon-14 (^{14}C) in contrast to the normal carbon-12 (^{12}C). If the sample synthesizes organic molecules from the labeled gas, the detector should show significant amounts of ^{14}C. Labeled release (center): Here it is assumed that biochemical reactions of Martian microorganisms require water, assimilate nutrients, and release gases containing some of the carbon from the nutrients. The detector then searches for ^{14}C in the gases as in the pyrolytic release experiment. Gas release (right): This experiment looks for changes in the composition of the gases in contact with the assumed living organisms. Possible products of metabolic processes are molecular hydrogen (H_2), molecular nitrogen (N_2), molecular oxygen (O_2), methane (CH_4), and carbon dioxide (CO_2). The detector searches for these molecules and for variations in time of their abundance. (NASA.)

mediate surroundings. Since these were stereo pictures, it was possible to judge the distance as well as the direction of an interesting clump of soil. Then the retractable sampler boom (Figure 15-44), a metal arm capable of reaching out to 3 meters from the lander, was used to scoop up the samples and dump them through a sifting device (to screen out large pebbles that might clog the instruments) into the spacecraft laboratory, where the material was distributed to the various instruments.

The biology investigations (Figure 15-45) consisted of three experiments. First, a pyrolytic release or "carbon assimilation" experiment simulated an environment in which the hypothetical Martian organisms would assimilate carbon dioxide from the atmosphere in a manner resembling the photosynthesis of plants on earth. A "gas exchange" experiment was designed to study changes in the atmosphere of a small chamber, containing a soil sample treated with a nutrient solution, to determine whether the gases were altered by biological activity (such as eating or reproduction) in the moistened sample. The "labeled release" experiment also involved nutrient solutions and was intended to search for evidence of growth and metabolism by another method.

These experiments were conducted several times with various soil samples in both Viking landers, and in some cases they were repeated after the same samples were heated and sterilized by equipment on the spacecraft. In several cases, the experimental results, as described in an official preliminary report, showed "some of the most general characteristics of known organisms." Thus, they were *consistent* with the possible presence of life on Mars. However, the instrument designed to detect organic molecules found that none were present in detectable amounts. It should be noted that organic molecules by themselves (as, for example, the ones that have been found in meteorites) are not necessarily *biogenic* (produced by living organisms). Thus, the finding of organic molecules by itself would not have been a *sufficient* condition to prove that there is life on Mars. However, such a finding is believed to be a *necessary* condition: where there is life there are biogenic organic molecules.

The Viking results show that not only are there fewer (if any) organic molecules on Mars than in the most nearly sterile portion of the earth (Antarctica), but there are even fewer organic molecules than we can find inside meteorites. This has led several scientists to suggest that the results of the biology experiments seeming to mimic the behavior of living organisms were actually due to unanticipated chemical reactions between the Martian soil and the laboratory substances that were added to it in the experiments. In particular, interest has concentrated on the possibility that the reactions involve hydrogen peroxide and similar compounds that

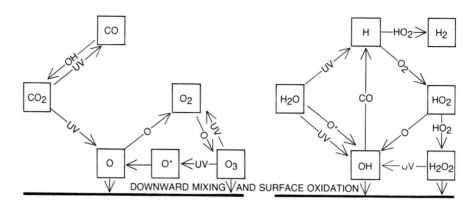

Figure 15-46. Schematic diagram of atmospheric chemical reactions on Mars. These are initiated when carbon dioxide (CO_2) and water vapor (H_2O) are dissociated by ultraviolet radiation (UV). The diagram shows the complexity of the situation; the effects of atmospheric mixing and of oxidation at the surface must be considered. It also shows that hydrogen peroxide (H_2O_2) is produced. (From *The Atmosphere of Mars* by C. B. Leovy. Copyright © 1977 by Scientific American, Inc. All rights reserved.)

Figure 15-47. Viking orbiter mosaic of meandering channels and islands, some of which are teardrop-shaped. (NASA.)

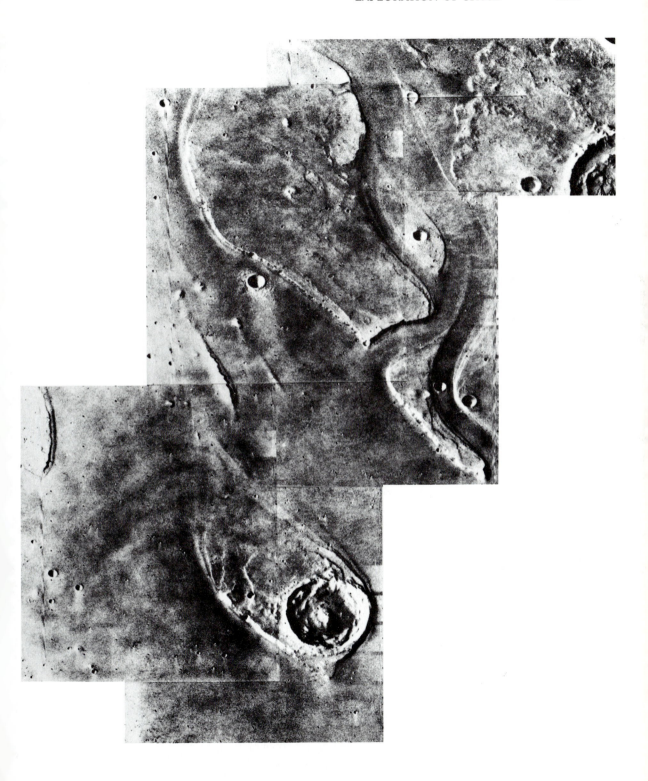

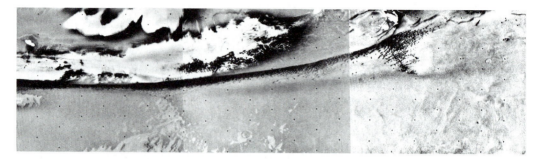

Figure 15-48. The largest belt of sand dunes in the solar system, as photographed by the Viking 2 orbiter near the north polar cap of Mars. (NASA.)

may have been produced in the soil of Mars (Figure 15-46) by reaction with oxygen and hydroxyl (OH) gas. This would truly be ironic if it is correct: at home we use hydrogen peroxide to kill microorganisms in order to sterilize minor wounds. Perhaps by adding liquid to the Martian soil, Viking produced a sterilizing solution during its search for life! However, all of these ideas are very controversial. The present consensus of scientists working in this field is that life has *not* been found on Mars, but that there are tantalizing indications that deserve to be followed up with future experiments, which we now can design on the basis of the chemical properties of the Martian soil that were discovered by Viking.

Among the results of the Viking orbiter observations (Figures 15-47– 15-51) were highly detailed views of the dry canyons that Mariner 9 had discovered. Although many astronomers had accepted the Mariner results as having proven the past existence of running water on Mars, others had disagreed, claiming that the channels might have been formed by powerful wind erosion or the flow of lava. However, the Viking pictures showed details of the channels, including teardrop-shaped islands that are strongly reminiscent of landforms carved by rivers on the earth. They leave little doubt that the analogy is correct. Intriguing views of the Martian satellites were also obtained (Figure 15-52).

MISSIONS TO THE INNER PLANETS

We still have little information about the surface of Venus. Radar observations tell us there are some large mountains, presumed to be volcanoes, and there may be craters, which we think are due to impacts. There is

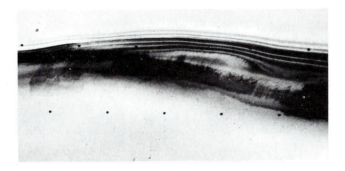

Figure 15-49.   Viking orbiter photograph of alternating layers of dust and ice near the north polar cap of Mars. (NASA.)

also a large valley, which may be a tectonic fracture such as Vallis Marineris on Mars and the Rift Valley on earth. However, these interpretations are obviously somewhat speculative. Two Soviet Venera landers have sent back a handful of excellent pictures (Figure 15-53), which show a rocky, desert surface, as one might expect at a temperature of 750 K (890 F)! Photos from the flyby of Mariner 10, discussed in Chapter 8, revealed details of the atmospheric circulation and clouds, but the planet itself remains very much a mystery. The results of the 1978–1979 Pioneer Venus mission and recent Soviet spacecraft should help to determine the detailed properties of Venus.

Mercury has been visited by only one spacecraft, Mariner 10, but this probe actually went into an orbit around the sun that took it past Mercury at close range on three occasions. Its most important finding was that Mercury has a global magnetic field like that of the earth, but a dead cratered surface (Figure 15-54) like that of the moon or the portion of Mars photographed by Mariner 4. It also has an enormous impact feature, called the Caloris Basin (Figure 15-55), that is 1,300 km (800 miles) in diameter. This feature resembles the lunar maria even to the extent of containing smooth lava plains. There is also a region of strange, broken terrain at the point on Mercury directly opposite (antipodal) the center of Caloris. It is thought to have been produced by the concentration of seismic waves traveling across and through the planet as a result of the great impact. Very similar features have been found on the back side of the moon, in positions antipodal to the centers of two of the large maria seen from the earth.

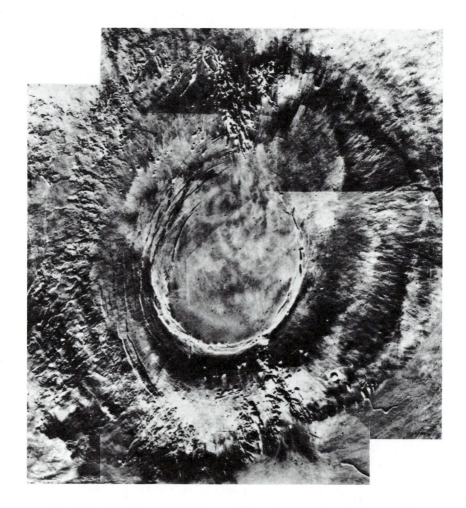

Figure 15-50. Mosaic of Viking orbiter photographs of the Martian volcano Arsia Mons. The central crater (caldera) is approximately 120 km in diameter. (NASA.) Aeronautics and Space Administration.)

Figure 15-51. Panoramic view of Mars from Viking 1 orbiter. (NASA.)

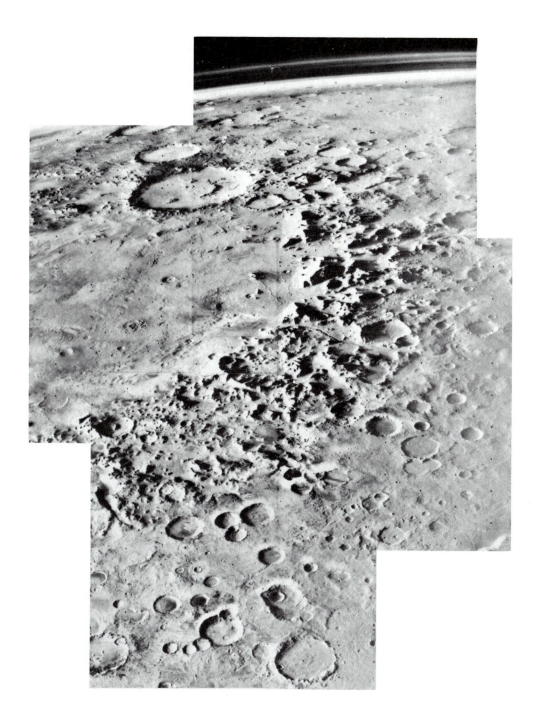

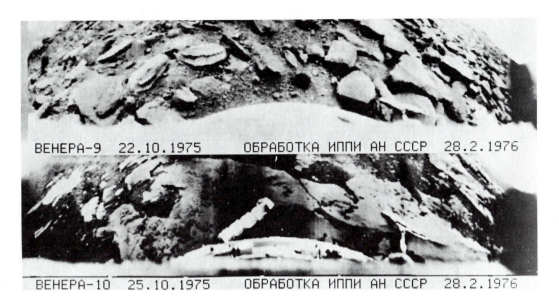

BEHEPA-9   22.10.1975   ОБРАБОТКА ИППИ АН СССР   28.2.1976

BEHEPA-10 25.10.1975   ОБРАБОТКА ИППИ АН СССР   28.2.1976

Figure 15-53. Photographs of the surface of Venus from the Soviet Union's Venera 9 and 10. The view is down 50° from the horizontal, and the field shown is 40° by 180° with the spacecraft at the center and the horizon at the edge. The lower, central part of each image shows part of the spacecraft. The scale near the spacecraft is indicated by the segmented piece of equipment in the lower image (just left of center), which is 10 by 40 cm. The Venera 9 photograph (top), taken on October 22, 1975, shows boulders with sharp edges. The Venera 10 photograph (bottom), taken on October 25, 1975, shows generally blunt and rounded boulders. The photograph bears the notation, "Processing by the Institute of Problems of Translation of Information, Academy of Sciences of the USSR," and the processing was completed on February 28, 1976. (Courtesy of National Space Science Data Center.)

Figure 15-52. A high-resolution photograph of Phobos taken from a range of 880 km by the Viking 2 orbiter. Features as small as 40 meters are shown. Notice the crater chains and the striations or grooves over much of the surface. These nearly parallel grooves are 150 to 200 meters wide and are even seen inside some large craters. Their origin is not understood, but some theories suggest that they are the result of stress, possibly caused by tides or by impact events. (NASA.)

Figure 15-54. Overall view of Mercury as obtained by Mariner 10. Note the many craters and the general resemblance to our moon. (NASA.)

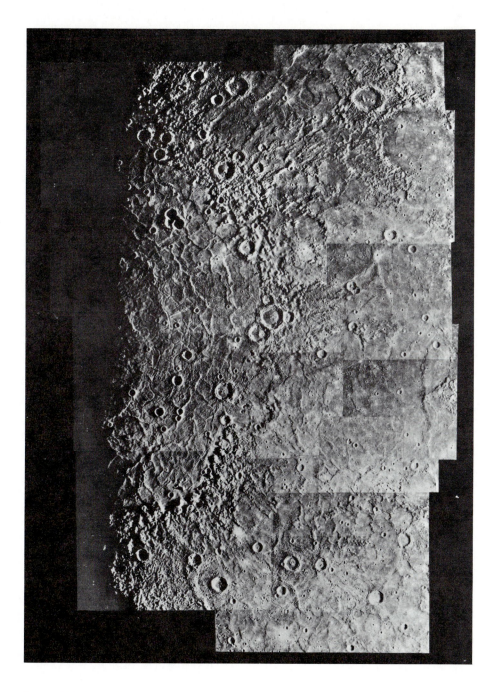

Figure 15-55. A large part of the Caloris Basin on Mercury as photographed by Mariner 10. (NASA.)

From Mariner 10, we learned that the density of Mercury as a whole is about the same as the corresponding quantity of earth. This means that Mercury is, on the average, made of *denser* material. One might think that if two planets have the same density, they must be made of material with the same density, but this is not so. The discrepancy is explained in this way: the earth is so much bigger than Mercury that material of a given natural density located deep inside the earth is squeezed down to higher ("artificial") densities by the weight of the overlying layers, whereas the same effect is insignificant on Mercury. Thus, considering the natural or uncompressed densities of the material inside the two planets, together with the observed overall planetary densities, we conclude that Mercury must contain a larger fraction of the denser elements.

The high density of Mercury, plus its magnetic field, are evidence that it has an iron core, but one that is much larger in proportion to the planet as a whole than the earth's core.[12] Recall that there is apparently no significant iron core inside our moon. With the cratered Mercurian surface, and the Caloris Basin considered as a kind of mare, the presence of the iron core explains why planetologist Bruce Murray has written: "Although Mercury is like the earth on the inside, it is like the moon on the outside."

Two aspects, however, in which the Mariner 10 photos show that the surfaces of Mercury and the moon are different are the spacing of secondary craters and the occurrence on Mercury of *lobate scarps* (Figure 15-56). Secondary craters are those smaller ones surrounding a large crater that are caused by the fall of material ejected from the main impact. They are noticeably less spread out from the primary craters on Mercury than they are on the moon. This is explained by the surface gravity of Mercury being twice as large as that of the moon; hence ejecta will fall back down more rapidly on Mercury, producing more tightly clustered secondary craters. The lobate scarps are described by Murray as "shallowly scalloped cliffs running for hundreds of kilometers." They are found all over the planet and are unlike any geological features found on earth or elsewhere in the solar system. Theorists claim that they may have been caused by an actual shrinkage of the outer layer of Mercury after its core formed and the lithosphere cooled. (It is thought that planetary core formation occurs when enough heat has been released by radioactivity within the planet to melt the central regions; then the heaviest substances, such as iron, sink towards the center.)

---

[12] The iron core of Mercury may be as large as the whole moon.

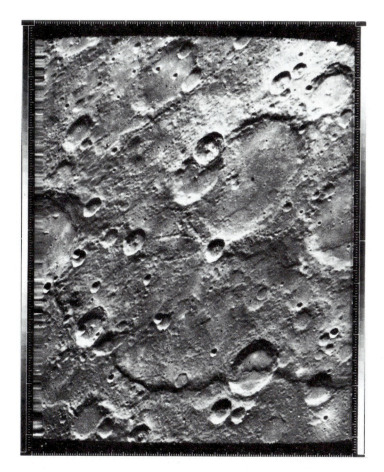

Figure 15-56. Mariner 10 view of Discovery Scarp on Mercury (top part of photograph). See text for a description of lobate scarps. (NASA.)

Nothing is known about the seismic properties of Mercury (and hence, little can be inferred about its interior, beyond the presence of a large, dense core), since no seismometer or other instrument has been landed on the planet. An ultraviolet spectrometer on Mariner 10 determined that Mercury does have an extremely thin atmosphere of hydrogen and helium, so thin that the mean free path (defined in Chapter 8) is larger than the planet itself. Collisions among the gas atoms are rare, and the atoms can readily escape from the planet. Thus there must be a constant resupply of the atmosphere, which may be furnished either by the solar wind or by outgassing from the interior.

PROBES OF THE OUTER PLANETS

So far, two spacecraft, Pioneers 10 and 11, have flown past Jupiter; results of these missions were summarized in Chapter 8. Pioneer 11 is also programmed to study Saturn, and two Voyager spacecraft are due to visit both of these planets on flyby missions that might bring one of them to Uranus as well. The spacecraft of Project Galileo will carry more advanced instruments to Jupiter in both an orbiter section and a probe to be dropped into the denser, unexplored layers of the atmosphere. The orbiter will study Jupiter and its larger satellites. At present, there are no prospects of visiting Neptune and Pluto.

THE ULTIMATE REQUIREMENT FOR SPACE TRAVEL

If civilization survives into the remote future, then space travel may become a supreme necessity. In about 6×10^9 years the sun will evolve into a red-giant star, and the earth will be uninhabitable. The requirement for liquid water, as opposed to ice or steam, on a planet like the earth demands a solar constant between 0.5 to 4 calories/square cm/minute. When the sun becomes a giant star, its luminosity will increase by 10^2 to 10^3 times. As the process begins, the sun's surface temperature will decrease, its surface will appear redder, and its radius will increase. On earth the oceans will boil away, and eventually the top layer of the atmosphere, the *exosphere*, will be so hot that all atmospheric gas molecules will escape in a relatively short time. The probable length of the sun's red-giant stage is thought to be several hundred million years. At its greatest extent the sun's surface will contain the orbits of Mercury and Venus and may even reach the orbit of earth. Our temperature requirements for survival at this time could be met on the satellites of Saturn, Uranus, and Neptune. Colonies could be established to ride out the sun's giant phase. Even if the earth remained afterward, there would be little point in returning to the charred cinder that was our birthplace, because the sun would be on the road to becoming a white dwarf, a star far too dim to supply our energy requirements.

Thus, the onset of the sun's red-giant phase will be, in effect, humanity's eviction notice from this solar system.[13] At that time, interstellar travel and the colonization of planets circling other stars will move from the realm of science fiction into a necessity for the survival of the species, if we are intelligent enough to last until that time.

THE COST OF ASTRONOMY AND SPACE RESEARCH

Since we have discussed the exploration of space in this chapter, and even the eventual need to move the population of earth to some planet of another star, it seems appropriate to discuss briefly the cost of space activities. The monies appropriated for science and space exploration are often discussed in connection with our many social and ecological problems. It would be surprising if the reader did not question the cost of the space program.

Let us begin by recalling the situation in the latter part of the nineteenth century. Some support of astronomy was undertaken by the government, such as the operation of the U.S. Naval Observatory in Washington, D.C., but much of astronomy was supported by private funds. In this matter the United States carried on the English tradition. Private donors supplied the cash to found Lick Observatory on Mount Hamilton near San Jose, California, in 1888. Similarly, funds were given for the Yerkes Observatory in Williams Bay, Wisconsin, in 1897. Since the prestige of the donor was involved, the quality of the telescope generally increased with each new observatory.

In the twentieth century, the Mount Wilson Observatory (near Pasadena, California), the Mount Palomar Observatory (north of San Diego, California), and the McDonald Observatory (near Fort Davis, Texas) were founded with private funds.

The situation began to change in the 1950s, when the creation of two major observatories was made possible by funds from the National Science Foundation (NSF). The Kitt Peak National Observatory with optical stellar and solar telescopes was built near Tucson, Arizona, and the Na-

---

[13]Perhaps not for everyone. Even very optimistic calculations suggest that the rate at which people could be launched into space is less than the rate at which they are presently being born on the earth.

tional Radio Astronomy Observatory was established at Green Bank, West Virginia. These two installations provide facilities for use by all qualified astronomers, including graduate students. Later, the Kitt Peak facility acquired a sister installation for observing southern skies, the Cerro Tololo Inter-American Observatory near La Serena, Chile, and the Arecibo Observatory in Puerto Rico was made a national radar and radio astronomy installation.

Other astronomical research was sponsored by branches of the military. In particular, the Air Force developed the Sacramento Peak Observatory, Sunspot, New Mexico, whose work is wholly in the domain of unclassified basic research on the sun; it was recently transferred to NSF control.

So far as astronomy is concerned, the transition from largely privately supported "little science" to largely federally supported "big science" was very much the result of the beginning of the space age in 1957. The launch of Sputnik 1 and the realization that the United States was not the leader in some vital aspects of space science came as a shock and, to many people, a challenge. Money was appropriated to develop programs to remedy the situation. The principal recipient of this money was the National Aeronautics and Space Administration (NASA), founded in 1958. Federal budgets are organized on the basis of a fiscal year, meaning the year ending on September 30. For example, the fiscal year 1977 ended on September 30, 1977. The NASA budget began in fiscal 1959 at 330 million dollars and rapidly increased to a peak of 5.25 billion dollars in fiscal 1965. Since then the NASA budget has decreased; the request for fiscal 1978 was down to 4.0 billion dollars. Obviously, most space science is supported through the NASA budget, but NASA is also the principal source of funds for astronomy as a whole.

What relationship does the NASA budget bear to the entire federal budget? The complexities of the federal budget cannot be underestimated, but for each year a summary is prepared by the federal Office of Management and Budget, entitled *The United States Budget in Brief*.[14] Figure 15-57 (based on this publication for fiscal 1978) shows the estimated spending by function. The total budget comes to about 450 billion dollars. Most of the items are fairly straightforward, but some clarification is desirable. The total budget includes substantial amounts of trust fund expenditures, such as social security, other retirement plans, health, and unemployment payments. Here the Government acts essentially as an

[14] Available from the Superintendent of Documents, U.S. Government Printing Office, Washington, D.C. 20402.

insurance company; these are not tax revenues. Trust funds in the fiscal 1978 budget come to approximately 146 billion dollars.

A very large slice of the budget goes for current national defense activities and the obligations due to past conflicts, as represented in the outlays for veterans' benefits and the interest on the national debt, which, until recently, resulted largely from deficit spending during war years. The large increases in the interest on the national debt during the peacetime years of 1975–1978 are not considered part of defense costs here. Viewed this way, 50 percent of the current tax dollar goes to pay for present and past defense costs, as compared to 70 percent in 1971. A discussion of the arguments pro and con concerning the various expenditures in the Federal Budget is beyond the scope of this book. Nevertheless, the figures presented in Figure 15-57 will serve as the beginning of any meaningful discussion in this area.

The breakdown of expenditures for General Science, Space, and Technology, as proposed for fiscal 1978, is given in Figure 15-58. The largest expenditures are in the category of "Space Flight, Earth Orbital Program"—that is, the manned Space Shuttle. The category "Space Sci-

Figure 15-57. Items in the U.S. Federal Budget as proposed for fiscal 1978, by function.

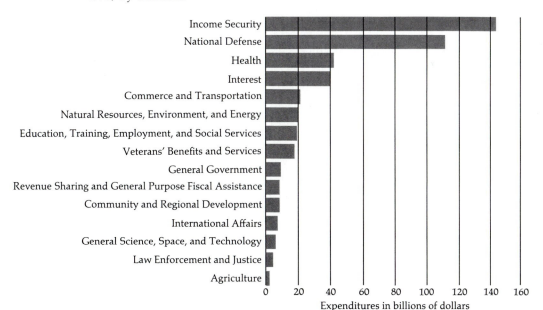

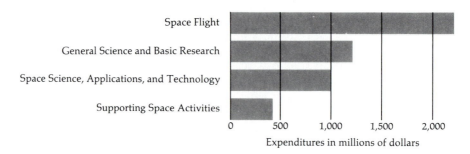

Figure 15-58. Breakdown of items for General Science, Space, and Technology in the U.S. Federal Budget, as proposed for fiscal 1978.

ence, Applications, and Technology" at 1.0 billion dollars pays for the orbiting Space Telescope to study the universe, the Galileo mission to study Jupiter, and further development in the Landsat series of satellites (used, for example, for global crop forecasting) as well as for space science and astronomy. The item "General Science and Basic Research" is for programs in the NSF and in the Energy Research and Development Administration (ERDA).

The National Science Foundation had a total estimated budget of 825 million dollars in fiscal 1978. The NSF sponsors the national observatories and a good deal of research at university observatories, but its funds must go to support many other scientific disciplines besides astronomy.

The opinions expressed here are our own, but the authors cannot claim to be unbiased. (We have taught in universities and have both worked for a national observatory and for NASA.) We believe that research in astronomy and the exploration of space are related, desirable, and important activities that transcend national boundaries and express much of the best accomplishments of humanity. They challenge thought and open new horizons for us today as they did for people in Galileo's time. It has become a cliche, and yet we feel bound to repeat it, that the Apollo 11 moon landing proved that we can commit ourselves to a seemingly impossible task and accomplish it. There is a great deal that needs to be done here on earth, and there are many competing priorities, human and environmental; nevertheless, we believe that there is an important place for astronomy and the exploration of space.

Problems in Modern Astronomy

We know a fair amount today about our local portion of the universe, the solar system. And we have some idea about the nature of stars and the makeup of our own galaxy. Some of the questions that dominated astronomical and philosophical thinking for a long time have apparently been answered; for example, we know roughly how old the earth is, and we know that its location is certainly not central to the scheme of things. We have good reason to believe that we exist in an expanding universe. On the other hand, many great problems remain, and these include questions that in some cases could not even have been asked before the development of modern astronomy, physics, and mathematics. Some of them have been singled out for presentation in this chapter; they are among the "hot topics" that have been vigorously debated at astronomers' conferences in the past few years. These problems are included here to give a taste of what lies beyond the elementary survey presented thus far—what is happening in astronomy today.

PULSARS AND THE CRAB NEBULA

Astronomers, like most people, have a great interest in something new. Indeed, it is the prospect of discovering a new comet, asteroid, or other object that inspired some of us to become astronomers. Others are more interested in investigating physical laws to derive a better understanding of the celestial phenomena recorded by the observers. Rarely, however, have scientists been fortunate enough (or clever enough!) to *predict* an entirely new *class* of objects that was subsequently actually found. Such, however, was the case in the matter of the neutron stars.

The story of neutron stars and pulsars begins in China in the year A.D. 1054 when Yang Wei-Tek recorded the presence of a brilliant "guest star" in the heavens. The object—a bright star where none had been seen before—was so luminous that it was visible in broad daylight for more than three weeks. Gradually, however, it faded, but it continued to be visible for some 650 days, after which it was too faint to be seen by the unaided eye. Today we identify the phenomenon observed by Yang as a *supernova*, the death of a star by an especially violent and rare means.

Centuries went by, and we have no information on the supernova in this interval, since it could not be seen, until finally the telescope was invented. Another century passed, and in 1731, an English physician, John Bevis, finally turned a telescope to the direction in which (unknown to him) the guest star had been observed and discovered a dim, fuzzy object. (Because of its appearance, it was later named the Crab nebula by the Earl of Rosse, who discovered its filaments; Plate 10.) During the next 235 years observational evidence about the nebula accumulated. By 1966 three of its main properties could be summarized as follows: (1) It consists of an irregular cloud of gas expanding at a rate of about 1,000 km/sec (presumably due to the original explosion). (2) Faint, probably short-lived luminous wisps, located near the center of the nebula, are moving at velocities much higher than that of the nebular expansion. (3) It is the source of optical, infrared, radio, and x-ray radiation of great intensity. Further, it had become clear that only one proposed theory could explain the continuous spectrum of the Crab nebula, that is, the relative intensities of radiation in the different wavelengths. According to this model, the light, radio waves, and x-rays were produced as *synchrotron radiation* by electrons moving at great speeds (approaching the speed of light) through a magnetic field in the nebula.

Strange Light at General Electric

At this point in the story of the Crab nebula, we will digress briefly from astronomy and talk about atomic accelerator machines. In 1944–1945, a number of frontier investigations in experimental physics centered on the bombardment of atoms with streams of energetic particles produced by such accelerator devices as the cyclotron. Just as astronomers keep hoping and asking for bigger and better telescopes, the physicists, then as now, always seem to need ever more powerful accelerators to advance the study of nuclear structure. At the time in question, the aim was to develop machines in the hundred MeV (million electron volt[1]) range and eventually break through to 1 BeV (billion electron volts). There is a basic limitation in the cyclotron that prevents it from reaching these high energies. More advanced instruments, the betatrons, were in existence, but the cost involved in upgrading them to higher energies became prohibitive.

The time was ripe for a basic improvement in accelerator design, and the design parameters for such a machine, the *synchrotron*, were described by Edwin M. McMillan of the University of California in a brief paper in the *Physical Review*. McMillan discussed a synchrotron that he planned to build at Berkeley. However, shortly after this paper was published, a group of scientists at the General Electric Company in Schenectady began work on a synchrotron. The specific purpose of their machine was to produce energetic electrons. When the synchrotron was built, the General Electric workers observed a fascinating effect through a glass window in its side; as the electrons moved along they emitted an intense light in the direction of motion. Appearing dull red in color when the synchrotron was idling at 30 MeV, the light brightened as the machine was gradually turned up; at 80 MeV, slightly above the design goal of the synchrotron, the radiation was a brilliant bluish-white. There was another interesting effect; viewing the light through a Polaroid filter, the researchers found that, although easily visible in one orientation of the Polaroid, the light became imperceptible as the filter was rotated through a 90 degree angle. Thus, it obviously was polarized.

The possibility of light emission by electrons moving at high speed in a powerful magnetic field, such as existed in the synchrotron, had been discussed half a century before by the physicist A. Lienard, and it had also occurred to McMillan as a possible side effect of his proposed machine. However, the strange light found at General Electric became known as *synchrotron radiation*. We mentioned in Chapter 5 that the path of a charged particle, such as an electron, will be deflected or bent by a magnetic field. In fact, in the case of the Crab nebula, the electrons actually spiral around the lines of magnetic force, and the synchrotron radiation that they emit is polarized in the direction perpendicular to the magnetic field lines (Figure 16-1). Synchrotron radiation may be emitted

---

[1]The electron volt or eV is a unit of kinetic energy often used by physicists. We say that the temperature of the atoms in the solar photosphere is 6,000 K; a physicist could equally well say that their average energy was about 0.5 eV.

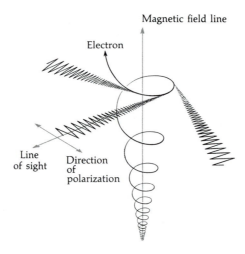

Figure 16-1. Schematic diagram showing production of polarized synchrotron radiation by electrons spiraling in a magnetic field. (Adapted from *Exploding Galaxies* by A. R. Sandage. Copyright © 1964 by Scientific American, Inc. All rights reserved.)

at any wavelength, depending on the velocity of the high-energy electron and the strength of the magnetic field.

The idea that the Crab nebula radiation was due to synchrotron emission was put forward by I. S. Shklovskii; it was soon confirmed when the nebula was observed with a polaroid filter.

Shklovskii's theory solved a problem—the radiation was produced by the synchrotron mechanism—but it eventually raised another one when x-rays from the Crab nebula were observed. The theory allowed one to calculate how long an electron of given energy would continue to emit radiation. In the case of x-ray emission, this *radiative lifetime* was only about a year. But the supernova explosion that produced the Crab was observed over 900 years ago. How then could electrons of radiative lifetime one year still be present to emit the x-rays? Presumably, some unknown energy source capable of producing fast electrons was still present in the nebula.

Neutron Stars

In the 1930s the Soviet physicist Lev Landau (and later the American J. Robert Oppenheimer) proposed a new state of matter that might exist in a previously unsuspected form: the neutron star. The matter in such a star

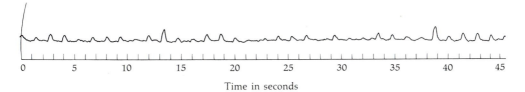

0   5   10   15   20   25   30   35   40   45

Time in seconds

Figure 16-2. The observed pulses from CP 1919, the first pulsar, showing pulses spaced by about 1.3 sec. Note that they are not of equal strength. (From *Pulsars* by Antony Hewish. Copyright © 1968 by Scientific American, Inc. All rights reserved.)

would be compressed to the enormous densities characteristic of the nuclear core of an atom. At the center of the neutron star, the density may be 10^{14} or 10^{15} times greater than that of water. As a result, all of the mass of an ordinary star like the sun, compressed to this enormously high density, would occupy a sphere with a diameter of only some 20 km (12 miles). Under certain circumstances, such a star might be formed in an *implosion* of the central region of a supernova, while an *expanding* gaseous remnant, like the Crab nebula, would also be generated. Most astronomers did not believe in this process, however, and calculations indicated that the neutron stars would be very hard to observe (due to their small size), so they were generally regarded as a physicist's dream. Although theoretical analyses of the properties of the hypothetical neutron stars were pursued by a few true believers (such as A. G. W. Cameron of Harvard University), the lack of observational evidence for the existence of these objects was discouraging, and this is where the problem of the neutron stars stood in 1967.

Pulsars

In July, 1967, radio astronomers at Cambridge University, led by Antony Hewish,[2] discovered a highly unusual object. Most previously known celestial sources of radio waves were relatively steady emitters, although a few (like the sun) produced bursts of radio waves at erratic times. However, the Cambridge source seemed to be emitting bursts (pulses) at perfectly spaced intervals (Figure 16-2), like the ticks of a clock. Although the

---

[2]Hewish later was awarded the Nobel Prize in Physics for the discovery and investigation of pulsars. He shared the Prize with another Cambridge radio astronomer, Sir Martin Ryle, who was honored for extragalactic research and innovations in radio telescope design. The first pulsar signals actually were recognized by Hewish's graduate student, Jocelyn Bell.

Table 16-1. Typical Parameters of Compact Stars

	White Dwarf	*Neutron Star*
Diameter	0.5 to 2 earth diameters	20 to 40 km
Mass	0.4 to 1 solar mass	0.5 to 4 solar masses
Central density*	10^5 to 10^9 gram/cm³	10^{14} to 10^{15} gram/cm³

*Density of ordinary water = 1 gram/cm³.

strength of the pulses varied, the observations indicated that the new radio source, called a *pulsar,* was keeping better time than most man-made clocks! This incredible discovery led to a great effort to discover more pulsars and to analyze their properties in an attempt to explain them. Among the first 100 pulsars to be found, the pulse-spacings (pulse periodicities) ranged from one-thirtieth of a second to over three seconds.

The extreme regularity of the pulse-spacings required an explanation. The three likely *clock mechanisms* that were generally considered were: (1) *pulsation* of a star, which might resemble the pulsations of a Cepheid variable star, but much faster; (2) *orbital motions* of a binary star; (3) *rotation* of a star. Two types of *clock stars* were considered for each of these theories: white dwarfs and neutron stars. These were selected because the shortness of the pulsar periods implied that small, dense objects were involved. (Typical parameters of white dwarfs and neutron stars are given in Table 16-1; also see Figure 10-19.) For example, an ordinary star cannot pulsate as fast as even the slowest pulsar; two stars like the sun cannot orbit with a pulsar period, because from Kepler's and Newton's Laws we know that, to orbit so rapidly, their separation would have to be less than their diameters; finally, an ordinary star cannot perform one rotation in a time as short as a pulsar period, because at such speeds matter would spin off at the equator, disrupting the star.

During the first few months after the pulsar discovery announcement, there was considerable debate over what both the clock mechanism and the clock stars might be. However, three additional discoveries convinced astronomers that the pulsars were rotating neutron stars: (1) two pulsars were found to be associated with supernova remnants; (2) the period of one of these two pulsars, namely, the object NP 0532[3] *in the*

---

[3] NP 0532 stands for "National Radio Astronomy Observatory Pulsar near Right Ascension 05 hours, 32 minutes." NP 0532 was discovered by David Staelin and Edward Reifenstein at the N.R.A.O.

Crab nebula, was extremely short: one-thirtieth of a second; (3) comparison of measurements over several months showed that many of the pulsar periods were not constant, but were gradually lengthening.

In connection with discovery (1) we recall that the hypothetical origin of a neutron star took place in a supernova event. Discovery (2) rules out white dwarfs, since the size and the distribution of matter in the interior of a white dwarf are incompatible with rotation, orbital motion, or vibration in so short a period as one-thirtieth of a second. Discovery (3) ruled out orbital motion or vibrations of neutron stars, because calculations showed that the periods would be expected to remain constant or to gradually *decrease.* The remaining possibility, which has survived all observational tests so far, is rotation of a neutron star. A typical neutron star, with properties as given in Table 16-1, could easily rotate 30 times per second without breaking up. Theorists are studying the ways in which the radiation might be beamed. Apparently a very strong magnetic field, perhaps even 10^{13} times that of the earth, may be involved. As the stars rotate, the beams sweep around like those of searchlights (Figure 16-3), and each time such a beam sweeps past the earth, we observe a pulse.

The pulsar NP 0532 in the Crab nebula is thus apparently a neutron star that is spinning at a rate of 30 rotations/sec. It has also been observed at optical wavelengths (Figure 16-4). The observations of its pulses show that the rotation period is slowly increasing. This slow-down of the spinning neutron star represents a loss of energy that must reappear in some other form, according to the law of conservation of energy. (For example, when an automobile is stopped, some of its kinetic energy appears in the brakes in the form of heat.) In fact, it has been calculated that the energy lost by the rotating neutron star in this way is somewhat larger than the total energy emitted by the Crab nebula in the form of electromagnetic radiation.

If the identification of pulsars as neutron stars is correct, then the evaluation of the energy loss involved in the slowing down of pulsar NP 0532 has solved the basic problem of the Crab nebula. The energy source of the nebula, which enables it to shine in the x-ray portion of the spectrum nine centuries after the supernova explosion, is the gradual slowing of a dense, rapidly rotating object, the neutron star. The spinning neutron star is losing its rotational energy, and the presence of a strong magnetic field allows the energy to reappear in the form of the kinetic energy of fast-moving electrons (precisely how this happens is also still being debated). The electrons encounter the magnetic field of the nebula and produce synchrotron radiation at all wavelengths, from the x-rays to the radio waves. Some of the fast particles produced by the neutron star may

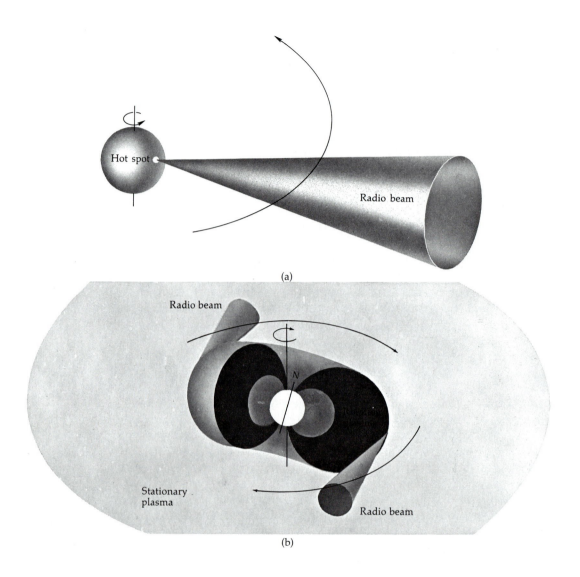

(a)

(b)

Figure 16-3.   Artist's impression of two beaming models for pulsars. (a) The pulsar radiation originates at a "hot spot" on the surface of the star. This might be the magnetic pole, and in the case drawn here, the magnetic pole would be at the star's equator. (b) The pulsar emission originates well out in the surroundings. In either case, as the rotating beam sweeps past the earth, we detect a "pulse." (After *Pulsars* by Antony Hewish. Copyright © 1968 by Scientific American, Inc. All rights reserved.)

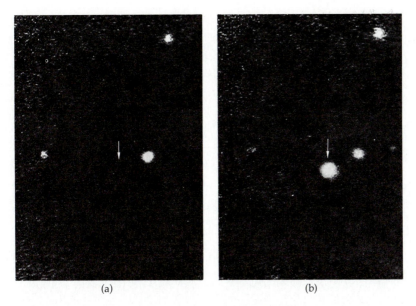

(a)	(b)

Figure 16-4. The Crab nebula pulsar (NP 0532) in optical wavelengths. (a) During the pulses. (b) Between the pulses. (Courtesy J. S.
Miller and E. J. Wampler, Lick Observatory.)

escape from the nebula, and thus pulsars like NP 0532 may represent a
source of cosmic rays.

RADIO GALAXIES

Most galaxies emit energy primarily in the form of visible light, the light
of their member stars and nebulae. However, there is a small but interesting fraction of known galaxies that produce huge amounts of energy in
the form of radio waves. The radio energy exceeds the visible light of
these *radio galaxies*, and, since the radio waves are not absorbed by dust in
space, we can use radio telescopes to observe these objects at far greater
distances than ordinary visible galaxies.

A fundamental characteristic of radio galaxies close enough for both
optical and radio study is that the bulk of the radio emission in each
object is produced in two optically invisible *radio lobes*, located on opposite
sides of the galaxy, often at huge distances from it. A typical separation of
a radio lobe from the associated galaxy might be as much as 100,000 parsecs, for example.

The problems posed by radio galaxies include: the origin of their
great radio energy, the nature and age of the radio lobes for a given

Figure 16-5. Cygnus A, showing the double nucleus. Compare with the photograph showing a much larger sky area in Figure 16-6. (Hale Observatories.)

galaxy, the relatively small sizes of the lobes, and the occurrence of specifically double-lobed structure. One clue comes from the fact that, in some cases, two or more lobe pairs are found along the same axis centered on a galaxy, but with different separations. This suggests that the lobes might be expelled from the galaxy by a process that occasionally repeats, so that the closer pair is the younger one. Additional evidence discussed below also implies that the process is repetitive.

The Energy Problem

The visible light emitted by a large normal galaxy, such as the Milky Way, is about 25 billion times the luminosity of our sun, but the Milky Way's *radio* energy output is only 25,000 times the solar luminosity. By comparison, a radio galaxy may generate radio waves with energy amounting to 250 billion times the solar luminosity! Although the visible light energy of normal galaxies can be attributed to the energy released by nuclear reac-

tions within their component stars (as well as to other much less significant processes), the source of the much greater energy of the radio galaxies
is yet to be explained. One of the earliest theories of radio galaxies ascribed the energy to the effects of collisions among galaxies, because the
first radio galaxy subjected to intensive study, known as Cygnus A (Figure 16-5), appeared on Palomar photographs to resemble a galaxy pair

Figure 16-6. Cygnus A, center of photograph, showing its huge size.
It is much larger than other galaxies in the cluster. Establish the difference in scale between the two photographs of Cygnus A by finding
the same circle of stars around the central region. In Figure 16-5, this
circle is approximately half the size of the entire photograph. In Figure
16-6, this circle is much smaller and found near the outer boundary of
the galaxy's image. (Courtesy of H. E. Smith and H. Spinrad, Kitt
Peak National Observatory 4-meter Telescope Photograph.)

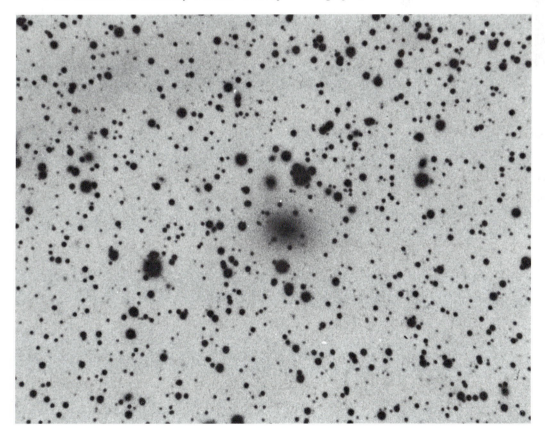

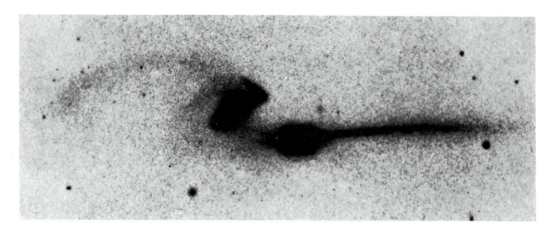

Figure 16-7. Photograph of two colliding galaxies, nicknamed "The Mice." (Courtesy of H. C. Arp, Hale Observatories.)

that indeed had collided. Since then, observations have shown that Cygnus A is a very large, single galaxy (Figure 16-6), perhaps with a double nucleus, which explains why the earlier observers thought that two galaxies were present. Furthermore, many other radio galaxies have been found that (when photographed) are clearly single galaxies, not colliding pairs. The final blow to the colliding galaxies theory comes from theoretical calculations which show that even when a collision does occur, it won't provide enough energy! In fact, some *genuine* galactic collisions (Figure 16-7) have now been observed, and they indeed are not strong radio sources.

Duration of the Radio Lobes

Suppose that the lobes of a radio galaxy are each 100,000 parsecs from the visible galaxy. Now, 100,000 parsecs is equal to about 326,000 light years. Thus, if the lobes were expelled at close to the speed of light (as assumed in some theories), then they are over 326,000 years old, whereas if they were expelled at one-tenth the speed of light, they are 3,260,000 years old, etc. Since the lobes presumably have been emitting radio waves at their present rate (or more) during the whole time since they were ejected, then the older they are, the more energy they have released, and thus the greater the energy problem. On the other hand, if they have moved at the highest possible speed and are just a few hundred thousand years old, then they are a short-lived phenomenon on the astronomical time scale of billions of years. (We rarely see lobes much more distant than one million parsecs from the associated galaxies, so presumably they fade out before reaching such large distances.)

If double-lobed radio galaxies are a short-lived phenomenon, however, then our chance of seeing one is very small. Let's say that the age of a galaxy is 15 billion (15×10^9) years. If the associated radio source lasts no more than one million (1×10^6) years, then there is only (1×10^6) ÷ (15×10^9) = $1/(15 \times 10^3$), or one chance in 15,000 that the radio source will be in existence when we observe the galaxy.[4] Thus, radio galaxies should be so rare that we hardly would detect any. But, in fact, if you study a rich cluster of galaxies (one with many galaxies, but nevertheless usually much less than 15,000), you have a fair chance of finding a radio galaxy among them, and in fact many thousands of radio galaxies have been observed in recent years. Thus, if double-lobed radio radio galaxies constitute a short-lived phenomenon, then the phenomenon must be recurrent, so that there are many times in the life of a galaxy when it produces a double radio source. But if this is true, then you must multiply the energy released in a single typical radio source by the number of radio sources that the galaxy produces over its lifetime to get the total radio energy that must be explained. Thus, whether the radio sources are long or short in endurance, the energy problem remains. And if radio sources are repetitively produced in the same galaxy, then we have to wonder what conditions exist inside the galaxy that keep building up to the point where such an extremely energetic event occurs.

The Confinement Problem

The emission from the lobes of a radio galaxy has the characteristics of synchrotron radiation, so it is presumably produced by high-speed electrons traveling through magnetic fields. Thus, one might think of the radio lobes as clouds of plasma and magnetic fields, perhaps ejected by some great explosion in the parent galaxy. But then one is faced with the *confinement problem:* When a cloud of gas or plasma is expelled into space, it must expand rapidly, just as a smoke ring soon expands to the point where it is dissipated in the air. By this standard, the lobes of a radio galaxy (Figure 16-8) are unexpectedly small, and, in particular, they sometimes are found to contain small, intense cores. What confines the lobes to such minute dimensions?

---

[4]There will be an even smaller chance if only *some* galaxies *ever* become radio galaxies. For example, if only ten percent of all galaxies ever become radio galaxies, then when we observe a randomly chosen galaxy, in the above example, there is only one chance in 150,000 that it will be a radio galaxy.

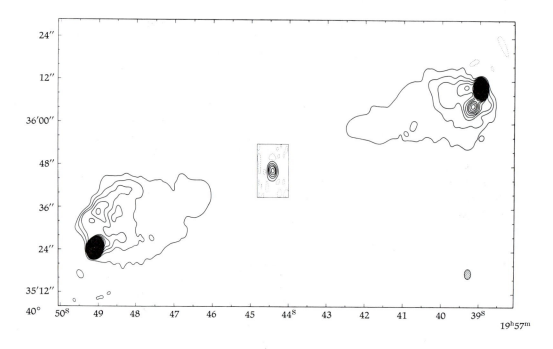

Figure 16-8.   Very high resolution radio map of Cygnus A. Notice the two conspicuous "hot spots" (shown black), one in each lobe. (Courtesy of Sir Martin Ryle, Cambridge University.)

One theory of radio galaxies, the *ram pressure model*, suggests that there is a thin, hot intergalactic gas retarding the growth of the lobes. Another idea, the *beaming model*, assumes that the lobes are not entities expelled from the galaxy after all. Instead, the authors of this theory imagine that there are powerful beams of energy (either in the form of electromagnetic waves or consisting of high-speed atomic particles) generated in the center of the galaxy. According to this concept, the radio lobes represent the regions where the beams interact with the intergalactic gas, producing the radio emission. The intense emission cores, also called *hot spots*, in the lobes might represent the actual regions where the beams strike the gas, whereas the surrounding lobes consist of radio-emitting plasma expanding from the hot spots. Still another much-discussed theory, called the *slingshot model*, supposes that a rapidly rotating, massive object (similar to a giant pulsar and known as a *spinar*) once existed in the center of the radio galaxy. For some reason, this object was disrupted, and fragments (each, in effect, a smaller spinar) were flung out in opposite directions. As each of the spinar fragments travels outward, it generates clouds of magnetized plasma which produce radio emission. Thus, the expelled spinars are located in the hot spots, according to this theory, and the radio lobes are the plasma clouds which they continuously generate. A given mass of plasma will dissipate into space, but the spinar is generating new plasma, so we continue to observe a relatively compact and bright radio lobe. (In an alternative version of the slingshot theory, it is not one giant spinar that broke apart, but rather an orbiting system of a few spinars that was disrupted, causing the individual components to fly outwards.) Schematic diagrams for the three models are given in Figure 16-9.

Objections to Theories

Serious objections have been made to each of the radio galaxy theories. The biggest problem is that each theory seems to invoke very crucial assumptions in areas where we are especially ignorant of the true facts. We have very limited knowledge of how much intergalactic gas there is, and, in fact, some of our ideas about this gas come from interpreting the observed sizes and shapes of the lobes and other radio-emitting structures in radio galaxies. But these interpretations are only valid if the ram pressure theory is correct, so the reasoning behind this theory could be regarded as circular. As to the slingshot model, it assumes the existence of large spinars, presumably somewhat like the rapidly rotating neutron star in the Crab nebula but much larger. A problem with the beaming model is that no one has proven that such intense energy beams actually can be

Figure 16-9. Schematic diagrams for three theories of radio galaxies. (a) Beaming model. (b) Ram Pressure model. (c) Slingshot model.

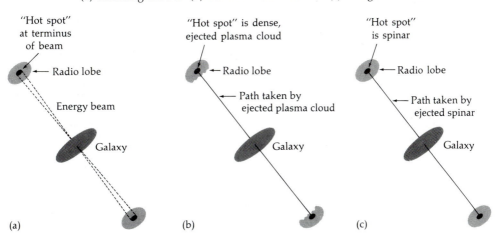

produced. In fact, there is a whole group of additional theories under discussion that try to explain what the central object is that must generate the beams! Is it a huge spinar, or a massive black hole, sucking in gas from surrounding space? Many of the theories seem to verge almost on science fiction at this point.

QUASARS

When radio telescopes were used to map the sky, it turned out that some of the strongest sources of cosmic radio waves were associated with very distant, optically inconspicuous galaxies. But at least they are recognizable as galaxies. In contrast, it was quite a problem to determine the nature of the objects associated with the radio sources now called *quasars*, and there is still considerable uncertainty about this question.

The story begins in the early 1960s, when a few radio sources, named 3C 48, 3C 196, and 3C 286,[5] were identified with faint stars (Figure 16-10), but when the spectra of the stars were photographed, they showed emission lines rather than absorption lines, and the patterns of the lines were unfamiliar. In fact, it was impossible to decide what the atoms were that produced these emission lines.

The breakthrough came in 1963, when radio astronomer Cyril Hazard in Australia observed the moon to occult (that is, eclipse) a hitherto unidentified radio source, 3C 273. Hazard's radio telescope was not ordinarily capable of making really precise position measurements, but it enabled him to record the exact times when the moon occulted 3C 273 and when 3C 273 emerged from behind the moon. Since the orbit of the moon is known with great precision, Hazard was able to calculate an accurate position of 3C 273 from his timings of the occultation. At Palomar, Maarten Schmidt photographed the spectrum of a faint blue star at this position (Figure 16-11). The emission line pattern of this star was recognizable. It was a well-known set of hydrogen lines, but it was remarkable. The lines had a very large red shift, amounting to 45,000 km/sec, or 15 percent of the velocity of light, as one would find in a galaxy at a distance of 450 megaparsecs. Thus, 3C 273 was apparently *not* a star in the Milky Way galaxy. With this clue in mind, the puzzle of the spectral line patterns of 3C 48, 3C 196, and 3C 286 was soon cleared up; they had been unrecognized because they had unexpectedly large red shifts. From the circumstance that these objects at first had seemed to be stars, they were named *quasi-stellar radio sources* or, more popularly, *quasars*.

---

[5] This is astronomical shorthand for the *Third Cambridge Catalogue of Radio Sources,* as mentioned in Chapter 6.

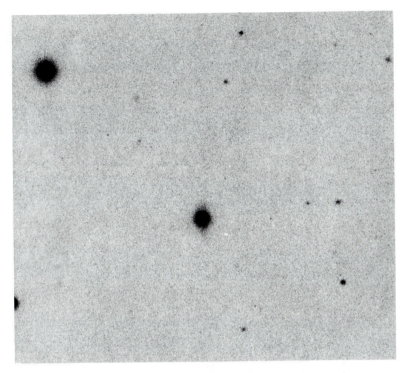

Figure 16-10. Quasar 3C 48 resembles a dim star; this is a negative print, so the stars appear as black dots. (Courtesy of the Hale Observatories.)

Today, more than 400 quasars are known, defined in the strict sense as radio sources identified with stellar-appearing objects whose optical spectra show large red shifts. Also known are radio-quiet quasars, or *blue stellar objects,* which appear to be virtually identical to quasars in their optical properties but are not strong radio sources. The celestial objects of greatest observed red shift are three quasars that appear to be receding at speeds of over 90 percent of the velocity of light, as deduced from spectrograms obtained with large telescopes.

The great red shifts of quasars imply that they are at great distances. But the quasars are also strong radio emitters in terms of *apparent* radio brightness as observed at the earth. These two circumstances taken together imply that the *absolute* radio brightnesses of quasars are immense. Indeed, they often are comparable to those of the most intense radio galaxies, although the quasar emission may arise in a vastly smaller volume.

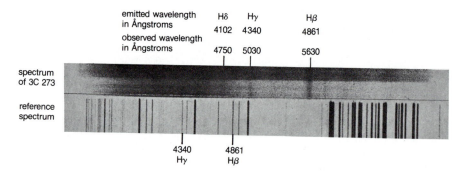

Figure 16-11. Laboratory reference spectrum (b) is compared with the spectrum of the closest known quasar, 3C 273 (a). The upper half of the quasar spectrum was given a longer exposure than the lower half, and thus appears darker on this negative reproduction. Note that although the laboratory spectrum shows the hydrogen line Hβ to be at wavelength 4,861 Ångstroms, this line appears at 5,630 Ångstroms in the spectrum of 3C 273, due to the red shift. (Courtesy of Maarten Schmidt, Hale Observatories.)

Improvements in radio astronomy techniques since Hazard's observations were made have allowed very high spatial resolution to be obtained in simultaneous measurements of the same quasar by radio telescopes in opposite hemispheres of the earth. These measurements have shown that the angular diameter of the emitting region of a quasar is usually very small. Combining these angular diameters with the distances computed from the red shifts and the Hubble constant, it is found that most of the radio emission of a typical quasar occurs in a region of less than a parsec in size. A spiral galaxy like the Milky Way has a diameter of about 30 kiloparsecs. In some cases, the radio emission of a quasar is a thousand times greater than the total emission at all wavelengths of a normal spiral galaxy, and yet it may be produced in a region of 30,000 times smaller diameter! Further, the visible light, infrared, and x-ray emission produced in the same small region of the quasar is far more energetic than the radio emission.

A great challenge to theoretical physicists has been to account for this enormous energy emission of a quasar, concentrated in such a small region. Some tried to get out of the problem by saying that the quasars were not at the great distances calculated from their red shifts and the Hubble constant. Instead, they suggested that the quasars were much closer, and that their red shifts were not due to the expansion of the universe but due to the *gravitational red shift*. (Einstein had shown that the gravity of a star would produce a slight red shift in its light; the *gravitational red shift theory* of quasars required that they have extremely strong gravitational fields in order to produce the observed high red shifts.) On the other hand, the *local theory of quasars,* extensively discussed by James Terrell of the Los Alamos Scientific Laboratory, suggested that the quasars were relatively close to the Milky Way. According to this theory, their huge red shifts were caused by ejection of the quasars at high velocity from an explosive event in the Milky Way or a nearby galaxy.

The arguments over the nature of the quasar red shifts have continued, but most astronomers believe that the *cosmological theory of quasars* is correct. This is the simple idea that the red shifts of quasars are due to the same effect as those of galaxies, namely, the expansion of the universe. One way of verifying this hypothesis would be to find quasars located in clusters of galaxies. At the California Institute of Technology, John Bahcall and James Gunn explored this possibility. They found several quasars and quasar-like objects that do appear to be associated with clusters of galaxies. Then, in 1978, Alan Stockton of the University of Hawaii found that, for a number of quasars, including 3C 273, there are adjacent galaxies with the same red shifts as the quasars. This implies that the quasars are associated with the galaxies as members of the same groups or clusters. Since galaxy red shifts are due to the expansion of the universe, the red shifts of these quasars must be due to the same effect. Therefore, the cosmological theory of quasars is almost surely correct.

The spectrum and polarization of the radio emission from many quasars have been measured, and it seems clear that the radio waves are produced by the synchrotron process, as in the Crab nebula. However, the question of the *source* of the enormous energy of a quasar is still unanswered. A great many speculative ideas have been put forth, and we mention only a few here to give the reader a feeling for the kind of thinking about quasars that is going on today.

The Swedish physicist Hannes Alfvén has revived the old idea of the possible existence of large amounts of *anti-matter* in connection with the energy source of quasars. Anti-matter consists of atoms resembling those found on earth but composed of particles with opposite charges—that is, the atoms have negatively charged nuclei and positively charged electrons. If a cloud of anti-matter encountered a cloud of ordinary matter, the result would be the annihilation of the atoms and the release of an enormous amount of energy.

Collisions between stars have also been widely discussed as an energy source for quasars. According to these theories, a quasar may consist of a large number of stars packed much more closely together than those of a globular cluster. Thomas Gold of Cornell University has calculated that collisions between these stars could supply the energy of a quasar. However, Stirling Colgate of the Los Alamos Scientific Laboratories has suggested that the most important effect of the collisions would be to cause smaller stars to coalesce into larger ones, which then evolve faster and become supernovae, according to the general ideas presented in Chapter 11. Thus, a large number of supernova explosions might account for the energy of a quasar. The *gravitational collapse* theory was proposed by the British astrophysicist Fred Hoyle and by William Fowler (California Institute of Technology). According to their ideas, hypothetical superstars with masses millions of times greater than that of the sun contract under the influence of their own gravitation, attaining great densities (as in a neutron star), and energy is released as the matter falls together. More recently, physicists have drawn analogies between quasars and pulsars. One suggestion is that a quasar derives its energy from the slowing down of a large rotating mass, and another idea is that many individual rotating magnetized objects (perhaps resembling the spinars mentioned in the section on radio galaxies) are found within a quasar and provide its energy.

Quasars, once thought to be stars in the Milky Way, are now regarded as among the most distant objects in the universe. They are also the most energetic phenomena known. They represent a challenge to the physicist, who must account for their enormous energies, and to the cosmologist, who must explain their significance in terms (perhaps) of a special stage in the evolution of galaxies or of the universe. As distant sources of intense radiation, they offer the possibility of studying intergalactic matter in terms of its effects on the transmission of quasar radiation, just as interstellar matter was studied through its reddening, polarizing, and line absorption effects on the light of stars. Indeed, many quasars have absorption lines with red shifts that, although large, are much smaller than those of the emission lines. According to one theory, these *intermediate red shift absorption lines* are produced by thin intergalactic gas, perhaps located in huge halos surrounding *intervening galaxies*, or galaxies much closer to us than the quasars. The halos must be far bigger than the galaxies, because these absorption lines are found in the spectra of so many quasars, although the galaxies themselves are not directly along the line of sight to the quasars. In fact, the intervening galaxies have not actually been located.

Finally, possibly the greatest challenge that quasars offer to the astronomer's imagination is the recent discovery of apparent *superluminal motions*. By means of interferometry using widely separated radio telescopes—literally on opposite sides of the earth—the positions of radio-emitting structures in several quasars have been measured with great precision. The measurements have been repeated at intervals over several years, and they indicate that the emitting structures are moving apart as seen on the plane of the sky—in other words, proper motion has been measured. When we calibrate these proper motions by adopting the distances implied by the quasar red shifts and the Hubble law, we derive tangential velocities. These velocities, in several cases, are well above the speed of light! Can you imagine something that moves at velocities greater than the speed of light, yet is produced by ordinary means, without the use of enormous energy or mysterious processes? Think about the spot of light projected from a rotating beacon of a lighthouse. How fast does it move across a ship a few miles away? How much faster does it move across another ship at twice that distance? Extrapolate!

BLACK HOLES

The *black hole* is a physical and mathematical concept that has captured the imagination of many astrophysicists. But do black holes exist? Many scientists have written theories about black holes and have invoked them to explain celestial phenomena that otherwise appear incomprehensible. A number of astronomers have even reached a consensus that one specific astronomical object actually is a black hole. However, both the reality of black holes and the identification of the first known example remain highly debatable.

What Is a Black Hole?

From one point of view, a black hole is nothing more than another kind of highly condensed dead or dying star. Stars of about 1 solar mass (see Table 16-1) eventually use up their nuclear fuel and contract to become white dwarfs. A star of 1.5 to 2 solar masses contracts beyond the white-dwarf state and becomes the much denser neutron star. In each case, a particular force or source of pressure comes into play to retard the contraction. However, for a contracting or imploding star of more than 3 solar masses, the force of gravity can become so powerful that no presently known force can balance or overcome it. Thus, according to calculations, such a star will contract *forever*.

Let us recall the concept of escape velocity, as mentioned in the discussion of space travel (Chapter 15). If we shoot a bullet directly up from the surface of the earth, it must travel at least 11 km/sec (7 miles/sec) to escape from the earth; otherwise, it will eventually fall back down. As listed in Table A-2, "Properties of the Planets," each planet has its own characteristic escape velocity, as calculated from the mass and diameter. The sun and other stars also have escape velocities that can be calculated. A black hole is an object with such powerful gravity that *its escape velocity exceeds the speed of light*. Thus, no ray of light emerges from the black hole, nor anything else. Matter can fall into the hole, but it cannot escape. Since we can't even see the matter, and can never hope to see it, even in principle, the contents of a black hole are sometimes said to have left the *observable* universe. Or, according to some of the more speculative or sensational theories, the contents of the black hole are said literally to have left the universe—whatever that means! It is even claimed that the material that has left our universe reemerges in some other universe, where it might be called a "white hole." Since it is very hard to explain either the enormous energy of quasars or the fact that the energy is concentrated in a very small region (as mentioned in the previous section), the claim is occasionally made that quasars are white holes in our universe. We don't really believe in white holes and other universes, but we don't know for sure, and neither, we suspect, does anyone else.

Do Black Holes Exist?

How might we determine if black holes actually exist? Can you think of any way to detect or study something that does not emit or reflect any light or particles of matter? Perhaps the first thing that comes to mind is that a dark object such as a black hole might pass in front of a bright object, such as a background star, and cut off its light.[6] This may or may not be a possibility, but expected black holes are so small (typically a few miles or less in size) and presumably are so far away that the chances of one eclipsing a background object, as seen from the earth, seem very small. And so are the chances that we would happen to be looking in the right place when it happened. And even if we were looking, would we be sure of what happened? How would we know that it wasn't some other dark object, such as a neutron star or a cold white dwarf, that did the eclipsing? So, looking for eclipses is not likely to help us find a black hole.

Current ideas on finding black holes depend on one property of a collapsed star that *can affect the observable properties of another star or of sur-*

---

[6] Actually, according to some ideas *(gravitational lens theory)*, a black hole passing in front of a distant galaxy might focus its light, so that it might appear briefly as a more intense spot of light. However, the arguments against our being able to observe and recognize such an event are equivalent to the arguments against our observing an eclipse of a background object by a black hole.

rounding matter, namely its gravity. If a massive star that happens to be part of a binary star system collapses and becomes a black hole, it can still remain as a member of the binary. Both the black hole and its companion star follow orbits around the center of mass of the system. We might not expect to see the black hole, but we can certainly observe the companion star. Observing its spectrum, we find that it is a single-lined spectroscopic binary, as defined in Chapter 10. If the spectrum of the visible star is recognizable, we can estimate the approximate mass of the star as the typical mass of an object with its spectrum and luminosity class (a blue giant, for example). Then we can estimate the mass of the unseen companion from the rate of motion of the visible star, as given by the Doppler shifts in the spectrum. (Remember that the visible star is in motion because it is accelerated by the gravitational attraction of its companion, the black hole.) Actually, this method gives a *lower limit* to the mass of the unseen companion. If such a calculation tells us that the unseen companion has a mass of at least (say) 8 solar masses, and if we can calculate that a normal star of that mass would be visible, then we must conclude that it is a collapsed star, too small to be seen. However, if it has more than three solar masses, the only kind of collapsed star that it can be is a black hole. This is the kind of reasoning that specifically led to the identification of Cygnus X-1 as a black hole.

Cygnus X-1

Cygnus X-1 is so named because it was the first x-ray source discovered in the constellation Cygnus. It was found during a brief survey with a rocket-borne x-ray telescope. Later, extensive studies were made with the UHURU satellite, and optical and radio telescopes located the object. A key finding was that the x-rays vary with a period of 5.6 days. So does the visible light of a single-lined spectroscopic binary star, known by its catalogue number, HDE 226868. Since this binary star is located in the region from which the x-rays were detected and has the identical periodicity, it was identified as the visible counterpart of Cygnus X-1.

Studies of the spectrum of HDE 226868 show the presence of emission lines that may be due to hot gas flowing from the visible star (a B0 supergiant) to its unseen companion. The spectral observations also indicate that the companion has a mass of at least 6 solar masses. Observations made from a later rocket flight revealed that the x-rays from Cygnus X-1 are flickering drastically, with the intensity changing by large factors in times as short as one millisecond, a thousandth of a second (Figure 16-13). Such rapid fluctuations imply that the x-ray source is an extremely small object. In fact, among supporters of the black hole theory,

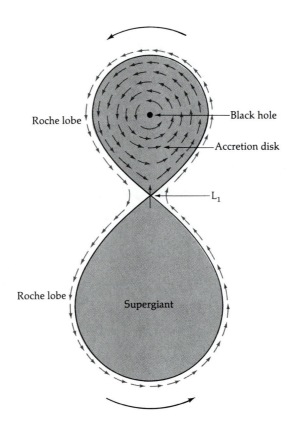

Figure 16-12. Schematic diagram, not to scale, showing a close binary system in which a supergiant star has expanded to fill its own Roche lobe. Flow of gas through point L_1 produces an x-ray emitting accretion disk around the companion star, a black hole. The same phenomena occur if a neutron star is in place of the black hole.

this finding has been regarded as the clincher, the definite indication that the unseen companion of HDE 226868 is a black hole. To understand why, we must say a few words about the theory of x-ray generation by collapsed stars.

Mass Accretion and X-Ray Emission

Suppose we have a binary system consisting of a supergiant star and a black hole. The supergiant may lose mass, some of which is attracted into the black hole. Two methods of mass loss are likely. One obvious method is the *stellar wind*, a process resembling the solar wind in which gas blows outward from the star's corona. Another method is called *Roche lobe overflow* (Figure 16-12). The Roche lobe is a mathematically defined surface that exists around each star in any binary system. Suppose we have a tiny particle and we place it somewhere in space in a particular binary system consisting of two stars, A and B. If the particle is placed inside the Roche lobe of A, it will orbit A or fall into it. Placed outside that lobe, it will orbit both stars, unless it is placed inside the Roche lobe of B. In the latter case, it will orbit or fall into B. Now imagine that A is a supergiant and B is a black hole. Further imagine that during the evolution of A, the star has grown bigger and bigger until finally its outer layer extends beyond the Roche lobe of A. Then the material in this layer is no longer part of A, and it will flow out into space. Some of the material passes through the point marked L+ in Figure 16-12 (L_1 is known technically as the *inner La-*

Figure 16-13. X-ray burst profile as recorded by a rocket instrument. This profile consists of data for several bursts combined from two flights. Large variation in the x-ray emission in times as short as a millisecond (msec) is clearly shown. (Courtesy of S. Holt, NASA-Goddard Space Flight Center.)

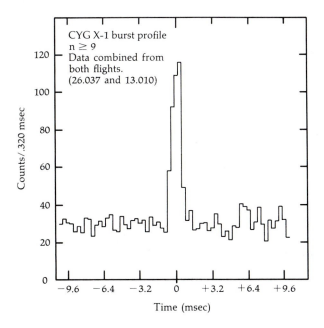

grangian point). The matter passing through L_1 is drawn into a spiral pattern, orbiting closer and closer until finally it falls into the black hole.

As the swirling gas approaches the black hole, orbiting it faster and faster, it assumes the shape of an accretion disk. (Recall the description of an accretion disk in the Star Formation section of Chapter 11.) Another important physical effect occurs. Note that the gas, as it approaches the black hole, is occupying a smaller and smaller region. In other words, it is being compressed. When a gas is compressed, it gets hot. As the accreting gas flows toward the black hole, it is heated. In fact, it gets so hot that, according to the Planck Law, it emits predominantly in the x-ray region of the spectrum. Of course, once the gas passes within the surface around the black hole where the escape velocity equals the speed of light (this surface is called the *event horizon* or *Schwarzschild limit*), x-rays no longer can escape. The theory predicts that conditions will be chaotic in the hot, turbulent gas just outside this limit, and that, for a black hole of around 6 solar masses, the last stable orbit for spiralling matter before it passes the event horizon will have an orbital period of around one millisecond.

Comparing observations with theory, we note two basic properties of the Cygnus X-1 binary system (Figure 16-13) that support the idea that a black hole is present: (1) powerful x-rays with one-millisecond fluctuations; (2) an unseen companion with a mass at least six times that of the sun. It should be emphasized that UHURU and other satellites have provided data on other x-ray-emitting binary systems in which the unseen companion has sufficiently low mass so that it is probably a neutron star. In addition, some theories of systems similar to Cygnus X-1 suggest that they evolve from the systems that contain a neutron star: as mass from the larger companion accretes onto the neutron star, the latter object grows. Eventually, it may exceed the upper limit of 3 solar masses for a stable neutron star. When that happens, it collapses into a black hole. At least, that's the idea.

The preceding arguments are just that, logical arguments for the presence of a black hole in Cygnus X-1.[7] There are some counterarguments, however. For example, a few astronomers believe that Cygnus X-1 might be a triple star rather than a binary, so that the presently unseen 6 solar masses might be divided among two stars: (1) a main-sequence star of a few solar masses, just faint enough so that it would not be detectable in the spectrum of HDE 226868 due to the much brighter light of the B0 supergiant, and (2) a neutron star on which the mass accretion would occur, generating the x-rays.

---

[7] A few other x-ray sources that recently were studied with HEAO-1 and other satellites resemble Cygnus X-1 and also may contain black holes.

Black Holes in Globular Clusters?

A globular cluster represents another case in which a black hole might affect the appearance of a visible star system. Some physicists have proposed that large black holes, with perhaps 1,000 solar masses or more, may exist at the centers of globular clusters. Such an object might have been created when a massive star or cloud of matter collapsed during the formative stage of the cluster. Then the black hole would have grown by accretion, sucking in interstellar gas from the vicinity and even swallowing up whole stars or pieces of them! The stars are so closely packed near the center of a globular cluster (in some cases only one-tenth of a parsec or less apart) that occasional collisions and many near misses must occur. When a star collides with a black hole, it is sucked in, adding to the mass and the gravitational field strength of the hole. Even when there is a near miss, the black hole's gravitational field may be sufficient to tear the passing star apart by tidal force, releasing huge amounts of gas into space, some of which may eventually end up in the hole.

What evidence exists for the fantastic scenario of the globular-cluster black hole theories? Three observational consequences can be predicted. First, one might expect bursts of x-rays when accumulations of matter fall into the hole. Second, calculations of star orbits in the cluster, as they evolve over time, show that the black hole should remain near the center of the cluster and that an abnormally dense concentration of stars should form at the center. As seen from earth, this concentration would resemble a small bright spot of light at the center, if it could be distinguished with available instruments. Third, spectroscopic observations, interpreted via the Doppler effect, should reveal greater orbital velocities than otherwise expected for stars near the center of a cluster. This would result from the huge gravitational force of the massive black hole.

X-ray bursts from the vicinities of a few globular clusters were discovered in 1975 and 1976, leading astronomers to speculate that they might actually represent evidence for large black holes in the clusters. Unfortunately, similar *bursting x-ray sources* were soon found to exist in regions of the galaxy where there are no globular clusters. We still don't know what these sources are (most of the experts think that they are some unusual kind of binary star), and so we cannot regard them as manifestations of black holes. Searches for unusually bright points of light at the centers of globular clusters have been hampered by the smearing effect of atmospheric seeing and have been inconclusive so far, with some astronomers under the impression that the concentrations of light have been found and others disputing the results. Atmospheric seeing also prevents suitable searches for the orbital velocity effect. As of now, there is no proof for black holes in globular clusters.

Other Uses for Black Holes

We saw in our discussion of the Crab nebula pulsar that it can be re-garded as a powerful source of stored energy, analogous to a giant flywheel. In the same way, one can compare mass accretion into a black hole with a hydroelectric dam—in each, energy is liberated because matter flows "down."

If black holes really do exist, mass accretion will surely be a powerful energy source. Therefore, theorists trying to explain such phenomena as the great energy of radio galaxies and quasars have suggested the existence of *supermassive black holes*. These hypothetical objects, with masses of a million solar masses or more, if located at the centers of galaxies or quasars, would accrete huge amounts of gas from surrounding space and thus release enormous energy. They should have observational consequences similar to those mentioned in the preceding section on Black Holes in Globular Clusters. Nobody knows if such huge black holes actually form.[8] However, nobody knows any better or more realistic way to explain the energy of quasars.

In conclusion, three basic families of black holes have been envisioned. The small or "normal" black holes have masses of roughly 3 to 10 solar masses and are believed to manifest themselves as members of some of the x-ray-emitting binary stars, with Cygnus X-1 as a fairly definite case. The large black holes, with masses of 1,000 to a few thousand solar masses, supposedly form in globular clusters. Supermassive black holes, with masses of a million solar masses or more, allegedly exist in quasars and radio galaxies. A person with a philosophical bent cannot help but comment that human beings, who have advanced by fashioning tools and otherwise converting elements of the natural world to their own ends, must have reached the ultimate expression of this tendency when they began to apply hypothetical objects, such as black holes, to their own use. That's exactly what is being done with black holes now—astrophysicists are using them to solve their own problems!

---

[8] Limited evidence was found in 1978 indicating the presence of a supermassive black hole in the nucleus of the giant elliptical galaxy M87.

17

Epilogue

The preceding chapters have outlined the continuing story of human probing and exploration of the universe. At every turn we have encountered new horizons. Here, we review these explorations and our changing world view.

THE EARTH MOVES IN MANY WAYS

The history of astronomy has involved a continuing realization that the earth is not the center of the universe, the earth is not fixed, and the heavens do not revolve around it. Just how many motions of the earth are there?

First of all, the earth rotates on its axis, thereby producing day and night and the apparent rising and setting of the stars. The axis of rotation performs a conical motion, tracing out a complete circle in the sky every 26,000 years, the phenomenon of precession.[1]

The center of mass of the earth-moon system travels around the sun in an elliptical orbit; the earth also performs a small orbital motion around this center of mass. The net result is that the earth's path around the sun is an ellipse with wiggles in it.

The earth also partakes of the sun's motion in the galaxy. The sun, traveling in a roughly circular orbit around the center of the Milky Way, is

---

[1]There is a slight wobble, called *nutation*, in this precession motion.

now moving toward a point in the constellation Hercules. Thus, the earth never returns to the same point in space; it does not trace out the same ellipse repeatedly, but follows a corkscrew path as seen from this point of view (Figure 10-4).

The earth and sun both partake of the motion of our galaxy, which is in orbit about the center of mass of a local group of galaxies, which includes the Andromeda galaxy (Plate 8) and the Magellanic Clouds (Figure 13-7). Finally, together with the local group, the earth participates in the general expansion of the universe, as revealed in the red shifts in the spectra of distant galaxies.

HOW LONG AGO?

Many aspects of our understanding of the universe can be summarized in a time sequence of events (Figure 17-1).[2] It begins with the explosion of the primeval fireball. In this expanding medium the galaxy and the oldest stars formed about 10 billion years ago. The sun condensed from a cloud of gas and dust after the galaxy had flattened into a disk-like configuration. About 4.5 billion years ago, the solid bodies of the solar system, including the earth, were formed. The oldest known rocks on earth formed about 3.7 billion years ago, or more than a half billion years after the oldest known lunar rocks. Already, extensive vulcanism had exhaled water vapor and atmospheric gases from the interior of the earth in massive quantities, and the oceans had formed by condensation of the water vapor. The primary broth arose over 4 billion years ago; life developed and important amounts of molecular oxygen (O_2) were injected into the atmosphere by the green plants. The atmospheric composition as we presently know it (pollutants excepted) was reached by 1 billion years ago.

Although life developed earlier, the fossil record and our generally accepted knowledge of geological time extend back only 600 million years. About 250 million years ago, the sun began the trip around the center of the Milky Way galaxy that it is now completing. The dinosaurs were developing at this time, and they were the dominant animals on earth for about 100 million years. The Rocky Mountains were formed about 80 million years ago. Nuclear reactions within the youngest stars now observed on the main sequence were ignited less than 3 million years ago. The development of man was under way at this time, and Advanced Australopithecus appeared about 3 or 4 million years ago.

---

[2] Also, review Figure 1-2 in connection with this section.

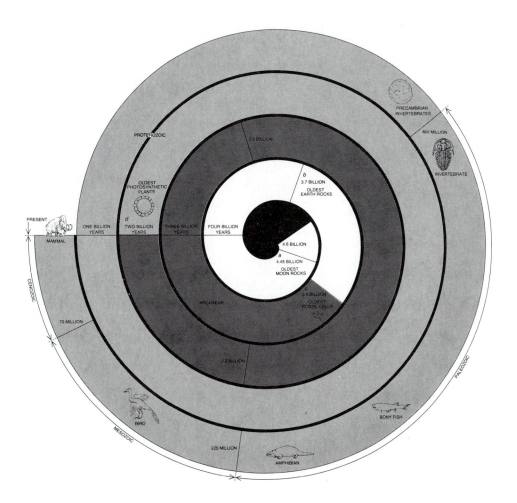

Figure 17-1. Spiral clock showing the passage of 4.6 billion years of earth history; each revolution of the hand takes one billion years. Some specific benchmarks are: (a) The age of the oldest known moon rocks at 4.45 billion years, according to some experts. Others have given ages of 4.6 billion years or greater for some moon rocks brought back by the Apollo astronauts. (b) The age of the oldest sedimentary rocks on earth, found in Greenland. (c) The oldest fossil cells (found in Swaziland), which may represent the planet's first flora. (d) The oldest photosynthetic plants, found as fossils in Canadian rocks. Human history spans a hair's breadth short of the line marked "present." (From *The Earth* by Raymond Siever. Copyright © 1975 by Scientific American, Inc. All rights reserved.)

Figure 17-2.   Spaceship Earth seen above lunar horizon from Apollo 8. (NASA.)

By 45,000 years ago the Cro-Magnon, virtually modern man, had arisen. About 26,000 years ago, the earth's rotation axis pointed to alpha Ursae Minoris; since then it has swept around a small circle in the sky and is now pointing roughly at that same star again. Man's written historical record dates from about 5,000 years ago, when the earth's axis pointed to the star Thuban (alpha Draconis), and Thuban was thus the North Star.

Almost 930 years ago, Yang Wei-Tek at Khaifêng, China observed a brilliant "guest star" in full daylight. Some 248 years ago Pluto was in the same location as now in its orbit around the sun; 22 years ago the sunspots were arranged on the sun roughly as they are now; 11 years ago the spots were also roughly as numerous and roughly at the same latitudes, but their magnetic polarities were opposite to the present situation. One year ago the earth was at the same position in its orbit around the sun. One month ago the moon was at about the same phase.

Light arriving at the earth from the sun left there about 8 minutes ago, while light from the moon left about 1.3 seconds ago. During the time that this light traveled from the moon to the earth, the neutron star at the heart of the Crab nebula, produced by the supernova event recorded by Yang, spun around 39 times.

THE FUTURE

Where do we go from here? The earth is the only place we know where an atmosphere exists that humans can breathe, although we have probed the universe with telescopes and are now beginning the direct exploration of the nearest astronomical bodies. The footprints of men are on the moon, and Mars may well be next. Hopefully, this exploration can continue in the name of people everywhere—people who have come to view themselves as integral parts of nature, every atom in their bodies having once been among or inside the stars, every substance they use having come from some spot in the total ecology of the globe and likewise destined to return to the environment of Spaceship Earth (Figure 17-2).

Study Guide

CHAPTER 1. INTRODUCTION

This chapter describes the philosophical approach taken to astronomy and the scope of this book. In reading the chapter, one should focus on the broad picture while being mindful of the uncertainties associated with some details.

Objectives

1. To understand the changing nature of astronomy as new information becomes available.
2. To begin to develop an appreciation for the interaction (sometimes conflicts) between astronomy and society.
3. To begin to consider the origin and evolution of the universe from the Primeval Fireball to the present day.

Key Words

Primeval Fireball   terra incognita
Homo sapiens

Exercises (Answers at End of Study Guide)

1. True or False. New information is constantly improving our astronomical picture of the universe.

2. True or False. The study of astronomy is unrelated to political or social institutions.

3. Our universe originated approximately _____ years ago.
   a. 20 thousand
   b. 20 million
   c. 20 billion
   d. 20 trillion
   e. none of the above

4. According to the best current evidence, modern man (*Homo sapiens*) has existed for approximately _____ years or more.
   a. 350 million
   b. 35 million
   c. 3.5 million
   d. 350 thousand
   e. 35 thousand

5. Humans have used some form of writing for approximately _____ years.
   a. 500
   b. 5,000
   c. 50,000
   d. 500,000
   e. none of the above

Questions and Assignments

1. Prepare a brief report on one of the three classical conflicts between science and established authority mentioned in this chapter.

2. Prepare a brief report on a contemporary conflict between science and authority.

3. How does the summary of the origin and evolution of the universe in Chapter 1 compare with your own ideas about the nature of the universe prior to beginning this course?

CHAPTER 2. HISTORICAL ASTRONOMY—SKY AND NO TELESCOPES

This chapter describes the historical development of astronomy before the invention of the telescope. Knowledge of this material should include understanding the objectives and the key words listed below.

Objectives

1. To understand the consequences of the moon's motion around the earth, a motion that produces eclipses and lunar phases.

2. To understand the origin and basis of the calendar.

3. To know the consequences of the earth's position, orientation, and motion in the solar system, such as a changing "North Star" and the cycle of seasons.

4. To understand the evolution of human thought about the earth's place in the solar system and the developments that caused the geocentric world-view to give way to the heliocentric world-view.

5. To know the basic ideas of gravitational attraction and why planets and satellites move in orbits.

6. To develop some insight into the scientific method—collecting and organizing facts, proposing a working explanation or hypothesis, and testing the hypothesis by new observations and measurements.

Key Words

constellations	geocentric hypothesis
planets	retrograde
phases	Ptolemy
plane of the ecliptic	epicycle
eclipse	deferent
umbra	primum mobile
penumbra	heliocentric world-view
diurnal motion	Copernicus
meridian	Tycho Brahe
solar day	Kepler
sidereal day	Kepler's Laws
year	ellipse
Julian calendar	astronomical unit (a.u.)
Gregorian calendar	Galileo
seasons	Newton
latitude	Newton's Laws
longitude	acceleration
right ascension	mass
declination	gravitation
Celestial Equator	orbit
Vernal Equinox	Halley
precession	aberration
magnitude	Bradley

Exercises (Answers at End of Study Guide)

1. When the moon passes directly between the sun and the earth,
   a. the layman calls it "full moon."
   b. nothing interesting occurs.

    c. the moon is only seen near midnight.

    d. solar eclipse occurs.

    e. lunar eclipse occurs.

2. True or False. The solar day and the sidereal day are identical.

3. True or False. The astronomical year is precisely 365 days long, which explains why a calendar year always has 365 days.

4. True or False. We have summer in the Northern Hemisphere because the earth is closest to the sun at that time.

5. True or False. Polaris is our eternal North Star.

6. Match the following:

Julius Caesar	epicycle
Copernicus	study of motion
Galileo	calendar reform
Ptolemy	gravitation
Newton	heliocentric world-view

7. The orbits of planets around the sun, according to Kepler's Laws, are _____.

8. True or False. The phases of the moon prove that one side of the moon is in perpetual darkness.

9. True or False. Gravitational theory enabled Halley to predict that his comet would return in 1758.

10. True or False. Certain stars may be hidden from our view, as they are always behind the sun as seen from earth.

Questions and Assignments

1. If a person has no telescope or technical training, but gazes at the sky most nights, what property of the planets might distinguish them from the stars?

2. Explain briefly the circumstances that produce the moon's phases. (Hint: A sketch should help.)

3. We do *not* have either a lunar or solar eclipse every month. Why?

4. Prepare a summary of differences between lunar and solar eclipses. Be sure to cover geometrical circumstances, duration, viewing area on earth, and the difference in visual spectacle as seen by a viewer with the unaided eye.

5. Explain how you can confirm (or rule out) that the seasons in the northern and southern hemispheres of the earth are caused by the sun being closer to the earth at one time of the year than at another.

6. Prepare a brief history of the attempts to approximate the uneven number of days in the year with a calendar scheme.

7. Make an improved calendar by devising a rule to be applied to the millennial years. Give the error.

8. Does the earth's axis of rotation always point to the same star? Explain your answer.

9. Sketch and explain the general arrangement of the solar system according to Aristotle and Ptolemy.

10. Name the five scientists largely responsible for our understanding of planetary motion, and state what each contributed.

11. According to Kepler, the orbits of the planets are ellipses. Draw one on a piece of cardboard, using the method described in the text. How would you arrange the nails to approximate the orbits of most planets?

12. Is the sun at the exact center of the earth's orbit? If so, what law does this prove? If not, tell why and state where the sun is located in the orbit.

13. Jupiter is approximately five times the earth's distance from the sun. Does it take a longer or shorter time to complete one orbit around the sun? Outline your reasoning.

14. Give everyday examples of Newton's three Laws of Motion, other than the ones mentioned in the text.

15. How does gravity differ from most forces that we encounter in our everyday lives?

16. Explain the physical situation that produces orbital motion.

17. The sun's mass can be determined from properties of the earth's orbit around it, and the earth's mass can be determined from the moon's orbital properties. Explain the basic idea, and name the quantities needed for these determinations.

18. What role did the appearance of Halley's comet in 1758 play in the history of the theory of gravitation?

19. Discuss the solar and sidereal days, taking into account the earth's rotation and orbital motion.

20. Explain what the aberration of starlight has in common with raindrops falling on a moving object. Assume you are jogging through Death Valley and cannot stop. There is a sudden rainstorm with no wind, so the raindrops fall vertically. What is the best way to tilt your head to get the most rain into your mouth as you run?

CHAPTER 3. OUR PLANET EARTH

This chapter describes the history and present condition of the earth's surface and interior. Knowledge of this material should include understanding the objectives and the key words listed below.

Objectives

1. To know the earth's basic properties (such as circumference, shape, age, and mass) and how they were determined.
2. To know the basic facts of the geologic time scale and how they were learned.
3. To understand the basic structure of the earth's interior and how it was inferred.
4. To understand the basic processes that produce change on the surface of the earth, such as continental drift, glaciation, and volcanism.
5. To develop an appreciation of how the principles and techniques of geology relate to the study of other planets and the moon.

Key Words

major planets	solar system
Aristotle	Lord Kelvin
Eratosthenes	Hutton
stadia	uniformitarianism
zenith	Lyell
equatorial bulge	P waves
neptunists	S waves
strata	epicenter
Becquerel	crust
Lord Rutherford	mantle
half-life	Mohorovičić
radioactive dating	Mohorovičić discontinuity
meteorites	core
fossils	inner core
permineralization	Xenophanes
superposition	fossil magnetization
Powell	continental drift
seismic waves	Bacon
surface waves	sea-floor spreading

crest   plates
trough   lithosphere
transverse waves   asthenosphere
shake waves   volcanoes
push-pull waves   glaciation
longitudinal waves   ice ages

Exercises (Answers at End of Study Guide)

1. True or False. Eratosthenes' method of estimating the earth's circumference was wrong in principle, but happened to give a fairly accurate answer.

2. The age of the earth is approximately _____ years.
   a. 4.5×10^1 d. 4.5×10^7
   b. 4.5×10^3 e. 4.5×10^9
   c. 4.5×10^5

3. True or False. Lord Kelvin obtained a good estimate of the earth's age by calculating the time that it took the earth to cool down to its present temperature.

4. True or False. Uniformitarianism was an interesting, but fundamentally incorrect, idea, which was replaced by neptunism.

5. Match the following:
   Eratosthenes   San Andreas fault
   San Francisco   uniformitarianism
   Bacon   crust-mantle boundary
   Lyell   earth's circumference
   Moho   continental drift

6. From fossils we learned the
   a. age of the earth.
   b. fact that older layers are higher.
   c. epicenter.
   d. geologic time scale.
   e. depth of the Grand Canyon.

7. True or False. The earth's age equals that of its oldest known rocks, as dated by measurements of radioactivity.

8. Which of the following is *not* a feature of the earth's interior structure?
   a. core   d. trough
   b. crust   e. mantle
   c. Moho

9. True or False. The theory of continental drift, in which blocks of material float on soft regions of the mantle, was disproved by studies of fossils on different continents.

10. True or False. Changes in the light of the sun did not cause climatic changes on the earth, because we are sure that no climatic changes were caused on other planets of the solar system.

Questions and Assignments

1. Summarize the early evidence for a spherical earth. Can you give a good example of modern evidence?

2. Carry out the principle of Eratosthenes' method by consulting a map to obtain distance and angular separation for two locations that are approximately on a north-south line.

3. Prepare a brief history of the attempts to determine the age of the earth.

4. Consult an appropriate library book and prepare a brief description of the process of radioactive decay.

5. What elements are the principal contributors to heating the earth's interior?

6. Prepare a brief summary of the geologic time scale.

7. Prepare a brief description of the formation of the Grand Canyon.

8. Summarize the structure of the earth's interior. Be sure to include the lithosphere and the asthenosphere.

9. Explain the physical principles of earthquake waves and how they are used to probe the earth's interior.

10. How has fossil magnetism helped to prove or disprove that the continents have drifted over the earth's surface?

11. Discuss several different ways that mountains can be formed.

12. Explain why there are so many earthquakes in California.

13. Give a plausible explanation for the lack of long chains of volcanoes on Mars.

14. Why is the date A.D. 1850 considered important in the study of the earth's climate?

15. Summarize the evidence for or against climatic change on Mars. Give your ideas about why such change does or does not occur.

CHAPTER 4. EVOLUTION AND LIFE—TERRESTRIAL AND EXTRATERRESTRIAL

This chapter describes the principles of the origin and evolution of life on earth and the possibility that it may exist elsewhere in the universe. Knowledge of this material should include understanding the objectives and the key words listed below.

Objectives

1. To know the basic ideas of biological evolution, such as natural selection, heredity, and the genetic code based on DNA.
2. To understand current scientific thinking about the origin of life on earth.
3. To develop some familiarity with the history of evolution as discovered from the fossil record.
4. To develop some familiarity with the evolution of man.
5. To stimulate your thinking about the possibility that life and even advanced civilizations may exist throughout the universe.

Key Words

Herodotus
Anaximander
fossils
Darwin
ecological niche
adaptive radiation
genetic code
nucleus
cytoplasm
DNA
panspermia
atmospheric evolution
primitive atmosphere
primary broth
prebiodonts

ecological balance
mutations
natural selection
heredity
genes
Oparin
chlorophyll
photosynthesis
fossil fuel
bipedalism
Neanderthal Man
Cro-Magnon Man
habitable zone
Kardashev

Exercises (Answers at End of Study Guide)

1. True or False. Darwin's study of finches on the Galapagos Islands was important in developing his ideas on natural selection.

2. True or False. Darwin's theory of natural selection was immediately accepted as a brilliant theory.

3. In the cell, the genetic code is carried in the chromosomes, which are found in the
   a. nucleus.
   b. cytoplasm.
   c. cell wall.
   d. all of the above.
   e. none of the above.

4. Match the following:
   Darwin   chemical origin of life
   Mendel   genetic code
   Oparin   natural selection
   DNA   advanced civilizations
   Kardashev   heredity

5. True or False. Laboratory experiments on the origin of life depend on knowledge gained from astronomy.

6. The basic process that produces carbohydrates from sunlight, carbon dioxide, and water is called
   a. bipedalism.
   b. radioactivity.
   c. adaptive radiation.
   d. photosynthesis.
   e. panspermia.

7. True or False. Our coal supply is often called fossil fuel because it is the fossil remains of extensive forests that flourished about 300 million years ago.

8. True or False. Dinosaurs, despite great size, never were a dominant life form on earth.

9. Modern man can be traced back 35,000 years to
   a. Proconsul.
   b. Cro-Magnon Man.
   c. Ramapithecus.
   d. Neanderthal Man.
   e. Archaeopteryx.

10. True or False. For life to occur on a planet of another star, it probably is necessary for water to exist in the liquid state.

Questions and Assignments

1. The voyage of H.M.S. *Beagle* reads like an adventure novel. Prepare a brief report on an aspect (other than his visit to the Galapagos) of Darwin's voyage.

2. What is the idea of ecological balance? Briefly describe a modern example of inadvertent upset of this balance.

3. Consult a biology or genetics book and explain why recessive traits (such as blue eyes) persist and do not die out.

4. The DNA molecule contains the genes as chains of nucleotides and has what other essential and remarkable property?

5. Refer to Question 4 and describe a mutation in terms of the DNA molecule. How do mutations occur? How often? Could astronomical phenomena affect mutations on earth?

6. Why are the panspermia hypothesis and the "garbage theory" unsatisfactory explanations for the origin of life?

7. Prepare a brief summary of the current view of the origin of life on earth.

8. What is the origin of most of the oxygen in the earth's atmosphere?

9. Give a simple reason for the fact that animal life was preceded by the development of plant life.

10. During what geological period did Man develop? What characteristic is shared by the principal fossils of the period?

11. Prepare a brief summary of the increase in human brain size during the last few million years. If this trend is still continuing, would you have predicted on a scientific basis that each U.S. President since George Washington would have been a little smarter than the previous one? If not, why not?

12. Does interest in astronomy extend back a long time in human history, or has it emerged only recently due to increased leisure time for the average person? What were some of the tasks carried out by the earliest known astronomers?

13. Why might you expect life to exist outside our solar system? Is there evidence for life on other planets within our solar system?

14. In terms of physical conditions, what is the significance of the habitable zone, and why is this situation important for the origin of life?

15. What bias inevitably affects our discussion of extraterrestrial life?

16. Prepare a short essay giving your *personal* views on the attempts to detect or even contact extraterrestrial civilizations.

17. On the Kardashev classification scheme, what type of civilization are we? How could Type III civilizations exist without our knowledge?

CHAPTER 5. MATTER, HEAT, AND LIGHT

This chapter describes the principles of physics that are used to interpret astronomical observations and measurements. Knowledge of this material should include understanding the objectives and the key words listed below.

Objectives

1. To understand the basic ideas of atomic theory.
2. To know the properties of the particles that make up the atom.
3. To be familiar with the processes that produce compounds and the different states of matter.
4. To understand the basic physical concepts of heat, temperature, and energy, and in particular the law of conservation of mass and energy.
5. To understand the basic properties of light.
6. To understand the concepts of elementary spectroscopy, including Kirchhoff's Laws.
7. To know the laws describing radiation from an ideal radiator, specifically Planck's Law and the Stefan-Boltzmann Law.
8. To understand the Doppler effect and its applications.

Key Words

Empedocles	Rutherford
atomism	nucleus
atoms	proton
Leucippus	neutron
Democritus	Chadwick
Epicurus	atomic mass
Boyle	tritium
Dalton	transuranium elements
atomic theory	Bohr
elements	Bohr atom

molecules
compound
isotopes
atomic weights
abundance
Newlands
Mendeleev
Periodic Table
atomic number
electron
cathode rays
photons
Thomson
Millikan
alpha rays
beta rays
gamma rays
calorie
absolute zero
Kelvin temperature scale
Brownian motion
pressure
pressure difference
kinetic energy
nuclear energy
chemical energy
work
potential energy
conservation of energy
Count Rumford
radiant energy
acoustical energy
nuclear energy
Einstein
conservation of mass and energy
electromagnetic radiation
electromagnetic spectrum
ether
interference

noble gases
short-range force
coulomb force
ion
electrovalent bond
covalent bond
van der Waals force
metallic bond
electron gas
crystalline solids
amorphous solids
conduction
convection
radiation
Fahrenheit scale
Celsius scale
centigrade scale
photomultiplier tube
Hertz
radiation pressure
spectroscopy
Newton
Angstrom unit
Angstrom
Bunsen
Kirchhoff's Laws
energy state
emission process
absorption process
quanta
Planck
spontaneous emission
free electrons
bound electrons
scattering
pure absorption
ideal radiator
Planck's Law
thermal radiation

linear polarization	Stefan-Boltzmann Law
diffraction grating	Doppler effect
photoelectric effect	velocity of light
Roemer	

Exercises (Answers at End of Study Guide)

1. True or False. Atoms are particles whose properties are not affected by chemical reactions.

2. Atoms of a given element may exist in several forms called
   a. electrons.   d. eka-silicons.
   b. molecules.   e. isotopes.
   c. compounds.

3. Which of the following is not a particle in the atom?
   a. electron   d. proton
   b. neutron   e. none of the above
   c. photon

4. Tritium is a rare isotope of hydrogen consisting of one proton, two neutrons, and one electron. Its atomic number is _____ and its atomic mass is _____ .
   a. 1   d. 4
   b. 2   e. 5
   c. 3

5. Absolute zero is
   a. $0°$ on the Fahrenheit temperature scale.
   b. $0°$ on the Celsius temperature scale.
   c. $0°$ on the Kelvin temperature scale.
   d. the definition of the calorie.
   e. none of the above.

6. Match the following:
   Dalton   nuclear energy release
   Mendeleev   spectroscopy
   Chadwick   thermal radiation
   Einstein   neutron
   Newton   atomic theory
   Planck   Periodic Table

7. True or False. The conservation of mass and energy is a useful concept, but one must be aware of numerous exceptions.

8. Which of the following is not part of the electromagnetic spectrum?
   a. gamma rays
   b. beta rays
   c. visible light
   d. x-rays
   e. radio waves

9. The total amount of light emitted from an ideal radiator is given by
   a. Dalton's Law.
   b. $E = mc^2$.
   c. Kirchhoff's Law.
   d. Planck's Law.
   e. Stefan-Boltzmann Law.

10. Red light has a _____ wavelength and _____ photon energy than blue light.

11. A visible-light emission line from an object moving away from the observer will be _____ compared with laboratory standard spectra.
    a. unchanged
    b. shifted toward the red
    c. increased in intensity
    d. shifted toward the blue
    e. decreased in intensity

Questions and Assignments

1. Prepare a brief history of atomism, starting with the early Greek philosophers.

2. Check one of Mendeleev's predictions (other than eka-silicon) by comparing the predicted and observed properties.

3. How was Thomson's "plum pudding" model of the atomic nucleus tested?

4. If opposite charges repel each other, how is the atomic nucleus held together?

5. How is the atomic weight of an element related to the atomic masses of its isotopes?

6. Explain (in terms of the Bohr atom) why the noble gases (for example, helium, neon, and argon) rarely form compounds.

7. Is potassium chloride a likely compound? If so, explain with a relevant diagram.

8. Briefly explain (in terms of the bonds between atoms and molecules) the differences among gases, liquids, and solids.

9. Give common examples (other than the ones mentioned in the text) for transport of heat by conduction, convection, and radiation.

10. Why do our hands become warm if we rub them together rapidly?

11. Why does the conversion of four hydrogen atoms into one helium atom result in the release of energy?

12. Give examples of light behaving as waves and as particles.

13. Summarize Kirchhoff's Laws in your own words.

14. Explain Kirchhoff's Laws on the basis of Bohr's model of the atom.

15. Explain the difference between scattering and pure absorption.

16. Explain how spectroscopic analysis enables us to identify elements in astronomical objects.

17. Describe the changes in the spectrum of thermal radiation from an ideal radiator as the temperature of the radiator increases.

18. Explain how the spectral changes described in Question 17 can be used to determine the surface temperatures of the sun and stars.

19. Explain how to determine the speed of a typical star toward or away from earth.

20. Prepare a brief history of the attempts to measure the velocity of light.

CHAPTER 6. TELESCOPES AND OBSERVATORIES

This chapter describes the instruments and facilities used in gathering astronomical observations. Knowledge of this material should include understanding the objectives and the key words listed below.

Objectives

1. To understand the basic principles of (and similarities among) the human eye, cameras, and telescopes.

2. To appreciate the many different kinds of telescopes, detectors, and observatories (including those in orbit above the earth's atmosphere) used to carry out astronomical observations.

3. To develop an awareness of the large number of observations, including sky surveys, necessary to carry out astronomical research.

4. To understand the astronomer's dependence upon and preoccupation with light and other forms of electromagnetic radiation, and to begin to understand the need to observe the universe in many different wavelengths.

Key Words

eye	retina
cornea	rods
pupil	cones
iris	camera
refracted	focal plane
lens	clock drive
Galileo	seeing
telescope	photometer
refractor	spectrograph
Newton	polarimeter
reflector	detector
scientific method	Daguerre
critical experiment	Draper
sunspots	prominences
radio telescope	photomultiplier
Giacconi	pulse counting technique
aperture	spectrometer
resolution	spectral scan
Lord Rosse	electronic image tube
Herschel	vidicon tube
Uranus	proportional counter
Schmidt camera	Germanium bolometer
correcting plate	sky survey
chromatic aberration	Argelander
Multi-Mirror Telescope	Bonner Durchmusterung
radio interferometer	Henry Draper Catalogue
angular resolution	Cannon
VLBI	Palomar Sky Survey
observatory	*Third Cambridge Catalog of*
light pollution	*Radio Sources*
Jansky	

Exercises (Answers at End of Study Guide)

1. True or False. Although the eye and camera perform the same function, they have little in common.

2. True or False. In photographing the stars with a telescope, the rotation of the earth must be compensated by means of a clock drive.

3. With one of the first telescopes, Galileo discovered
   a. some features of the lunar landscape.
   b. that sunspots were solar features, and that the sun rotates.
   c. what seemed to be appendages of Saturn.
   d. that the Milky Way was composed of many faint stars.
   e. all of the above.

4. Today, the world's largest optical telescope is located on
   a. Mt. Palomar in California.
   b. Kitt Peak in Arizona.
   c. Mt. Pastukhov in the Soviet Union.
   d. Cerro Tololo in Chile.
   e. Mt. Hopkins in Arizona.

5. Match the following:
   lens   photography
   mirror   reflecting telescope
   correcting plate   refracting telescope
   Daguerre   sky survey
   Argelander   Schmidt camera

6. Radio telescopes are generally much larger than optical telescopes, because the larger size is necessary to obtain usable
   a. angular resolution.   d. chromatic aberration.
   b. interferometry.   e. polarization.
   c. sensitivity to x-rays.

7. Observatory locations are usually chosen for good weather and clear skies. In the United States, the preferred locations are in the
   a. desert Southwest and Hawaii.   d. Middle-Atlantic states.
   b. Gulf Coast states.   e. Pacific Northwest.
   c. Midwest.

8. True or False. The photographic plate is still valuable as a recording device in astronomical research.

9. Which of the following is an astronomical sky survey?
   a. Bonner Durchmusterung   c. Palomar Sky Survey
   b. *Third Cambridge Catalog of*   d. all of the above
      Radio Sources   e. none of the above

10. True or False. Observing the sky at different wavelengths is often important because the appearance in, for example, optical and radio wavelengths, can be quite different.

Questions and Assignments

1. Two astronomical instruments are based on the fact that light is bent when passing from air to glass. Describe these instruments and give the additional principle utilized in one of them.

2. Prepare a brief report on the history and use of the first telescopes or on a historically significant telescope.

3. Why can reflecting telescopes be made over 200 inches in diameter, whereas refracting telescopes do not exceed 40 inches in diameter? You may wish to supplement your answer with a drawing.

4. Astronomers are interested in constructing telescopes of larger and larger aperture. Explain why.

5. Is the tendency to construct optical telescopes with larger and larger primary mirrors likely to continue? Why? If not, what are some alternatives?

6. What are some of the solutions adopted by radio astronomers to the need for angular resolutions higher than can be obtained with a single dish?

7. Both optical and radio telescopes can be adversely affected by pollution. What are the specific problems for these two cases?

8. Observatories are found in many different locations, some on the earth's surface and some in orbit around the earth. Pick a location and specify the functions that the observatory must supply for successful operation of the telescopes.

9. Prepare a brief report on the construction of a well-known modern observatory.

10. The popular notion of astronomical observations often consists of an astronomer looking through the telescope and making notes. Is this still true? If not, explain why.

11. Briefly describe several different instruments that one might use at the telescope, and indicate the type of observation carried out with each instrument.

12. Prepare a brief report describing the operation of a specific detector, and indicate the types of observations that one might make with it.

13. Prepare a brief report describing an observing system for use outside the visible-light range.

14. Who was Friedrich Argelander? Describe and indicate the importance of his major work.

15. Who was Annie Jump Cannon? Describe and indicate the importance of her major work.

16. What is the Palomar Sky Survey? With what instrument was it made? Why is it so important?

17. As a library research project, prepare a brief report on a sky survey made in wavelengths other than visible light.

CHAPTER 7. THE SOLAR SYSTEM—AN OVERVIEW

This chapter presents the history of discoveries of the objects in the solar system. The properties of some of these objects are discussed here also. Knowledge of this material should include understanding the objectives and the key words listed below.

Objectives

1. To develop some understanding of the circumstances and the processes of the telescopic discovery of solar-system objects, starting with Uranus in 1781.

2. To know the properties of the asteroids, comets, meteoroids, meteors, meteorites, and satellites.

3. To appreciate from recent discoveries—for example, the rings of Uranus and the satellite of Pluto—that there is still much to learn about the solar system.

4. To understand the general properties of the solar system, which must be explained by any successful theory of the solar system's origin.

Key Words

solar system	Adams
planets	Leverrier
astronomical unit	Neptune
Bode's Law	Lowell
Uranus	blink comparator
Herschel	Pickering

Tombaugh
Pluto
asteroid
Piazzi
Ceres
Pallas
Juno
Vesta
asteroid belt
Chiron
Kowal
Eros
comet
Tycho Brahe
Halley's comet
Halley
Clairaut
nucleus
coma
hydrogen cloud
tail
Delsemme
icy-grain halo
solar wind
Oort cloud
Oort
perihelion

aphelion
meteor
meteoroid
micrometeorite
meteorite
meteoroid stream
meteor shower
sporadic meteoroid
meteor crater
satellite
Io
Europa
Ganymede
Callisto
Deimos
Phobos
Hall
Shklovskii
Nicholson
Janus
Dollfus
Titan
Saturn's rings
Nereid
Triton
Uranus' rings
Charon

Exercises (Answers at End of Study Guide)

1. Which of the following is *not* a major planet?
   a. Mercury
   b. Earth
   c. Jupiter
   d. Vulcan
   e. Saturn

2. The distances of most planets from the sun are reasonably given by a sequence of numbers described by
   a. Kepler's Law.
   b. Bode's Law.
   c. Newton's First Law.
   d. Kirchhoff's Law.
   e. Halley's Law.

3. Match the following:

Herschel	Pluto
Adams and Leverrier	Ceres
Tombaugh	comet
Piazzi	Deimos and Phobos
Halley	Uranus
Hall	Neptune

4. True or False. Adams and Leverrier predicted the existence of a planet in order to explain perturbations in the orbit of Uranus.

5. True or False. The mass of Pluto clearly is responsible for the perturbations in the orbit of Neptune.

6. The modern value for the astronomical unit (a.u.) is
   a. 1.496×10^8 km.
   b. 1.496×10^9 km.
   c. 1.496×10^{10} km.
   d. 1.496×10^{11} km.
   e. 1.496×10^{12} km.

7. The most famous comet, Halley's comet, should return in about
   a. 1986.
   b. 1991.
   c. 2001.
   d. 2010.
   e. none of the above.

8. Which of the following is not a component of comets?
   a. coma
   b. hydrogen cloud
   c. rings
   d. tail
   e. nucleus

9. Meteors are a light phenomenon produced when a _____ enters the earth's atmosphere at high speed.
   a. comet tail
   b. crater
   c. satellite
   d. meteoroid
   e. none of the above

10 True or False. Now that Pluto's moon has been found, all the major planets are known to have one or more satellites.

Questions and Assignments

1. Make up your own mnemonic device for remembering the order of the planets in distance from the sun.

2. Prepare a brief report on the discovery of Uranus, Neptune, or Pluto. Alternatively, prepare a scientifically sound, fictional account of the future discovery of a tenth planet.

3. Briefly summarize the known properties of the asteroids.

4. Prepare a brief report on the discovery of the asteroid Chiron. (Hint: Consult periodicals rather than books.) Why is it thought to be unusual?

5. Why is the determination of the astronomical unit (a.u.) important?

6. Make a sketch showing the principal parts of a comet. Briefly describe the processes (if known) affecting each part as the comet approaches the sun.

7. Summarize the historical circumstances of some famous apparitions of Halley's comet.

8. What is the difference between planetary orbits and the orbits of most comets? Use a sketch if necessary.

9. What causes a comet to approach the sun for the first time? Where does such a comet come from?

10. Describe what happens when a *very large* interplanetary rock enters the earth's atmosphere. Use the terms *meteoroid, meteor, meteorite,* and *meteor crater,* as appropriate.

11. What is the source of material for some, if not all, meteor streams?

12. What is a meteor shower? Summarize an account of a major shower such as the Perseid shower, which occurs each year about August 12.

13. Explain the observation that more meteors are seen after midnight than before midnight.

14. Summarize the important reasons for studying meteorites.

15. Prepare a brief account of the discovery of the Martian satellites, Deimos and Phobos.

16. Prepare a brief account of the discovery of Uranus' rings. Compare them with the rings of Saturn.

17. Summarize what was learned about Pluto as a result of the discovery of its moon, Charon. Describe the appearance of Charon in the sky as seen respectively from several different locations on Pluto.

18. List the fundamental properties of the solar system that must be explained by a satisfactory theory of its origin.

CHAPTER 8. THE PLANETS AND THEIR ATMOSPHERES

This chapter presents the general properties of the planets, with emphasis on their atmospheres. Knowledge of this material should include understanding the objectives and the key words listed below.

Objectives

1. To know the basic properties of the planets.
2. To understand the application of radar to the study of planetary surfaces and rotation.
3. To understand the physical principles that determine planetary temperatures.
4. To understand the origin and evolution of the earth's atmosphere and why it differs so much from the atmospheres of Mars and Venus.
5. To know the basic properties of the atmospheres of Mars, Venus, and the Jovian planets.

Key Words

planetology	Doppler spreading
interior planet	sol
exterior planet	albedo
terrestrial planet	greenhouse effect
Jovian planet	planetesimals
solar nebula	evaporation
radar astronomy	mean free path
Schiaparelli	escape speed
ozone layer	great red spot
ionosphere	red spot hollow
electron density	Burke
trade winds	Franklin
horse latitudes	nonthermal radiation
jet stream	Shain
hydrologic cycle	Rayleigh scattering
magnetosphere	zonal circulation
bow waves	Rasool
metallic hydrogen	

Exercises (Answers at End of Study Guide)

1. True or False. In general, a planet can be termed either *terrestrial* (dense bodies and low-mass atmospheres) or *Jovian* (low-density bodies and massive atmospheres).

2. The rotation period of Venus was accurately determined by
   a. viewing the surface markings as the planet turned.
   b. the greenhouse effect.   d. evaporation.
   c. planetesimals.   e. radar astronomy.

3. Some of the earth's atmosphere can escape through the process of
   a. Rayleigh scattering.   d. trade winds.
   b. evaporation.   e. photosynthesis.
   c. Doppler spreading.

4. True or False. The largest constituent of the earth's atmosphere is ozone.

5. The atmospheres of Mars and Venus are composed primarily of
   a. argon.   d. oxygen.
   b. hydrogen.   e. unknown gases.
   c. carbon dioxide.

6. Match the following:

   Rasool   planetary rotation
   Venus   radio bursts
   Jupiter   greenhouse effect
   Io   atmospheric evolution
   Doppler spread   great red spot

7. True or False. The contribution of volcanoes to the earth's atmosphere has been negligible, as their eruptions are so infrequent.

8. The atmospheric region responsible for long-distance radio propagation is produced by solar radiation and is called the
   a. lithosphere.   d. jet stream.
   b. magnetosphere.   e. mean free path.
   c. ionosphere.

9. Which of the following are found in the atmosphere of Jupiter?
   a. molecular hydrogen   d. helium
   b. methane   e. all of the above
   c. ammonia

10. True or False. The radio waves from Jupiter indicate that intelligent life may be there, since such waves must be produced by electronic equipment.

Questions and Assignments

1. Summarize and contrast the properties of terrestrial and Jovian planets.

2. Use the information in Question 1 to present a plausible scheme for the respective origins of the terrestrial and Jovian planets.

3. Explain the technique used to determine the rotation periods of Mercury and Venus. Why are these periods somewhat unusual?

4. Describe some actual and potential methods for mapping the surface of Venus.

5. Explain the physical principles that determine the temperature of a planetary surface. How good are the predictions of this simple theory?

6. Prepare a broad outline or flow chart for the origin and evolution of the earth's atmosphere.

7. If oxygen production were stopped today, approximately how long would the existing oxygen last, and what would happen to it?

8. Explain the connection between limestone and the evolution of the earth's atmosphere.

9. What does the ozone layer do for us now? Did it have a role earlier in the earth's history?

10. Give a simple physical explanation for the blue color of the sky.

11. Consult a transcontinental airline schedule and derive your own rough estimate of the jet stream's speed.

12. Prepare a brief summary of the hydrologic cycle.

13. Why is the surface temperature of Venus so high?

14. Describe the atmospheric circulation on Venus.

15. Describe the differences and similarities of the atmospheres of Venus and Mars.

16. Explain Rasool's theory of the differences in the atmospheres of Venus, Earth, and Mars.

17. Compare the appearance and atmospheric composition of Jupiter and Saturn.

18. Summarize present knowledge on the appearance and atmospheric composition of Uranus and Neptune.

19. Describe the radiation environment of Jupiter.

CHAPTER 9. OUR SUN

This chapter describes the properties of the sun. Knowledge of this material should include understanding objectives and the key words listed below.

Objectives

1. To understand basic solar properties and phenomena.
2. To understand the energy source that supplies the sun's luminosity.
3. To develop an awareness of some of the unsolved problems of solar physics, such as solar neutrinos and the missing sunspots.
4. To appreciate studies of the sun and how they reveal phenomena that cannot be studied directly on other stars.

Key Words

sunspot	pore
Galileo	penumbra
sunspot maximum	umbra
sunspot minimum	sunspot groups
sunspot cycle	quiet sun
Schwabe	Fraunhofer
Carrington	chromosphere
differential rotation	flash spectrum
butterfly diagram	Wollaston
photosphere	continuous absorption
granulation	H^- ion
arch filaments	continuum
solar luminosity	solar activity
solar constant	center of activity
limb darkening	calcium filtergrams
Mayer	flare
Helmholtz-Kelvin	condensation
proton-proton chain	prominences
carbon cycle	filaments
convective region	Zeeman
corona	Zeeman effect
plasma	magnetogram
solar wind	x-ray bursts
solar neutrinos	Maunder minimum
Davis	

Exercises (Answers at End of Study Guide)

1. True or False. By comparing observations of sunspots on different

days, Galileo concluded that the sun rotated on its axis with a period of roughly one month.

2. Which of the following are regions of the sun?
   a. photosphere
   b. chromosphere
   c. corona
   d. convective region
   e. all of the above

3. Continuous absorption in the solar photosphere is provided by
   a. an unknown source.
   b. The H⁻ion.
   c. the sodium D lines.
   d. the solar luminosity.
   e. limb darkening.

4. The solar luminosity is determined by measuring the _____ _____ .

5. True or False. Physicists know of several different sources that could supply the solar luminosity, but they do not know which one is the most important.

6. From solar limb darkening, we can deduce that the temperature in the photosphere
   a. is constant.
   b. decreases inward.
   c. increases inward.
   d. is undetermined.
   e. none of the above.

7. Match the following:
   Galileo   magnetic fields
   Carrington   sunspots
   Fraunhofer   solar neutrinos
   proton-proton chain   solar spectrum
   Davis   differential rotation
   Zeeman   solar luminosity

8. True or False. The outer solar atmosphere extends well beyond the earth's distance from the sun.

9. Which of the following is an aspect of solar activity?
   a. flares
   b. great red spot
   c. core of the sun
   d. all of the above
   e. none of the above

10. True or False. The sunspot cycle has faithfully repeated every eleven years since A.D. 1700.

Questions and Assignments

1. Prepare a brief report on the solar discoveries of the astronomer R.C. Carrington.

2. What is deceptive about single photographs of the sun?

3. Describe the development of a single sunspot.

4. Discuss the lines in the solar spectrum (named after J. Fraunhofer) in terms of Kirchhoff's Laws.

5. What is the flash spectrum? How do you explain it in terms of Kirchhoff's Laws?

6. Explain the origin of lines in the solar spectrum in terms of the concepts of pure absorption and pure scattering.

7. Explain two methods for determining the photospheric temperature. Should the values be the same?

8. Explain the Helmholtz-Kelvin hypothesis. Why is it inadequate to explain the sun's luminosity? Does it have any other application?

9. Describe the principal process for supplying the solar luminosity, according to present knowledge. Explain why there is or is not doubt about this theory.

10. What is the direct evidence for the convective region below the photosphere?

11. The temperature above the photosphere increases to a high value in the corona. Explain how this occurs. Can the corona ever be observed without the aid of special instruments?

12. What is the material that constantly streams away from the sun? Describe its effects in the solar system.

13. What is so remarkable about the neutrinos from the sun that physicists have gone to great efforts to detect?

14. Is there a method for studying the sun's interior structure that has been used to study the internal structure of other solar-system bodies?

15. Sunspots appear dark when compared with the surrounding photosphere. Are they really cool areas on the sun? Explain your answer.

16. How can the solar magnetic field be measured?

17. Prepare a summary of events that usually occur in a large "center of activity."

18. What physical property seems to be common to all aspects of a "center of activity"? Discuss its role.

19. What effects do solar flares produce on earth?

20. What was the Maunder minimum? Why are we rather sure that it occurred?

CHAPTER 10. THE STARS—A CENSUS

This chapter summarizes the properties of the stars. Knowledge of this material should include understanding the objectives and the key words listed below.

Objectives

1. To know the nature and properties of stars and how they were learned.
2. To develop an awareness of the great range of such stellar properties as temperature, size, and luminosity.
3. To appreciate the data on stellar properties as the foundation for theories of the origin and evolution of stars.

Key Words

Proxima Centauri	light period
parallax	RR Lyrae stars
Bessel	spectroscopic parallax
Struve	space velocity
Henderson	statistical parallax
light-year	proper motion
parsec	radial velocity
trigonometric parallaxes	Barnard
intrinsic brightness	Barnard's star
apparent brightness	tangential velocity
Cepheid variable	white dwarf
Herschel	visual binary
relative brightness	Michell
standard stars	Goodricke
magnitudes	light curve
UBV	van de Kamp
color index	field stars
bolometric magnitudes	multiple stars
absolute magnitude	associations
spectral class	open clusters
Saha	globular clusters
giant stars	pulsating star
binary stars	Shapley

center of mass
spectroscopic binary stars
double-lined spectroscopic binary
single-lined spectroscopic binary
eclipsing binary
mass-luminosity relation
main-sequence stars
Hertzsprung
Russell
H-R Diagram
supergiant

Eddington
Leavitt
period-luminosity relation
Classical Cepheid
W Virginis star
long period variable
electron degenerate matter
stellar winds
contact binaries
Struve

Exercises (Answers at End of Study Guide)

1. The standard unit of distance at which a star would have a parallax of one second of arc is called the
   a. astronomical unit.
   b. absolute magnitude.
   c. parsec.
   d. apparent magnitude.
   e. light-year.

2. Match the following:
   Bessel   pulsating star
   Barnard   parallax
   Herschel   white dwarf
   Cepheid   proper motion
   Sirius B   sun's motion

3. The star with the largest known proper motion moves about _____ seconds of arc per year.
   a. 1   d. 1000
   b. 10   e. none of the above
   c. 100

4. True or False. Distances to some stars can be determined if the star pulsates and the period of light variation is known.

5. True or False. The absolute magnitude of a star is the magnitude the star would have at a distance of ten parsecs, if there were no absorption of its light by interstellar matter between the star and the earth.

6. True or False. The names *supergiant, giant, main sequence,* and *white dwarf* come from old myths and astrology and have no connection with the real nature of stars.

7. The most difficult basic property of a star to determine is its
    a. radius.   d. mass.
    b. temperature.   e. color.
    c. luminosity.

8. True or False. Most stars occur in pairs or in multiple systems.

9. Classical Cepheids, W Virginis stars, and RR Lyrae stars are all
    _____ stars.

10. True or False. White dwarfs have normal stellar temperatures, but
    their sizes are roughly that of the earth.

Questions and Assignments

1. Summarize the different methods of parallax determination.

2. Are there stars within 1 parsec of the sun?

3. The measured trigonometric parallax of a star is so small that the uncertainty is comparable to the parallax itself. What do we know about
   the distance to this star?

4. Briefly review the measurements and steps necessary to determine
   the total space velocity of a star relative to the sun.

5. How can the temperature of a star be estimated from its color index?

6. What corrections must one make to observed magnitudes in order to
   estimate a bolometric magnitude?

7. What does a star's luminosity tell us about its interior?

8. Think up an original mnemonic device for the sequence of spectral
   classes from O to M.

9. What single factor accounts *primarily* for the sequence of spectral
   classes? Name one factor that clearly does not.

10. Give some examples of the great range of stellar sizes. How are most
    stellar radii determined?

11. Describe one method for determining the masses of stars.

12. Astronomers state that a certain star is a binary system. However,
    even through the most powerful telescope, only one star is seen. Its
    brightness is steady, so obviously it's not an eclipsing system. How
    did they find out that the star is a binary?

13. What is the mass-luminosity relation and how is it determined?

14. Compare the sun's mass, radius, temperature, and luminosity with
    the range of stellar values.

15. Draw a simplified version of the Hertzsprung-Russell Diagram, and label the areas occupied by the various kinds of stars.

16. How can one determine the diameters of the two stars in an eclipsing binary?

17. How might dark companions of nearby stars be detected?

18. Briefly summarize the properties of the different kinds of star clusters.

19. Why are Cepheid variable stars so valuable to astronomers?

20. Describe an example of stars so close together that they are distorted and/or transfer mass.

CHAPTER 11. BIRTH AND DEATH OF STARS

In this chapter we present the evidence for and our current view of stellar evolution—the birth, life, and death of stars. Knowledge of this material should include understanding the objectives and the key words listed below.

Objectives

1. To understand the basic argument that stars must be evolving.

2. To understand the uses of H-R Diagrams of star clusters.

3. To be familiar with the current theory of the birth, life, and death of stars.

4. To understand the consequences of the interaction between stars and the interstellar medium.

5. To develop an appreciation for the aspects of stellar evolution that predict an ultimate end to an inhabitable solar system for humankind.

Key Words

composite H-R Diagram
cluster parallax
Sandage
turnoff points
protostars
Larson
core
Becklin

"embryo star"
accretion disk
envelope
Herbig-Haro objects
T-Tauri stars
cocoon stars
Neugebauer
heavy elements

Cameron	"metals"
Hayashi	stellar populations
helium flash	Population II
horizontal-branch star	nebular hypothesis
white dwarf	Kant
black dwarf	Laplace
central stars of planetary nebulae	conservation of angular
nova	momentum
supernova	Helmholtz-Kelvin contraction
supernova remnant	Kraft
Crab nebula	collisional hypothesis
Veil nebula	tidal hypothesis
neutron star	Bode's Law
black hole	dynamical relaxation

Exercises (Answers at End of Study Guide)

1. True or False. Most of the main-sequence stars, including O stars and our sun, were born about the same time.

2. True or False. The similarities of some parts of cluster H-R Diagrams permit the determination of cluster parallaxes.

3. Stars spend most of their life in what region of the H-R Diagram?
   a. red-giant region
   b. main sequence
   c. blue-supergiant region
   d. white-dwarf region
   e. cluster region

4. The _____ _____ appears to stop the evolution of red-giant stars, which had been proceeding up and to the right on the H-R Diagram.

5. True or False. Although supernovae can produce spectacular supernova remnants, they almost never exceed 10^4 suns in luminosity.

6. Match the following:

Kant-Laplace	heavy elements
Kraft	stellar populations
turnoff points	rotation speeds
Baade	nebular hypothesis
supernova remnants	cluster ages

7. True or False. Nuclear reactions in stars occur when elements of high atomic weight are burned to produce an ash of elements of low atomic weight, just as heavy coal, when burned, becomes light ash.

8. Which of the following is not considered a possible final state of stellar evolution?

   a. white or black dwarf   d. red giant
   b. neutron star   e. none of the above
   c. black hole

9. True or False. The material that formed the sun may previously have been in about two other stars.

10. True or False. Thanks to nuclear physics, our understanding of the formation of stars and particularly of the planets is well established and complete.

Questions and Assignments

1. Carry through the calculation that shows ages of a few million years for O stars. What is the obvious conclusion from this exercise?

2. Sketch an H-R Diagram for two hypothetical open clusters at different distances from the sun. If the distance to one of the clusters is known, explain how the distance to the other would be determined.

3. Sketch your own composite H-R Diagram for the open clusters shown in Figure 11-3a. Determine the age of a cluster and explain the reasoning behind the method.

4. Explain the general features of the sketch prepared for Question 3.

5. Prepare your own brief outline of stellar evolution for a star of 1 solar mass.

6. How should the outline prepared in Question 5 be modified if the star is much more massive than the sun?

7. Briefly describe the phenomena of novae and supernovae.

8. What is the evidence for the statement that supernovae eject material?

9. What class of observed objects appears to correspond to the theoretically predicted neutron stars?

10. Where are the elements heavier than hydrogen and helium produced?

11. The solar spectrum is rich in iron lines. What is the source of the sun's iron?

12. Briefly summarize the concept of "population" as introduced by W. Baade.

13. Describe the nebular hypothesis for formation of the sun and planets.

14. Discuss a major problem associated with the nebular hypothesis.

15. Why are the tidal and collisional hypotheses no longer considered viable?

16. Does Bode's Law necessarily bear directly on the formation of the solar system? Explain your answer.

17. Explain why star clusters are so important in the study of stellar evolution.

18. What phenomenon of stellar evolution will cause the earth to be uninhabitable in the future? When will this happen?

CHAPTER 12. THE MILKY WAY GALAXY

In this chapter we describe the properties of our galaxy. Knowledge of this material should include understanding the objectives and the key words listed below.

Objectives

1. To understand the structure and dimensions of our galaxy and, in particular, the sun's location in it.

2. To know the basic properties of the interstellar medium and appreciate its importance in galactic studies.

3. To understand the galactic rotation and how it is studied.

4. To know the basic model of the Milky Way galaxy and understand its organization into arm, disk, and halo populations.

Key Words

Milky Way galaxy	nebula
galactic coordinate system	H II regions
galactic plane	nebulium
galactic equator	Bowen
galactic latitude	forbidden lines
galactic longitude	collisional excitation
Kapteyn	fluorescence
star density	recombination radiation
Kapteyn universe	Strömgren

Shapley
dark nebulae
Horsehead nebula
interstellar medium
intrinsic colors
Trumpler
Hiltner
Hall
Jansky
Reber
synchrotron radiation
21-centimeter line
van de Hulst
Shklovskii
Ewen and Purcell
interstellar clouds

Orion nebula
giant molecular clouds
thermal radiation
radio recombination lines
interstellar molecules
spiral arms
Oort
galactic rotation curve
high velocity stars
Lindblad
arm
disk
halo
protogalaxy
density-wave theory

Exercises (Answers at End of Study Guide)

1. True or False. The Kapteyn universe is a generally accepted model of our galaxy.

2. True or False. Shapley determined the distance and direction to the galactic center by studying RR Lyrae stars in globular clusters.

3. Interstellar absorption dims the light from a star and _____ it.

4. Which of the following molecules has *not* been detected in the interstellar medium?
   a. ammonia
   b. formaldehyde
   c. methyl alcohol
   d. dimethyl ether
   e. none of the above

5. Match the following:
   Shapley
   Bowen
   Strömgren
   Trumpler
   van de Hulst
   Oort

   21-cm line
   H II regions
   galactic rotation
   nebular lines
   interstellar absorption
   galactic dimensions

6. True or False. The interstellar gas is more strongly concentrated in a ring between 4 and 7 kiloparsecs from the galactic center.

7. The sun takes approximately _____ years to complete a revolution around the galactic center.

a. 5 million
b. 25 million
c. 250 million

d. 3 billion
e. 5 billion

8. The mass of the galaxy can be estimated from the rotation curve and is approximately ——————— suns.
   a. 3×10^7
   b. 3×10^8
   c. 3×10^9

   d. 3×10^{10}
   e. 3×10^{11}

9. True or False. The concept of stellar populations developed by W. Baade is applicable to models of the Milky Way galaxy.

10. True or False. The stars in a spiral arm are *not* trapped there, but pass slowly through.

11. The space between the stars contains
    a. bright gas clouds.
    b. dark gas clouds.
    c. dust.

    d. all of the above.
    e. none of the above.

Questions and Assignments

1. Use a sketch to explain the relation between the band of light called the Milky Way and the Milky Way galaxy.

2. Make your own sketch of the galactic coordinate system. Is it centered on the sun? Why?

3. Summarize the methods used by Kapteyn and Shapley to investigate the size of the galaxy. Why did only one method work?

4. What is so unusual about many of the emission lines normally observed in bright nebulae?

5. Trace the origin of the energy that appears in the emission lines of bright nebulae.

6. Briefly explain how the size of an H II region depends on its exciting star and the gas density.

7. What everyday phenomenon is physically analogous to the changes in stellar colors produced by the interstellar dust? Explain.

8. Summarize Trumpler's argument for interstellar absorption in your own words.

9. As a library research project, prepare a report on the work of Grote Reber in the pioneering days of radio astronomy.

10. Why was the discovery of the 21-cm line of hydrogen so important?

11. Summarize the currently accepted model of the distribution of interstellar gas in the galaxy.

12. Briefly summarize the problems associated with the discovery of complex molecules in the interstellar gas.

13. Briefly describe a method for showing that the galaxy rotates.

14. What are the so-called high velocity stars, and how were they named?

15. Summarize the galactic model in terms of the arm, disk, and halo populations.

16. Use the summary developed for Question 15 to describe briefly how the galaxy probably evolved to its present state.

17. Use a different analogy from the one given in the text to explain the density-wave theory of spiral arms in everyday terms.

CHAPTER 13. GALAXIES AND THE UNIVERSE

This chapter describes the properties of galaxies and the present state of cosmology, the study of the nature and origin of the universe. Knowledge of this material should include understanding the objectives and the key words listed below.

Objectives

1. To know the basic properties of galaxies.

2. To understand the observational facts of cosmology and how well the Big Bang theory can explain them.

3. To develop an appreciation for the possible future fates of the universe.

Key Words

Shapley	protogalaxies
Curtis	Cepheids
spiral galaxies	distance indicators
Messier	clusters of galaxies
barred spiral galaxies	Slipher
irregular galaxies	red shift
Magellanic Clouds	Hubble constant
elliptical galaxies	Penzias and Wilson
dwarf elliptical galaxies	Gamow
Baade	cosmic background radiation
stellar populations	Olbers' Paradox
Olbers	Bondi, Gold, and Hoyle

cosmology	oscillating universe
Big Bang	critical density
Primeval Fireball	open universe
Lemaître	closed universe
Hubble time	Steady State theory

Exercises (Answers at End of Study Guide)

1. True or False. In the National Academy of Sciences debates, Shapley was correct on the size of the Milky Way galaxy and the sun's location in it, and Curtis was correct in his view that the spirals were star systems like the Milky Way.

2. Which of the following is *not* a known type of galaxy?
   a. barred elliptical
   b. spiral
   c. elliptical
   d. dwarf elliptical
   e. none of the above

3. True or False. The stars Baade resolved in the central region of M31 were quite similar to the stars in the outer regions resolved by Hubble years earlier.

4. Match the following:

Olbers	red shift
Lemaître	critical density
Hubble constant	dark night sky
future of universe	irregular galaxies
Magellanic Clouds	Big Bang

5. Which of the following are used as distance indicators in the study of galaxies and cosmology?
   a. Cepheids
   b. supergiants
   c. size of H II regions
   d. brightest galaxy in cluster
   e. all of the above

6. True or False. Most galaxies occur in groups called clusters of galaxies.

7. True or False. Most galaxies are moving away from us at about the same speed, or "universal expansion."

8. True or False. Olbers' Paradox is not a problem on the Big Bang model.

9. The age of the universe can be calculated from the Hubble constant, and the current value is
   a. 40 million years.
   b. 250 million years.
   c. 1 billion years.
   d. 4.5 billion years.
   e. 10 to 20 billion years.

10. The parameter that determines whether the universe is open or closed is the ——————— ———————.

Questions and Assignments

1. How was the distance to M31 determined and confirmed?

2. What type of galaxy is similar to the Milky Way?

3. Briefly summarize the properties of the different types of galaxies.

4. Summarize the known range of masses of galaxies, and briefly indicate the methods used to determine them.

5. As a library research project, prepare a report describing in some detail Baade's work on stellar populations and galaxies in the 1940s.

6. What probably determines the shape of a galaxy?

7. Describe a property of spiral galaxies that shows variation among different galaxies. Cite some examples.

8. What would you use as a distance indicator for a remote cluster of galaxies with unknown red shift?

9. Describe the three basic facts that must be explained by any acceptable theory of cosmology.

10. Use a graph to explain how the red shift depends on distance.

11. Explain how the recession speeds of galaxies permit an age determination for the universe. How does this age compare with other ages, such as those of the solar system and the oldest star clusters in our galaxy?

12. If the distant galaxies appear to be receding from us in all directions, can we conclude that the Milky Way is near the center of the universe? Explain.

13. Describe the Big Bang theory and, in particular, how it explains the three facts of cosmology described in Question 9.

14. Briefly describe the Steady State theory and the so-called oscillating universe. If these theories can be ruled out, cite the evidence.

15. If the universe is decelerating, how would the graph prepared for Question 10 appear? As an astronomer, how would you test for deceleration?

16. Why can't we simply determine the average density of the universe, compare it with the critical density, and thereby determine whether the universe is open or closed?

CHAPTER 14. SPACE AGE ASTRONOMY

This chapter describes the need for, techniques of, and some results from astronomical observations made above the earth's atmosphere. Knowledge of this material should include understanding the objectives and the key words listed below.

Objectives

1. To understand the difficulties and limitations of observations made from the earth's surface.
2. To appreciate fully that understanding the universe requires observations in essentially all regions of the electromagnetic spectrum.
3. To be familiar with the several types of orbiting observatories and some important results obtained from them.

Key Words

seeing
scattering
absorption
airglow
Low
Kuiper Airborne Observatory
Project Stratoscope
UHURU
supernova remnants
x-ray bright points
Orbiting Astronomical
    Observatory (OAO)
Code

x-ray binaries
transient x-ray sources
Aerobee
Astrobee
orbiting observatory
Orbiting Solar Observatories
    (OSOs)
isothermal plateau
SKYLAB
Spitzer
Copernicus
High Energy Astrophysical
    Observatory (HEAO)
hydrogen cloud

Exercises (Answers at End of Study Guide)

1. Which of the following does *not* affect observations from the earth's surface?
    a. seeing
    b. dark globules
    c. absorption
    d. airglow
    e. scattering

2. True or False. Observations of the same object in several different regions of the electromagnetic spectrum are generally equivalent, hence only one is necessary.

3. True or False. Observations from a high-altitude airplane showed that the galactic center is a powerful infrared source.

4. True or False. The basic problem with observations from rockets is that the observing time is normally only a few minutes.

5. Match the following:

SKYLAB   isothermal plateau
HEAO   Copernicus
Spitzer   x-ray bright points
OSO   rocket
Aerobee   x-ray light curves

6. True or False. Ultraviolet instruments on the early solar research satellites could not resolve separate portions of the solar disk.

7. True or False. Temperatures in a solar flare can briefly exceed the temperature in the center of the sun.

8. True or False. Photographs from HEAO showed that the corona occasionally ejects huge bubbles of solar material.

9. Studies of comets made with satellite instruments led to the discovery of the cometary

   a. nucleus.   d. hydrogen cloud.
   b. coma.   e. icy-grain halo.
   c. tail.

10. True or False. Studies with Copernicus showed that many of the heavier elements, such as iron, had the same abundance relative to hydrogen in the interstellar gas as in the atmospheres of stars.

Questions and Assignments

1. Briefly discuss each of the four limitations of ground-based observations, and cite a significant observation compromised or prohibited by each limitation.

2. Discuss in detail the absorption of the different regions of the electromagnetic spectrum by the earth's atmosphere. Mention the atmospheric constituent that is primarily responsible in each case.

3. Summarize the arguments in favor of observations in many different wavelength regions.

4. As a library research project, prepare a report on the location and conditions at a well-known high-altitude (mountain-top) observatory.

5. What types of observations are made from aircraft?

6. What aspect of solar research was pioneered by studies from balloons?

7. Describe briefly the three classes of x-ray sources discovered by instruments carried on rockets.

8. What are the x-ray bright points? How were they discovered?

9. What was the purpose of the Celescope experiment?

10. Describe the structure of the interstellar gas as determined from data obtained from Copernicus.

11. According to results obtained from UHURU data, what is thought to be the origin of x-rays from clusters of galaxies?

12. According to results obtained from HEAO data, what is a possible origin of the diffuse background of x-rays?

13. As a library research project, prepare a report on a scientific discovery from an earth-orbiting observatory *not* discussed in the text. Examples are the long-wavelength radio waves studied by the Radio Astronomy Explorer and the gamma-ray bursts studied by the Vela satellites.

CHAPTER 15. EXPLORATION OF SPACE

This chapter briefly describes the techniques of space exploration and presents the major scientific results obtained to date. Knowledge of this material should include understanding the objectives and the key words listed below.

Objectives

1. To know the history and basic physics of rocketry.

2. To understand the basic strategies of space exploration and their relative strengths and weaknesses.

3. To know the important discoveries of space physics, particularly those about interplanetary space, the moon, and the planets.

4. To appreciate how techniques developed for the study of the earth are modified and extended to investigate other objects in the solar system.

5. To develop an awareness of the costs of astronomy and space research within the larger contexts of costs of scientific research in general and government spending for other purposes.

Key Words

Newton
Sputnik 1
escape speed
Tsiolkovsky
Goddard
Van Allen Belts
magnetopause
magnetic tail
Interplanetary Monitoring
    Platform
bow wave
dipole field
aurora borealis
aurora australis
auroral electrons
cosmic rays
sector structure
libration points
acoustic detection
time-of-flight detectors
mascons
maria
highland areas
craters
Surveyors
rills
rays
domes
meteoroid impact
volcanic origin
crater chains
drowned craters
albedo

Oberth
Saturn 5
magnetosphere
mirror points
Explorer 1
Armstrong
Aldrin
Collins
regolith
lunar bedrock
igneous rocks
breccia
moonquakes
foci
rubble
moon rocks
spring tides
neap tides
equatorial bulge
angular momentum
flyby
orbiter
soft-lander
lander-rovers
Schiaparelli
canali
canals
Lowell
wave of darkening
Mariner
Viking
Olympus Mons
Vallis Marineris

Luna 9	macrobes
Apollo 11	Caloris Basin
sampler boom	lobate scarps
channels	*The Budget in Brief*
Venera	

Exercises (Answers at End of Study Guide)

1. The escape speed from the earth's surface is approximately _____ km/sec.
   a. 11   d. 17
   b. 13   e. 19
   c. 15

2. Which of the following men was *not* involved in the development of useful rockets?
   a. Goddard   d. Wells
   b. Oberth   e. none of the above
   c. Tsiolkovsky

3. Match the following:

solar wind	Jupiter
Mariner 9	sector structure
Mariner 10	radiation belts of earth
Lowell	Mercury
Pioneer 11	Mars
Van Allen	canals

4. Which of the following is not a feature of the lunar surface?
   a. maria   d. craters
   b. mascons   e. rays
   c. highland areas

5. True or False. The most likely explanation for most craters on the moon is meteoroid impact.

6. True or False. Moonquakes have been found to be quite similar to earthquakes.

7. True or False. The Mariner and Viking investigations of Mars have provided evidence for extensive volcanism and the presence of liquid water at some time in the past.

8. True or False. The biological experiments on the Viking landers produced no conclusive evidence for life on Mars.

9. True or False. Because of the thick clouds on Venus, we still have no photographs of its surface.

10. A geological feature found on Mercury but on no other solar-system body is the ———————— ————————.

Questions and Assignments

1. Explain the physical basis of the escape speed and the rocket.

2. Sketch the earth's magnetosphere, and show its major features. What energetic particles are trapped in the magnetosphere?

3. Are aurorae caused by Van Allen Belt particles raining down on the earth's atmosphere? Explain your answer.

4. What are the properties of interplanetary dust at the earth's distance from the sun? How were they determined?

5. Describe the surface appearance of the moon. Mention the principal features.

6. What is the evidence for volcanism on the moon?

7. Is the lunar bedrock exposed on the moon? What two basic types of rock are found on the moon?

8. Describe the properties of natural and man-made moonquakes.

9. What model of the lunar interior was formulated from the studies of moonquakes and related data? Where do the moonquakes occur?

10. Summarize the history of the lunar surface.

11. Does the length of the day on earth remain constant? Explain the cause of the situation.

12. Briefly describe the usual steps in a program of space exploration of a planet.

13. Is there good evidence for canals on Mars? Explain the current view.

14. Describe the major features of the Martian surface as determined by data from the Mariner and Viking spacecraft.

15. What have we inferred about the Martian interior?

16. Is there water on Mars now? Was there more or less water in the past? Justify your answer.

17. Describe *one* of the Viking biology experiments. Summarize the results concerning the existence of primitive life from all three experiments.

18. Describe the surface of Venus as revealed by the Venera landers. What basic reservation should you have concerning this answer?

19. Mercury has been described as having an interior like the earth and a surface like the moon. What is the evidence for this viewpoint?

20. What are lobate scarps? How might they have been formed?

21. Several planetary missions are planned or in progress. Prepare a brief report on the new results as they become available.

22. Summarize the terrestrial effects of the sun becoming a red giant.

23. What are the major elements of the United States Federal Budget? What percentage of the total supports "General Science, Space, and Technology"?

CHAPTER 16. PROBLEMS IN MODERN ASTRONOMY

In this chapter we present some of the "hot topics" in astronomy—exciting problems that have attracted great interest. Knowledge of this material should include understanding the objectives and the key words listed below.

Objectives

1. To reinforce the fact that astronomy is a constantly changing science.

2. To understand the history of pulsar discoveries and current ideas about pulsars.

3. To understand the observational facts about and the difficulties in understanding radio galaxies and quasars.

4. To know the basic physics of black holes and the evidence for their existence.

Key Words

neutron stars	Oppenheimer
pulsars	Hewish
supernova	pulsar
Crab nebula	clock mechanisms
synchrotron	radio galaxies
McMillan	radio lobes
Shklovskii	confinement problem
radiative lifetime	quasars

Landau
synchrotron radiation
Schmidt
blue stellar objects
gravitational red shift
Terrell
Bahcall
Gunn
superluminal motions

Hazard
black hole
Cygnus X-1
stellar wind
Roche lobe overflow
Schwarzschild limit
bursting x-ray sources
supermassive black holes

Exercises (Answers at End of Study Guide)

1. True or False. The bright "guest star" observed by Chinese astronomers in A.D. 1054 was probably the supernova that produced the Crab nebula.

2. The light from the Crab nebula results from high-speed electrons moving in a magnetic field and is called _____ _____.

3. Which mechanism accounts for the extreme regularity of the pulse spacings in pulsars?
   a. rotation of a star
   b. pulsation of a star
   c. orbital motion of a binary star
   d. magnetic changes like the solar cycle
   e. regular starquakes

4. True or False. The radio luminosity of a radio galaxy equals that of a large normal galaxy.

5. True or False. At present, there is no satisfactory explanation for the confinement of the emitting plasma to the two lobes of a radio galaxy.

6. Match the following:
   McMillan
   Shklovskii
   pulsars
   Schmidt
   Cygnus X-1

   quasars
   synchrotron
   black hole
   Crab nebula
   neutron stars

7. True or False. Quasars take their name from the time when they were believed to be stars and were named quasi-stellar radio sources.

8. True or False. We now know that the interpretation of quasar red shifts being caused by the expansion of the universe is incorrect.

9. According to theoretical calculations, a contracting, dying star becomes a black hole if its mass is
   a. greater than 3 M$_\odot$.   d. about 0.1 M$_\odot$.
   b. in the range 1.5 to 2 M$_\odot$.   e. less than 0.1 M$_\odot$.
   c. close to 1 M$_\odot$.

10. True or False. To search for black holes, it makes sense to study binary stars.

Questions and Assignments

1. Briefly describe the history of observations of the Crab nebula.

2. What is synchrotron radiation?

3. Shklovskii's suggestion that the Crab nebula radiation was synchrotron radiation solved one problem but created another. Explain this situation.

4. What are the properties of a neutron star? How might one be formed?

5. What is the observational evidence that neutron stars exist?

6. How did the study of a pulsar solve the problem of energy supply for the Crab nebula?

7. Summarize the properties and problems of radio galaxies. Can they be explained as collisions of galaxies?

8. Which solution to the confinement problem for radio galaxies do you prefer? Why?

9. Summarize the observed properties of quasars.

10. Compare the evidence for and against the two theories on the nature of quasar red shifts—the local theory and the cosmological theory.

11. What are superluminal motions?

12. Briefly summarize the properties and the conditions that lead to each form of dead star.

13. Why can't we receive light from a black hole?

14. Can we receive light from the *vicinity* of a black hole? If so, outline a situation that produces it.

15. Summarize the evidence for the existence of a black hole in Cygnus X-1.

16. Is it likely that large black holes exist at the centers of globular clusters? Justify your answer.

17. What "use" have astronomers proposed for the hypothetical supermassive black holes?

CHAPTER 17. EPILOGUE

This short chapter presents a review of the new horizons discussed in this book in terms of the earth's motion in space and a chronological summary of events. Knowledge of this material should include understanding the objective and comparing the summaries with the detailed treatment in earlier chapters.

Objective

1. To review briefly the grand scheme of cosmic evolution from the Big Bang to the current state of life on earth.

Assignment

Colonies in space—near earth, elsewhere in the solar system, and ultimately in more distant parts of the galaxy—should eventually become possible. Prepare a brief report on one such project. What are your personal views on space colonization?

Answers to Exercises

CHAPTER 1

1. True 2. False 3. c 4. e 5. b

CHAPTER 2

1. d 2. False 3. False 4. False 5. False 6. Julius Caesar—calendar reform; Copernicus—heliocentric world-view; Galileo—study of motion; Ptolemy—epicycle; Newton—gravitation 7. ellipses 8. False 9. True 10. False

CHAPTER 3

1. False 2. e 3. False 4. False 5. Eratosthenes—earth's circumference; San Francisco—San Andreas fault; Bacon—continental drift; Lyell—uniformitarianism; Moho—crust-mantle boundary 6. d 7. False 8. d 9. False 10. False

CHAPTER 4

1. True 2. False 3. a 4. Darwin—natural selection; Mendel—heredity; Oparin—chemical origin of life; Kardashev—advanced civilizations; DNA—genetic code 5. True 6. d 7. True 8. False 9. b 10. True

CHAPTER 5

1. True 2. e 3. c 4. a, c 5. c 6. Dalton—atomic theory; Mendeleev—Periodic Table; Chadwick—neutron; Einstein—nuclear energy release; Newton—spectroscopy; Planck—thermal radiation 7. False 8. b 9. e 10. larger, smaller 11. b

CHAPTER 6

1. False 2. True 3. e 4. c 5. lens—refracting telescope; mirror—reflecting telescope; correcting plate—Schmidt camera; Daguerre—photography; Argelander—sky survey 6. a 7. a 8. True 9. d 10. True

CHAPTER 7

1. d 2. b 3. Herschel—Uranus; Adams and Leverrier—Neptune; Tombaugh—Pluto; Piazzi—Ceres; Halley—comet; Hall—Deimos and Phobos 4. True 5. False 6. a 7. a 8. c 9. d 10. False

CHAPTER 8

1. True 2. e 3. b 4. False 5. c 6. Rasool—atmospheric evolution; Venus—greenhouse effect; Jupiter—great red spot; Io—radio bursts; Doppler spread—planetary rotation 7. False 8. c 9. e 10. False

CHAPTER 9

1. True 2. e 3. b 4. solar constant 5. False 6. c 7. Galileo—sunspots; Carrington—differential rotation; Fraunhofer—solar spectrum; proton-proton chain—solar luminosity; Davis—solar neutrinos; Zeeman—magnetic fields 8. True 9. a 10. False

CHAPTER 10

1. c 2. Bessel—parallax; Barnard—proper motion; Herschel—sun's motion; Cepheid—pulsating star; Sirius B—white dwarf 3. b 4. True 5. True 6. False 7. d 8. True 9. pulsating 10. True

CHAPTER 11

1. False 2. True 3. b 4. helium flash 5. False 6. Kant-Laplace—nebular hypothesis; Kraft—rotation speeds; Sandage—composite H-R Diagram; turnoff points—cluster ages; Baade—stellar populations 7. False 8. d 9. True 10. False

CHAPTER 12

1. True 2. a 3. False 4. Olbers—dark night sky; Lemaître—Big Bang; Bowen—nebular lines; Strömgren—H II regions; Trumpler—interstellar absorption; van de Hulst—21-cm line; Oort—galactic rotation 6. True 7. c 8. e 9. True 10. True 11. d

CHAPTER 13

1. True 2. a 3. False 4. Olbers—dark night sky; Lemaítre—Big Bang; Hubble constant—red shift; future of universe—critical density; Magellanic Clouds—irregular galaxies 5. e 6. True 7. False 8. True 9. e 10. critical density

CHAPTER 14

1. b 2. False 3. True 4. True 5. Aerobee—rocket; OSO—isothermal plateau; SKYLAB—x-ray bright points; Spitzer—Copernicus; HEAO—x-ray light curves 6. True 7. True 8. False 9. d 10. False

CHAPTER 15

1. a 2. d 3. solar wind—sector structure; Mariner 9—Mars; Lowell—canals; Van Allen—radiation belts; Mariner 10—Mercury; Pioneer 11—Jupiter 4. b 5. True 6. False 7. True 8. True 9. False 10. lobate scarp

CHAPTER 16

1. True 2. synchrotron radiation 3. a 4. False 5. True 6. McMillan—synchrotron; Shklovskii—Crab nebula; pulsars—neutron stars; Schmidt—quasars; Cygnus X-1—black hole 7. True 8. False 9. a 10. True

Powers of Ten Notation

There is an enormous range in the scale of objects studied in astronomy. For example, the wavelength of the hydrogen line used by radio astronomers to map the spiral pattern of the galaxy (page 358) is 21 centimeters or 8.3 inches; the diameter of the earth is 6,400 kilometers or 640,000,000 centimeters (250,000,000 inches); the distance to the sun, one astronomical unit (page 187), is 15,000,000,000,000 centimeters (5,900,000,000,000) inches); the distance to the nearest star beyond the solar system is 1.31 parsecs (page 282) or 4,300,000,000,000,000,000 centimeters (1,600,000,000,000,000,000 inches); and the distance to the Virgo cluster of galaxies (page 386) is about 11,000 kiloparsecs[1] or 34,000,000,000,000,000,000,000,000 centimeters (13,000,000,000,000, 000,000,000,000 inches). Very small quantities also are encountered in astronomy. For example, the radius of the helium atom is about 1 Angstrom or 0.00000001 centimeter (0.000000004 inch).

The range in sizes described above is so great that it is very inconvenient to use any one unit, such as the centimeter, for every kind of length—there are just too many cases in which an annoying number of zeros has to be written down. It is also troublesome to use so many different units (centimeter, kilometer, astronomical unit, or parsec) and to

---

[1]This value assumes a Hubble constant of 100 kilometers per second per megaparsec. If the Hubble constant is 50 kilometers per second per megaparsec, the distance would be twice as large, or 22,000 kiloparsecs.

keep referring to Appendix 2 for the respective values of these units. A technique that allows us to minimize the number of zeros involved in using a common unit over a large range is the scientific notation or powers of ten method of representing numbers. It also happens to be very handy for doing simple calculations.

The power of ten simply locates the decimal point as shown in the following listing:

$$
\begin{aligned}
10^{10} &= 10{,}000{,}000{,}000 \\
10^{9} &= 1{,}000{,}000{,}000 \\
10^{8} &= 100{,}000{,}000 \\
10^{7} &= 10{,}000{,}000 \\
10^{6} &= 1{,}000{,}000 \\
10^{5} &= 100{,}000 \\
10^{4} &= 10{,}000 \\
10^{3} &= 1{,}000 \\
10^{2} &= 100 \\
10^{1} &= 10 \\
10^{0} &= 1 \\
10^{-1} &= 0.1 \\
10^{-2} &= 0.01 \\
10^{-3} &= 0.001 \\
10^{-4} &= 0.0001 \\
10^{-5} &= 0.00001 \\
10^{-6} &= 0.000001 \\
10^{-7} &= 0.0000001 \\
10^{-8} &= 0.00000001 \\
10^{-9} &= 0.000000001 \\
10^{-10} &= 0.0000000001
\end{aligned}
$$

Note that one thousand is simply $10 \times 10 \times 10$, or 10^3. One hundredth is $1 \div (10 \times 10)$, or 10^{-2}. Obviously, the listing could be extended to any large or small number. A number that is not an even multiple of ten is also easily represented by this method. For example, 365 is simply 3.65×100, or $3.65 \times 10 \times 10$, or 3.65×10^2.

This technique conveniently allows us to rewrite some of the cumbersome numbers quoted at the beginning of this appendix. The distance to the Virgo cluster of galaxies can be written as 3.4×10^{25} cm; the radius of the helium atom can be written as 1×10^{-8} cm.

Powers of ten notation allows easy multiplication and division, using the rules that *the powers add when multiplying* and *the powers subtract when dividing*. For example,

Multiplication: $(200) \times (3,000)$ $= (2 \times 10^2) \times (3 \times 10^3)$
$= (2 \times 3) \times 10^{2+3} = 6 \times 10^5$

Division: $\dfrac{5,000}{200}$ $= \dfrac{5 \times 10^3}{2 \times 10^2} = \dfrac{5}{2} \times 10^{3-2} = 2.5 \times 10^1 = 25$

Division: $\dfrac{200}{4,000}$ $= \dfrac{2 \times 10^2}{4 \times 10^3} = \dfrac{2}{4} \times 10^{2-3} = \dfrac{1}{2} \times 10^{-1}$

$= 5 \times 10^{-1} \times 10^{-1} = 5 \times 10^{-2} = 0.05$

Division: $\dfrac{600}{20,000}$ $= \dfrac{6 \times 10^2}{2 \times 10^4} = 3 \times 10^{2-4} = 3 \times 10^{-2} = 0.03$

A little practice with these ideas will enable the reader to make power of ten calculations with ease. Even if you do not become a scientist, powers of ten may help you to understand other things of interest; for example, the total revenue of the U.S. Government from personal income taxes in 1968 was 7.82×10^{10} dollars, and the population was about 2×10^8. Thus the average tax per person was

$$\frac{7.82 \times 10^{10}}{2 \times 10^8} = \text{about } 3.9 \times 10^2 \text{ dollars}$$

Units

LENGTH, AREA, AND VOLUME

The scientific units of measurement are often confusing to citizens of the United States and other English-speaking countries. However, we are a minority in this respect, and in fact the "scientific" or metric system of units is used in everyday affairs in a great many countries. It is rather difficult to defend our own system, where the thickness of a thumb once determined the inch, the length of someone's foot determined the unit "foot," and the distance from the nose to the end of the thumb of Henry I of England was the basis of the yard. Some of these origins may be legendary, but the units are certainly not fundamental, nor is conversion from one unit to another convenient.

The *metric system* is based on the *meter*, which was defined as one-millionth of the distance from the North Pole to the equator along the arc shown in Figure A-1. The distance from Barcelona, Spain, to Dunkirk, France, was surveyed between 1792 and 1799, and the length of the entire arc was calculated from astronomical observations. One meter is about 1.1 yards. In the metric system, units are converted by changing the power of ten. Common examples are:

> 1 kilometer $= 10^3$ meters (0.6 miles)
>
> 1 centimeter $= 10^{-2}$ meters (0.39 inches)
>
> 1 millimeter $= 10^{-3}$ meters
>
> 1 Angstrom $= 10^{-10}$ meters or 10^{-8} cm

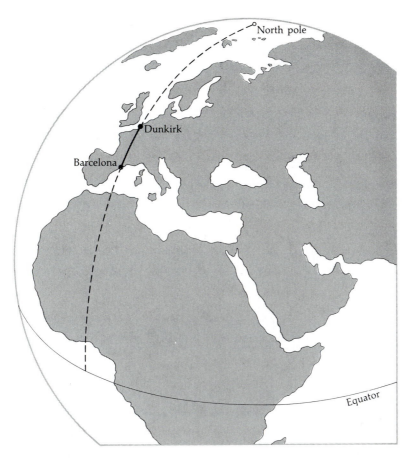

Figure A-1. The meter was based on the quadrant of the earth's meridian passing through Dunkirk and Barcelona, as shown. (From *Standards of Measurement* by A. V. Astin. Copyright © 1968 by Scientific American, Inc. All rights reserved.)

The centimeter is the most frequently used unit of distance in scientific work. However, in astronomy and space science, several larger units of length are often used. These include:

earth's radius $= 6.4 \times 10^8$ cm

sun's radius $= 7.0 \times 10^{10}$ cm

astronomical unit $= 1.5 \times 10^{13}$ cm

light-year $= 9.5 \times 10^{17}$ cm

parsec = 3.1×10^{18} cm

kiloparsec = 3.1×10^{21} cm

megaparsec = 3.1×10^{24} cm

To be precise, the meter is defined as 1,650,763.73 wavelengths of a specific atomic line of an isotope of the gas krypton emitting light under standard conditions. (Emission lines produced by gases are discussed in Chapter 5.)

Surface area is usually given in square centimeters or square meters.

1 square centimeter = 0.16 square inches

1 square meter = 10.8 square feet

Volume is usually given in cubic centimeters or cubic meters:

1 cubic centimeter = 0.061 cubic inches

1 cubic meter = 35 cubic feet

A specific version of the metric system frequently used in physics is the *c.g.s.* system, in which the basic units of length, mass, and time are the centimeter, gram, and second. In the *m.k.s.* system, these basic units are the meter, kilogram, and second.

MASS

For practical purposes, the basic unit of mass, the gram, can be defined as the mass of 1 cubic centimeter of water, as measured at a temperature of 4 degrees Celsius. (The Celsius temperature scale is virtually the same as the formerly used centigrade scale.) The kilogram (10^3 grams) is another common unit. A mass of 1 kilogram weighs about 2.2 pounds on the earth's surface.

TIME

The basic unit of time is the second. It can be defined as a certain fraction of the rotation period of the earth. Greater precision comes through the use of *atomic clocks*, devices that provide a time source based on the frequency of an optical or radio emission line. There are about 3.16×10^7 seconds per year.

ANGLE

Almost everyone knows that the circle is divided into 360 degrees. This convention may possibly be traced back to an early, inaccurate determina-

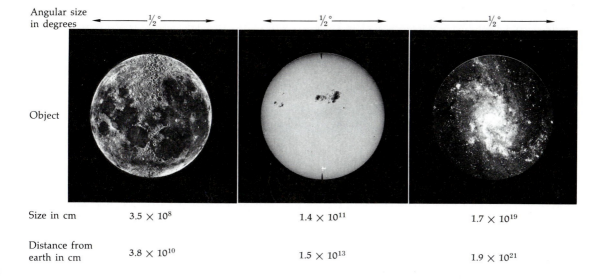

Angular size in degrees	← —— ½° —— →	← —— ½° —— →	← —— ½° —— →
Object			
Size in cm	3.5×10^8	1.4×10^{11}	1.7×10^{19}
Distance from earth in cm	3.8×10^{10}	1.5×10^{13}	1.9×10^{21}

Figure A-2. The concept of angular size. (Bottom) If the angle in the example were about one-half degree, the moon, sun, and the galaxy M33 (which have vastly different linear sizes) would fit into it as shown. (Top) Photographs printed to the same angular scale show the effect cited. (Courtesy of the Lick Observatory, Hale Observatories, and Hale Observatories, respectively.)

tion of the number of days in a year; if there were 360 days per year, and if the earth's orbit were a circle, the earth would move around the orbit at the rate of one degree per day. The degree is further subdivided into 60 minutes of arc (60′), and the minute into 60 seconds of arc (60″). Thus, there are $60 \times 60 = 3,600$ seconds of arc in one degree.

The concept of *angular size* of an object is used throughout astronomy, because it constitutes an apparent size that we can measure. On the other hand, in order to know the true (or "linear") size, as measured in length units (for example, kilometers), we have to know how far away the object is; note that objects of entirely different true sizes can have the same angular size as seen in the sky; this effect is shown in Figure A-2.

APPENDIX *3*

Additional Reading

An older book written with the same general philosophy as ours is:

THROUGH SPACE AND TIME, by Sir James Jeans. London: Macmillan, 1934.

Books with the same general philosophy include:

KNOWLEDGE AND WONDER, by Victor F. Weisskopf. New York: Doubleday, 1966.

MATTER, EARTH AND SKY, by George Gamow. Englewood Cliffs, N.J.: Prentice-Hall, 1965.

INTELLIGENT LIFE IN THE UNIVERSE, by I. S. Shklovskii and Carl Sagan. San Francisco: Holden-Day, 1966.

Two standard astronomy texts to refer to are:

DRAMA OF THE UNIVERSE, by George Abell. New York: Holt, Rinehart, and Winston, 1978.

THE UNIVERSE UNFOLDING, by Ivan R. King. San Francisco: W. H. Freeman and Company, 1976.

To learn or refresh your knowledge of basic mathematical concepts, read:

MATHEMATICS, A HUMAN ENDEAVOR, by Harold R. Jacobs. San Francisco: W. H. Freeman and Company, 1970.

We think you will also enjoy:

THE IMMENSE JOURNEY, by Loren Eiseley. New York: Random House, 1957.

Well-illustrated accounts of the latest developments in astronomy and space research appear in SCIENTIFIC AMERICAN, SKY AND TELESCOPE, NATURAL HISTORY, SMITHSONIAN, and SCIENCE NEWS.

For the history of great advances in astronomy, told in the words of the masters themselves, browse through:

A SOURCE BOOK IN ASTRONOMY, edited by Harlow Shapley and Helen E. Howarth. New York: McGraw-Hill, 1929.

SOURCE BOOK IN ASTRONOMY, 1900–1950, edited by Harlow Shapley. Cambridge, Mass.: Harvard University Press, 1960.

A useful publication is the American Association for the Advancement of Science (AAAS) quarterly review journal, SCIENCE BOOKS AND FILMS. Also see THE AAAS SCIENCE BOOK LIST SUPPLEMENT.

Basic Astronomical Equations

Included in this Appendix are the simple mathematical expressions that are relevant to the content of this book—for readers who wish to know and use them.

KEPLER'S THIRD LAW

The average distance of a planet from the sun, d (in astronomical units), and its period of revolution, p (in years), are related by

$$p^2 = d^3$$

This equation was used in constructing Figure 2-21, page 37.

NEWTON'S SECOND LAW

The force, F, the mass, m, and the acceleration, a, are related by

$$F = ma$$

The acceleration is in the same direction as the force. Recall that an acceleration is a change in velocity and that no acceleration means a constant velocity. Hence, Newton's First Law of Motion (page 41) is really a special case of his Second Law. In the c.g.s. system, the units are dynes (F), grams (m), and centimeters per second per second (a). The dyne is defined as the amount of force required to accelerate a mass of one gram by one centimeter per second per second.

NEWTON'S LAW OF UNIVERSAL GRAVITATION

The gravitational force, F, between two masses, m and M, that are separated by a distance, r, is

$$F = \frac{MmG}{r^2}$$

The units of force, mass, and distance are dynes, grams, and centimeters, respectively, in the c.g.s. system. G is the constant of gravitation, which has a numerical value of about 6.7×10^{-8}. See page 42.

EINSTEIN'S EQUATION FOR MASS-ENERGY EQUIVALENCE

The energy equivalent, E, of a mass, m, is given by

$$E = mc^2$$

The units of E and m are ergs and grams, respectively (c.g.s. system), and c, the speed of light, is 3×10^{10} centimeters per second. See page 122.

RADIATION LAWS

The total energy, E, emitted by the surface of a black body or ideal radiator is given by the area under the Planck curve. This is called the Stefan-Boltzmann Law.

$$E = \sigma T^4$$

In the c.g.s. system, the constant σ is about 5.67×10^{-5} erg/cm^2 sec deg^4, and T is in degrees Kelvin.

The *wavelength* (λ) at which the maximum energy is emitted by an ideal radiator is given by Wien's Law.

$$\lambda = \frac{0.29}{T}$$

The units of λ are centimeters, and T is the temperature in degrees Kelvin.

PARALLAX EQUATION

A star's parallax, p (in seconds of arc), is related to its distance, d (in parsecs), by

$$d = \frac{1}{p}$$

HUBBLE'S LAW

The distance, D, of a galaxy is related to its red-shift velocity, V, by

$$D = \frac{V}{H}$$

Here H is Hubble's constant, roughly 100 kilometers per second per megaparsec, and the units of D and V are megaparsecs and kilometers per second, respectively. Other modern estimates of H are as low as 50 km/sec/mpc. See page 388.

THE MAGNITUDE EQUATION

If two stars have brightnesses b_1 and b_2, their magnitudes, m_1 and m_2, are related by

$$m_1 - m_2 = 2.5 \log_{10} \left(\frac{b_2}{b_1}\right)$$

Thus, if star 2 is 100 times brighter than star 1, $\log_{10} b_2/b_1$ equals 2, and the magnitude difference, $m_1 - m_2$, is 5, as required by the definition of magnitude on page 25. This equation can be applied to the brightnesses and magnitudes in a specific wavelength interval, such as that of the B or blue magnitudes, page 287.

THE DOPPLER FORMULA

The change in wavelength of a spectral line produced by a moving object is given by:

$$\Delta\lambda = \frac{\lambda V}{c}$$

Here, $\Delta\lambda$ is the doppler shift; λ is the *rest wavelength* of the spectral line, that is, its wavelength as measured in a stationary laboratory light source; V is the radial velocity of the moving object, that is, the speed with which it is moving toward or away from the observer, and c is the velocity of light. When the object is approaching the observer, the Doppler shift is toward shorter wavelengths *(blue shift)*; a receding motion produces a shift toward longer wavelengths *(red shift)*. Typically, in research on visible-light spectroscopy of astronomical objects, $\Delta\lambda$ and λ are each expressed in Angstroms when this formula is used, and V and c are each expressed in kilometers per second ($c = 3 \times 10^5$ km/sec).

Summary Tables

Table A-1. Properties of the Sun

Mass	2×10^{33} grams, or 3×10^5 earth masses
Diameter of the photosphere	1.4×10^6 km, or 109 earth diameters
Temperature of the photosphere	5,750 K
Temperature of the corona	2×10^6 K
Rate at which mass streams away in the solar wind	2×10^{12} grams/sec
Rate at which solar energy is released in photons (luminosity of sun)	4×10^{26} watts, or 4×10^{33} ergs/sec
Rate at which mass is converted into energy by nuclear reactions in the core of the sun	5×10^{12} grams/sec
Rate at which solar energy is received by the earth (solar constant)	2 calories/cm²-min or 1.4×10^6 ergs/cm²-sec
Estimated age	5×10^9 years
Distance from the center of the galaxy	10,000 parsecs
Speed of motion through the galaxy	250 km/sec

Table A-2. Properties of the Planets

Property		Mercury	Venus	Earth	Mars	Jupiter	Saturn	Uranus	Neptune	Pluto
Average distance from sun	(km)	0.58×10^8	1.08×10^8	1.50×10^8	2.28×10^8	7.78×10^8	14.2×10^8	28.7×10^8	44.9×10^8	58.9×10^8
	(a.u.)	0.39	0.72	1.00	1.52	5.20	9.52	19.2	30.0	39.4
Average orbital speed (km/sec)		48	35	30	24	13	9.6	6.8	5.4	4.7
Orbital period (sidereal)		88 days	225 days	365 days	1.88 years	11.9 years	29.5 years	84 years	164 years	247 years
Radius (km)		2,440	6,050	6,378	3,397	71,600	60,000	25,900	24,750	1,500
Mass (grams)		3.3×10^{26}	4.9×10^{27}	6.0×10^{27}	6.4×10^{26}	1.9×10^{30}	5.7×10^{29}	8.7×10^{28}	1.0×10^{29}	1.4×10^{25}
(earth masses)		0.055	0.82	1.00	0.11	318	95	15	17	.002
Escape velocity (km/sec)		4.2	10	11	5.0	59	36	21	23	1
Average axial rotation period (sidereal)		58.6 days	243 days	$23^h\ 56^m$	$24^h\ 37^m$	$9^h\ 56^m$	$10^h\ 14^m$	16^h ?	18^h ?	6.4 days
Tilt of equatorial plane to orbital plane		0°	3°	23.5°	25°	3°	27°	98°	29°	90° ?

Question marks indicate uncertainties.

Table A-3.   Satellites of the Planets*

Satellite	Average Distance from Planet (km)	Orbital Period (sidereal days)	Direction of Orbit (with respect to planet's rotation)	Tilt of Orbit to Planet's Equator	Radius (km, where known)	Mass (grams, where known)
EARTH						
Moon	384,000	27.3	direct	23½° ± 5°‡	1,738	7.3×10^{25}
MARS						
Phobos	9,380	0.32	direct	1°	13§	—
Deimos	23,460	1.26	direct	3°	7§	—
JUPITER						
I Io	422,000	1.77	direct	0°	1,818	8.9×10^{25}
II Europa	671,000	3.55	direct	0°	1,533	4.8×10^{25}
III Ganymede	1,071,000	7.15	direct	0°	2,608	1.5×10^{26}
IV Callisto	1,884,000	16.69	direct	0°	2,445	1.1×10^{26}
V Amalthea	182,000	0.50	direct	0°	120	—
VI Himalia	11,487,000	251	direct	28°	85	—
VII Elara	11,747,000	260	direct	25°	40	—
VIII Pasiphae	23,510,000	739	retrograde	35°	—	—
IX Sinope	23,670,000	758	retrograde	27°	—	—
X Lysithea	11,861,000	264	direct	29°	—	—
XI Carme	22,540,000	692	retrograde	16°	—	—
XII Ananke	21,250,000	631	retrograde	33°	—	—
XIII Leda	11,094,000	239	direct	27°	—	—
SATURN						
I Mimas	185,500	0.94	direct	2°	200	4×10^{22}
II Enceladus	238,000	1.37	direct	0°	275	7×10^{22}
III Tethys	294,700	1.89	direct	1°	524	6.2×10^{23}
IV Dione	377,400	2.74	direct	0°	412	1.2×10^{24}
V Rhea	527,000	4.52	direct	0°	788	2.7×10^{24}
VI Titan	1,222,000	15.95	direct	0°	2,912	1.4×10^{26}
VII Hyperion	1,484,000	21.28	direct	0°–1°	112	1.1×10^{23}
VIII Iapetus	3,562,000	79	direct	15°	800	1.5×10^{24}
IX Phoebe	12,960,000	550	retrograde	30°	100	—
X Janus	168,700	0.82	direct	0°	—	—
URANUS†						
I Ariel	191,000	2.52	direct	0°	300	1.3×10^{24}
II Umbriel	266,000	4.14	direct	0°	200	5×10^{23}
III Titania	436,000	8.71	direct	0°	500	4×10^{24}
IV Oberon	583,000	13.46	direct	0°	400	2.5×10^{24}
V Miranda	130,000	1.41	direct	3°	100	1×10^{23}

Table A-3. Satellites of the Planets* *(continued)*

Satellite	Average Distance from Planet (km)	Orbital Period (sidereal days)	Direction of Orbit (with respect to planet's rotation)	Tilt of Orbit to Planet's Equator	Radius (km, where known)	Mass (grams, where known)
NEPTUNE						
I Triton	355,550	5.88	retrograde	20°	2,200	1.3×10^{26}
II Nereid	5,567,000	360	direct	28°	100	3×10^{22}
PLUTO						
Charon	19,300	6.4**	direct	0°	600	—

*Unlisted planets have no known satellites.

†Note that the rotation of Uranus is retrograde.

‡Tilt varies from 18° to 29°.

§Oblong shape; this "radius" equals one-half of the long dimension.

**Charon appears to be in a "Pluto-synchronous" orbit, that is, Charon's orbital period equals Pluto's rotation period. Thus, Charon is always above the same spot on Pluto's surface.

Index